FOOD PROCESS
ENGINEERING

FOOD PROCESS ENGINEERING
Theory and Laboratory Experiments

Shri K. Sharma

Steven J. Mulvaney

Syed S. H. Rizvi

Cornell University
Ithaca, New York

WILEY-INTERSCIENCE

A John Wiley & Sons, Inc., Publication

New York / Chichester / Weinheim / Brisbane / Singapore / Toronto

This book is printed on acid-free paper. ∞

Copyright © 2000 by John Wiley & Sons, Inc. All rights reserved.

Published simultaneously in Canada.

No part of this publication may be reproduced, stored in a retrieval system or transmitted in any form or by any means, electronic, mechanical, photocopying, recording, scanning or otherwise, except as permitted under Sections 107 or 108 of the 1976 United States Copyright Act, without either the prior written permission of the Publisher, or authorization through payment of the appropriate per-copy fee to the Copyright Clearance Center, 222 Rosewood Drive, Danvers, MA 01923, (978) 750-8400, fax (978) 750-4744. Requests to the Publisher for permission should be addressed to the Permissions Department, John Wiley & Sons, Inc., 605 Third Avenue, New York, NY 10158-0012, (212) 850-6011, fax (212) 850–6008, E-Mail: PERMREQ@WILEY.COM.

For ordering and customer service call 1-800-CALL-WILEY.

Library of Congress Cataloging-in-Publication Data

Sharma, S. K. (Shri Kamal)
    Food process engineering : theory and laboratory experiments /
  Shri K. Sharma, Steven J. Mulvaney, Syed S. H. Rizvi.
      p. cm.
    ISBN 0-471-32241-5 (paper : alk. paper)
    1. Food industry and trade.   I. Mulvaney, Steven J.   II. Rizvi, Syed S. H.   III. Title
TP370.S423 2000   99-28987
664—dc21   CIP

Printed in the United States of America

10 9 8 7 6 5 4 3 2 1

This work is dedicated to
all our loved ones

CONTENTS

of cans in retort. Calculation of lethality by the general method (graphical and numerical techniques).

Evaluation of thermal processing of cans in a retort using Ball's formula. Calculating processing times for foods heated in a conduction and convection mode in different can sizes.

Blanching of fruits and vegetables and its effect on product quality. Thermodynamics of food freezing. Estimation of freezing times. Effect of freezing on product quality.

UHT processing of low- and high-viscosity foods. Comparison of UHT processing with cans processing of foods in a retort. Estimation of the overall heat transfer coefficient and effect of UHT on product quality.

Concentration of liquid foods using membrane processing. Effect of flow, transmembrane pressure, membrane resistance and cake layer formation on the permeate flux rate.

Use of an evaporator for the concentration of maple syrup/milk. Engineering and product quality considerations.

Concept of experimental design, treatment factors, treatment levels, interactions among factors, effect of blocks on experimental design, true replication in product development experiments. Use of analysis of variance and estimating the significant effect of treatment factors.

Engineering and processing concepts of spray drying and drum drying of concentrated skim milk; effect of process conditions on the overall quality of nonfat dried milk (NFDM).

PREFACE

This book is the culmination of several years of collaborative teaching among the authors at Cornell University, in which we attempted to differentiate the principles of food process engineering from those of food engineering alone. This is harder than it might seem at first. Food processing, or food manufacturing as we prefer to call it, is an all-encompassing endeavor in practice, which makes it difficult to teach. Some food processing courses focus on raw materials (i.e., a commodity approach) and track a raw material from start to finish, such as, milk to cheese. Others may focus on the engineering aspects of food processing (typically referred to as a unit operations approach). However, the latter approach may not be too different from a *food engineering* course if the main emphasis is on heat and mass transfer aspects of unit operations. Our approach has been to develop a core course in the area of food process engineering that focuses on developing specific quantitative skills, which would be generally useful in a wide variety of food processing or manufacturing environments. Our general view is that a course in the principles of food engineering with its own stand-alone lab exercises emphasizing engineering properties and transport processes would be a prerequisite for this course. This course could also be considered as the introduction to a more advanced food process engineering course. However, it is important to us that all (or nearly all) graduates with a B.S. in food science have the ability to systematically analyze a food manufacturing process based on its underlying physics and chemistry.

In practice, this involves a combination of unit operations type of lab exercises supplemented with lab exercises in the linear programming, experimental design, and rheological properties of fluid and solidlike foods. In addition, the actual unit operation labs used are selected to demonstrate the interplay between engineering, chemistry, and the microbiological and sensory aspects of food manufacturing. The variety of lab exercises also allows for a broad exposure to different foods, such as the extrusion of cereals, spray drying of milk, frying of chips, or retorting of vegetables. Emphasis has been placed wherever possible on relating a physical aspect of food quality such as color, texture, or material properties, density or porosity, and so forth to various process conditions. In part, this is because of our interest in this aspect of food processing; however, these tests are generally rapid enough to be done as part of a single laboratory period, and they tend not to be taught in a separate stand-alone food science course.

A unique aspect of this book is the inclusion of considerable background information on the relevant physics and chemistry for a particular topic with a lab exercise to demonstrate these principles. Thus, students have all the relevant background information at hand to understand a lab exercise. It is our experience that as a practical matter, upper-level students do not refer to their freshman chemistry or physics books to complete assignments. Our book is even more useful for students without a food science background, such as engineering majors with an interest in foods. The book is also convenient for instructors who teach a process-oriented course with essentially one textbook, rather than having to assign chapters from several texts to complete their class syllabus. Since there are 19 chapters in this book, instructors can easily tailor a one-semester course to meet their specific needs, interests, and available process equipment and analytical instruments.

The authors take great pleasure in acknowledging the contributions of the many teaching assistants and graduate students who have helped develop these lab exercises, along with the many students who have taken these courses over the past 10 years. We cannot name them all, but you know who you are. We are also deeply indebted to George Houghton, who served as a teaching support specialist to us for several years prior to his retirement. In particular, the material on linear programming and experimental design is essentially George's work. George also prepared the vast majority of original figures throughout the text. If this book stimulates an increased interest in "quantitative food processing," then we will have accomplished our objectives.

Shri K. Sharma
Steven J. Mulvaney
Syed S. H. Rizvi

Ithaca, New York

1

PRODUCT FORMULATION AND PROCESS OPTIMIZATION USING LINEAR PROGRAMMING

1.1 BACKGROUND

Many foods consist of a mixture of ingredients that must be blended together. For the resulting product to be satisfactory, it must usually meet specifications for levels of fats, protein, water, and other ingredients. Frequency, you will find that many formulations meet these specifications. Under those circumstance, you may wish to select the formulation with the lowest cost. Linear programming is a mathematical technique for optimizing some function such as cost while meeting a set of specifications or constraints.

1.2 LINEAR PROGRAMMING EXAMPLE

The following simplified example will serve as an introduction to linear programming.

1.2.1 The Problem

You are in charge of formulating a dogfood that consists of three ingredients: Woof Meal, Fido Bits, and a nonnutrient filler. One hundred lb of the finished product must contain at least 10 lb of protein, 6 lb of fat, and 15 lb of fiber. The ingredients have the following levels of these components: Woof Meal contains 10% protein, 12% fat, and 75% fiber and Fido Bits contains 50% protein, 15% fat, and 20% fiber. This morning's quotations indicate that Woof Meal is selling for $2.50 per hundred lb and Fido Bits is selling for $3.00 per hundred lb. What is the least-cost formulation that will meet specification? Assume that the filler is so cheap that it can be ignored.

1. *Problem variables.* This problem is solved by specifying the values of two variables:

$$W = \text{lb of Woof Meal in 100 lb of product}$$
$$F = \text{lb of Fido Bits in 100 lb of product}$$

2. *The objective function.* Since the costs of the ingredients are $2.50 and $3.00 per hundred lb or $0.025 and $0.030 per lb, the cost of 100 lb of product is given by the equation

$$C = 0.025W + 0.030F \qquad (1.1)$$

Your objective is to find the values of W and F that minimize this cost, that is, minimize C. The equation to be minimized or maximized is called the objective function of the problem.

3. *Nonnegative constraints.* Since it makes no sense for ingredients to have negative values, the solution must meet the following constraints:

$$W \geq 0 \qquad F \geq 0 \qquad (1.2)$$

4. *Combined weight constraint.* In addition, since we want the formulation for about 100 lb of dogfood, we add the constraint

$$\text{Combined weight:} \quad W + F \leq 100 \qquad (1.3)$$

If $F + W$ is less than 100, the balance is made up with the filler.

5. *Other constraints.* The problem states that the total protein must exceed 10% or 10 lb per hundred lb. Since Woof Meal is 10% protein and Fido Bits are 50% protein, this requirement can be expressed by the following constraint:

$$\text{Protein:} \quad 0.10W + 0.50F \geq 10 \text{ lb} \qquad (1.4)$$

Similarly, the fat and fiber requirements can be expressed with the following constraints:

$$\text{Fat:} \quad 0.12W + 0.15F \geq 6 \text{ lb} \qquad (1.5)$$
$$\text{Fiber:} \quad 0.75W + 0.20F \geq 15 \text{ lb} \qquad (1.6)$$

6. *Problem statement.* With these constraints and objective function, the problem then can be restated in formal mathematical terms as follows. Find $W \geq 0$ and $F \geq 0$ so

that

$$\text{Protein:}\quad 0.10W + 0.50F \geq 10$$
$$\text{Fat:}\quad 0.12S + 0.15F \geq 6$$
$$\text{Fiber:}\quad 0.75W + 0.20F \geq 15$$
$$\text{Weight:}\quad W + F \leq 100$$

and so that $C = 0.025W + 0.030F$ is minimized.

1.2.2 Graphic Solution

For simple problems with only two variables, a linear programming problem can be solved graphically. In addition, finding a graphic solution provides insight into linear programming. The following describes how this is done:

1. *The problem space.* Begin by drawing the Cartesian plane shown in Figure 1.1 with two axes representing the problem variables, namely the weight of Woof Meal (W) and the weight of Fido Bits (F). This plane is called the "problem space" for this problem and extends infinitely far in all directions. Any point in this plane represents the weights of the ingredients for one particular formulation of the product.

2. *Potential solutions.* The solution to this problem is a pair of values, one for W and one for F. But these quantities are coordinates of a point in the plane, so the solution to the problem can be represented by a point in the problem space. Figure 1.1 shows five possible solutions. The entire problem space contains infinitely many solutions.

3. *Strategy.* The strategy for finding a solution to a linear programming problem is to use the constraints to limit the part of the problem space in which a solution can be found. The part of the space that meets all constraints is called the feasible region. We then use the objective function to find one optimal solution within this feasible region.

4. *Half-planes.* Notice that the plane represented in Figure 1.1 contains negative values of both W and F. As we pointed out above, negative weights are mean-

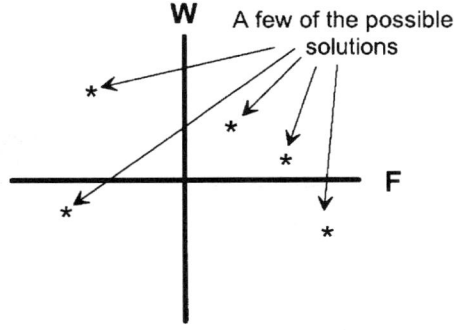

Figure 1.1 Woof Meal (W) and Fido Bits (F) in a 2-D plane.

ingless and only areas of the plane that meet the nonnegativity constraint should be considered. For example, the constraint in Eq. (1.2) implies that $F \geq 0$ limits consideration to areas of the plane that include the vertical axis and the area to the right of the vertical axis. This is the shaded areas in Figure 1.2A. Note that this constraint eliminates half the problem space from further consideration. We call the region allowed by this constraint a half-plane. Similarly, Figure 1.2B is shaded to indicate the half-plane corresponding to the constraint shown in Eq. (1.3) as $W \geq 0$. This constraint also eliminates half the entire problem space.

5. *Intersection of planes.* Since both constraints

$$F \geq 0 \qquad \text{and} \qquad W \geq 0$$

must hold, the area of the plane that can contain the solution is reduced to the intersection of the two half-planes as shown by the shaded area in Figure 1.2C. Therefore, we are left with only a quarter of the problem space to consider.

6. *Graphing nutrient constraints.* Every constraint define a half-plane. Consider the constraint for fat (Eq. 1.5) as shown below:

$$0.12W + 0.15F \geq 6$$

Unlike the first two constraints, the boundary of this one is not along an axis but along the line whose equation is

$$0.12W + 0.15F = 6 \qquad\qquad (1.7)$$

To draw this line, we first locate two points along the line. A point can be located by substituting any value, for one variable and solving for the other variable. Although any value can be used, let's keep things simple and substitute 0 for F, thus,

$$0.12W + 0.15(0) = 6, \qquad W = \frac{6}{0.12} = 50$$

This tells us that the line for this equation passes through the point (0,50), where 0 is the F coordinate of the point and 50 is the W coordinate. This also tells us that a mix containing 50 lb of Woof Meal and no Fido Bits would meet the fat constraint.

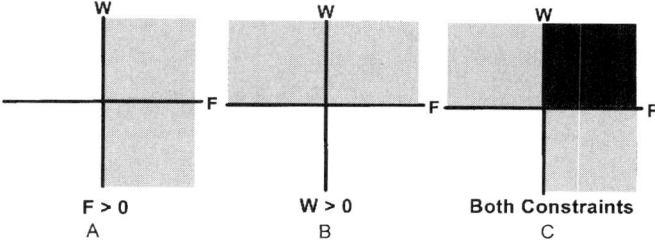

Figure 1.2   Weight constraints (both Woof Meal and Fido Bits weights must be positive).

For the second point, let's substitute 0 for W and solve for F:

$$0.12(0) + 0.15F = 6, \qquad F = \frac{6}{0.15} = 40$$

This tells us that the line also passes through the point (40,0), or that a mix containing 40 lb of Fido Bits and no Woof Meal will also meet the fat constraint. If we draw a line through both these points, the half-plane that fits the fat constraint will contain the line and, because the constraint contains a "greater than" relation, the entire area above and to the right of the line. This half-plane is shaded in Figure 1.3A. Any point in this half-plane meets the fat constraint. Similarly, the protein constraint is bounded by a line through (0,100) and (20,0), whereas the fiber constraint is bounded by a line through (0,20) and (75,0). These constraints are shown graphically in Figure 1.3B and 1.3C, respectively. (The reader is encouraged to verify the correctness of these half-planes.)

7. *The combined weight constraint.* Finally, the constraint that the weights of Woof Meal and Fido Bits must be less than or equal to 100 lb (Eq. 1.3) defines a half-plane to the left of a line through (100,0) and (0,100). This is shown in Figure 1.4.

8. *The feasible region.* Each of the constraints defines a half-plane. Since the solution must meet all the constraints, it must lie within all these half-planes and, hence, within the intersection of all these planes. This intersection is called the feasible region for the problem and is shown as the shaded region in Figure 1.5. The solution must be a point in the feasible region.

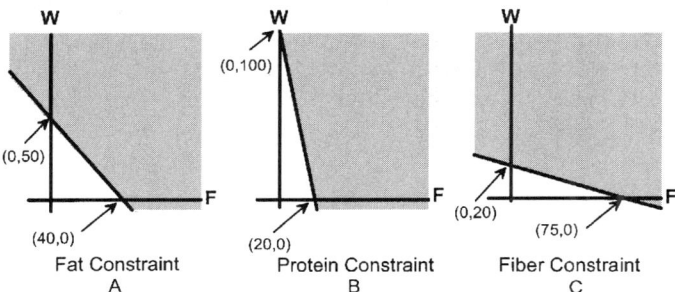

Figure 1.3 Nutrient constraints.

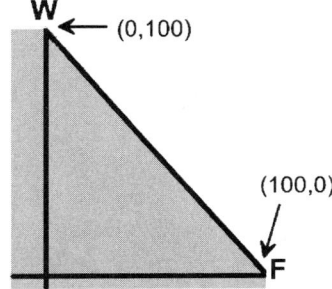

Figure 1.4 Total weight constraints.

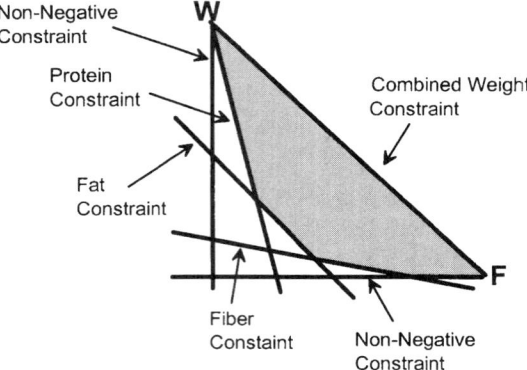

Figure 1.5   The feasible region.

9. *Extreme points.* Notice that the feasible region is bounded by polygon consisting of a set of straight-line segments. The line segments intersect at vertices that we call extreme points as shown in Figure 1.6. As you will see shortly, the least-cost solutions to linear programming problems always occur at an extreme point. In some specific cases, the solution may include two adjacent extreme points and the line segment between them. Thus, we have reduced the dogfood problem from infinitely many formulations to just five.

10. *Graphing the objective function.* If there were no further conditions, any point within the feasible region would be a valid solution to this problem. The point (60,40), for example, falls in the feasible region. This point correponds to using 60 lb of Woof Meal and 40 lb of Fido Bits in the formulation. This solution meets all constraints. It is not however, likely to be the least cost-solution. To find this, we must locate one or more points in the feasible region that minimizes the value of the objective function (Eq. 1.1):

$$C = 0.025W + 0.030F$$

where C is the total cost of the two major ingredients per hundred lb of product. Let us start by searching for the minimum value by picking an arbitrary value for C and finding all points that will have this value. For example, to find all solutions

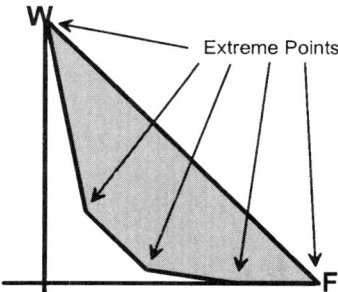

Figure 1.6   Extreme points.

that cost $2.00 per 100 lb, substituting 2.00 in the objective function, we obtain

$$2.00 = 0.025W + 0.030F$$

This is the equation of a straight line so we need to find two points along the line. Substituting 0 for W gives

$$2.00 = 0.025(0) + 0.030F, \qquad F = \frac{2.00}{0.030} = 66.7$$

so that the line for $C = 2.00$ passes through (66.7,0). Substiting 0 for F tells us that the line also passes through (0,80). This line is shown in Figure 1.7. Any point along this line represents a formulation that will cost $2.00 for these two ingredients. Where this line passes through the feasible region, it contains feasible solutions that cost $2. A line representing all formulations that cost $1.00 per hundred lb of product is also shown in Figure 1.7. None of these formulations falls within the feasible region, so none of them meets all constraints.

11. *The least-cost solution.* Notice that the two lines in Figure 1.7 have the same slope. In fact, all lines generated by the same objective function will have the same slope. Therefore, to locate the least-cost solution, simply plot any line that fits the objective function, then move it without changing its slope until you find the least-cost line that falls within the feasible region. In Figure 1.8, we see that the line representing a cost of $1.21 per hundred lb of product touches the feasible region at

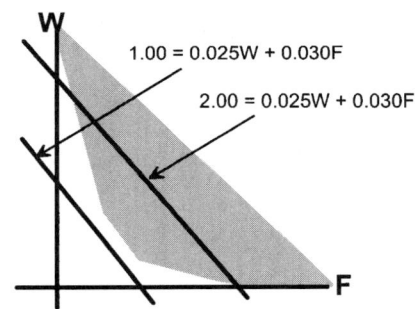

Figure 1.7   Values of the objective function.

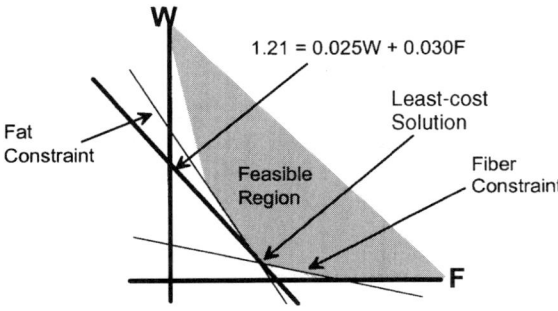

Figure 1.8   The intersecting lines showing the solution to the problem.

just one point. This point represents the solution to the problem. Any lower cost would move the line to the left, out of the feasible region, so this is the lowest cost that will meet all constraints.

12. *Solution values.* In this example, the solution falls at the intersection of the lines defined by Equations (1.4) and (1.5):

$$\text{Fat:}\quad 0.12W + 0.15F = 6$$
$$\text{Fiber:}\quad 0.75W + 0.20F = 15$$

The values of W and F at this intersection can be read directly from the graph as shown in Figure 1.9 or can be obtained by solving these equations simultaneously. You can solve these equations with algebra or restate these equations in matrix form and use a computer program that handles matrices. Using ordinary algebra, you might proceed as follows:

- First, change the coefficient of F in the Fat equation to 1 by multiplying by $1/0.15$:

$$\frac{1}{0.15}(0.12W + 0.15F) = \frac{1}{0.15}(6), \qquad 0.8W + 1F = 40$$

- Next, change the coefficient of F in this equation to 0.2 by multiplying by 0.2:

$$0.2(0.8W + 1F) = 0.2(40), \qquad 0.16W + 0.2F = 8$$

- Now eliminate F by subtracting this equation from the Fiber equation:

$$(0.75 - 0.16)W + (0.2 - 0.2)F = 15 - 8, \qquad 0.59W = 7$$

- Solve for W:

$$W = \frac{7}{0.59} = 11.9$$

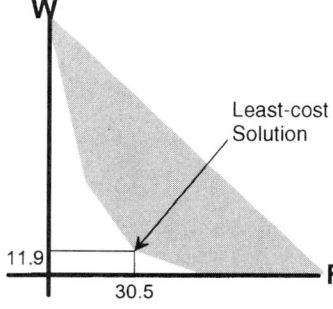

Figure 1.9   The point showing the least-cost solution.

- Substitute this solution in one of the original equations and solve for F:

$$0.12(11.9) + 0.15F = 6, \qquad F = \frac{6 - 0.12(11.9)}{0.15} = 30.5$$

For those familiar with matrix methods, the equations for Fat and Fiber are

$$\begin{pmatrix} 0.12 & 0.15 \\ 0.75 & 0.20 \end{pmatrix} \begin{pmatrix} W \\ F \end{pmatrix} = \begin{pmatrix} 6 \\ 15 \end{pmatrix} \tag{1.8}$$

The solution to this matrix equation, which can be obtained with Microsoft$^{\circledR}$ Excel 97, is

$$\begin{pmatrix} W \\ F \end{pmatrix} = \begin{pmatrix} -2.260 & 1.695 \\ 8.475 & -1.356 \end{pmatrix} \begin{pmatrix} 6 \\ 15 \end{pmatrix} = \begin{pmatrix} 11.9 \\ 30.5 \end{pmatrix}$$

Either way, the cheapest formulation that meets the constraints requires 11.9 lb of Woof Meal and 30.5 lb of Fido Bits.

13. *Solution cost.* The cost of this formulation is obtained by substituting these values in the objective function (Eq. 1.1):

$$C = 0.025(11.9) + 0.030(30.5) = \$1.21 \text{ per hundred lb}$$

14. *Total weight.* If we substitute the solutions in Eq. (1.3), the total weight of ingredients is

$$11.9 + 30.5 = 42.4 \text{ lb}$$

From this, we conclude that we need $100 - 42.4 = 57.6$ lb of filler to make a bag of mix. In a more realistic problem, the filler would have a cost and be included as a variable. For example, if the cost of filler is \$0.05/lb, the objective function would become

$$0.12W + 0.15F + 0.05N = C$$

and the constraint on total weight would change to the equality

$$W + F + N = 100$$

This, however, would make it a three-dimensional problem that cannot easily be solved graphically.

1.2.3 Modifying the Problem

1. *Slack.* Substituting the solution values into constraint equations (1.4, 1.5, and 1.6) shows that the solution formula has the following composition:

$$\text{Protein:} \quad 0.10(11.9) + 0.50(30.5) = 16.4 \text{ lb}$$
$$\text{Fat:} \quad 0.12(11.9) + 0.15(30.5) = 6.0 \text{ lb}$$
$$\text{Fiber:} \quad 0.75(11.9) + 0.20(30.5) = 15.0 \text{ lb}$$

Notice that all nutrients are present in amounts greater than or equal to the specified limits and so meet the constraints. Furthermore, the fat and fiber levels exactly meet the minimum constraints. On the other hand, the protein level exceeds the minimum by the amount

$$16.4 - 10.0 = 6.4 \text{ lb}/100 \text{ lb} \tag{1.9}$$

This excess is called "slack" and is usually present in solution to linear programming problems.

2. *Two-sided constraints.* In some cases, you may judge the slack to be too large. Excess slack can be taken care of by setting a two-sided constraint on some property. For example, you may wish to keep protein between 10 and 15 lb per 100. This constraint can be stated as

$$10 \le 0.10W + 0.50F \le 15 \tag{1.10}$$

Two-sided constraints must be handled as two separate constraints, thus

$$
\begin{aligned}
0.10W + 0.50F &\ge 10 \quad \text{(the original protein constraint)} \\
0.10W + 0.50F &\le 15 \quad \text{(the added protein constraint)}
\end{aligned}
\tag{1.11}
$$

The new constraint passes through the points (10,100) and (30,0). (The reader should verify these values.) Figure 1.10A shows how the new constraint reduces the feasible region. Figure 1.10b illustrates the new solution at

$$W = 16.7 \text{ lb of Woof Meal}$$
$$F = 26.7 \text{ lb of Fido Bits}$$

The slack in this new solution is in the fiber since

$$[0.75(16.7) + 0.20(26.7)] - 15 = 2.87 \text{ lb fiber}$$

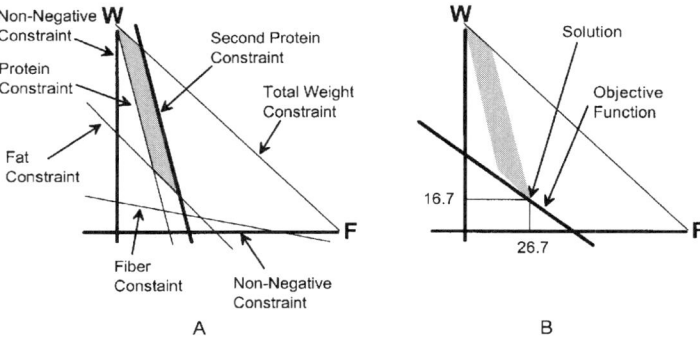

Figure 1.10 The feasible region with two constraints on protein.

1.2.4 Ratio Constraints

Sometimes constraints take the form of ratios. For example, we might require that the lecithin (an emulsifier) content of the dogfood be at least 2% of the fat content. In other words,

$$\frac{\text{Lecithin}}{\text{Fat}} \geq 0.02 \qquad \text{or} \qquad \frac{\text{Fat}}{\text{Lecithin}} \leq \frac{1}{0.02} = 50 \tag{1.12}$$

(Notice that when the two sides of an inequality are inverted, the direction of the inequality changes.)

Suppose Woof Meal contains 0.4% lecithin are Fido Bits 0.1%. How do we write the ratio constraint into this problem? The trick is to convert it to the same form as all other constraints as follows:

1. Rearrange the ratio

$$\text{Fat} \leq 50(\text{Lecithin}) \tag{1.13}$$

2. Recall that Woof Meal has 12% fat and Fido Bits has 15% fat, substituting this in Eq. (1.13):

$$0.12W + 0.15F \leq 50(\text{Lecithin}) \tag{1.14}$$

Furthermore, we have just stated that Woof Meal has 0.4% lecithin, whereas Fido Bits contains 0.1% lecithin. Substitute this in Eq. (1.14):

$$0.12W + 0.15F \leq 50(0.004W + 0.001F) \tag{1.15}$$

3. Rearrange Eq. (1.15) to put the variable terms on the left and the constant on the right. In this example, we find no constant term so the right-hand side becomes 0.

$$[(0.12 - 50(0.004)]W + [(0.15 - 50(0.001)] F \leq 0$$
$$-0.080W + 0.10 F \leq 0 \tag{1.16}$$

Equation (1.16) is the constraint we will use to limit the fat-to-lecithin ratio. There might also be a constraint specifying upper and/or lower limits on lecithin, but in this particular example, there is only this ratio constraint. Equation (1.16) defines a half-plane bounded by a line with the equation

$$-0.080W + 0.10F = 0 \tag{1.17}$$

4. Substituting 0 for W in Eq. (1.17) makes $F = 0$ so this line passes through the origin (0,0). Substituting 50 for W makes $F = 40$ so this constraint is bounded by a line through (0,0) and (40,50) as shown in Figure 1.11.

5. To determine which side of this line fits the constraint, note that in inequality (Eq. 1.16), the left-hand side can be made less than 0, either by increasing W or

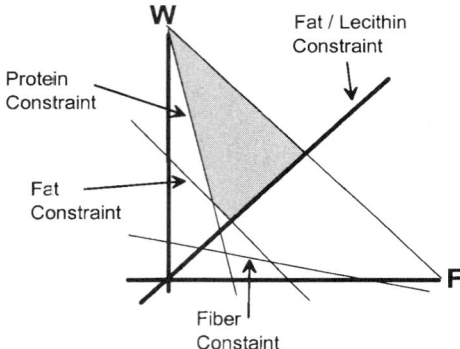

Figure 1.11   A ratio constraint.

decreasing F. This tells us that the feasible region must be above (larger W) and to the left (smaller F) of the line as shown in Figure 1.11.

6. The solution will now be at one of the extreme points of the new feasible region illustrated in Figure 1.11. (Can you locate the solution in this figure?)

1.2.5   Degenerate Problems

Sometimes not every set of constraints allows you to find a solution. For example, suppose you wanted the salt level in the dogfood to be below 5%. If the salt content of Woof Meal is 10% and that of Fido Bits is 15%, this leads to the following constraint:

$$\text{Salt:}\quad 0.10W + 0.15F \le 5$$

This constraint defines a half-plane below a line through (0,50) and (33.3,0) as shown in Figure 1.12A. Since the feasible region must satisfy all constraints, it must consist of the intersection of this half-plane and the previously defined feasible region. As illustrated in Figure 1.12B, these two regions do not overlap, so there is no intersection. Thus, adding this constraint eliminates the feasible region and there is no solution to the problem. We refer to this as a degenerate problem. To find a solution, you must either relax some constraints (allow less fat, e.g.) or look for other ingredients.

1.2.6   A Spreadsheet Solution

The graphic method described above is a convenient method of illustrating linear programming. However, it only works with problems of two variables that can be represented by a two-dimensional space. When a formulation requires more than two ingredients, a nongraphic method must be used. For example, a spreadsheet such as Microsoft® Excel 97 as shown in Table 1.1.

1. *Variables.* Table 1.1 shows a spreadsheet that can be set up to solve the dogfood problem. Variables are named in cells B1 and C1. Values for these variables are placed in cells B2 and C2. Initially, any value will do for these variables. We will use a value of 1 for each. These will be replaced by the solutions.

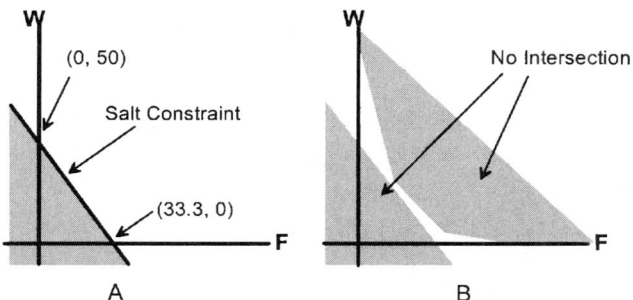

Figure 1.12 A degenerate problem.

Table 1.1 **Spreadsheet to solve the dogfood problem**

	A	B	C	D	E	F	G
1		W	F	Total			
2	Weight	1	1	2			
3	Cost	0.025	0.030	0.055	Min		
4						Limits	Slack
5		1	0	1.00	>=	0	1
6		0	1	1.00	>=	0	1
7	Weight	1	1	2.00	<=	100	−98
8	Protein	0.10	0.50	0.60	>=	10	−9.4
9	Fat	0.12	0.15	0.27	>=	6	−5.73
10	Fiber	0.75	0.20	0.95	>=	15	−14.05

2. *Sum.* Although it is not needed to solve the problem, it is frequently useful to sum the values of the variables. In this example, the sum will tell us the total weight of the ingredients. The following formula is entered in the indicated cell:

$$\text{Microsoft}^{\text{R}} \text{ Excel:} \quad D2: \quad = \text{SUM(B2:C2)}$$

You can test this formula by entering different values in cells B2 and C2 checking the sum in D2.

3. *Costs.* The costs, profits, or other values that are coefficients in the objective function are placed in cells B3 and C3. In this example, these values are the ingredient costs per lb.

4. *The objective function.* The objective function is placed in cell D3. This function multiplies each variable in row 2 by the coefficient in row 3 and adds the results like this:

$$\text{Excel:} \quad D3: \quad = \$B\$2 * B3 + \$C\$2 * C3$$

The constraints are going to take the same form, so you will be copying this expression into each constraint. Since every copy will use the same variables in row 2, references to these variables have *dollar signs* to prevent changes in these addresses when they are copied. The [F4] function key can be used to insert the dollar sign. Test these formulas by entering different values in cells B2, C2, B3, and C3 and checking the results in D3.

5. *Constraints.* The nutrient constraints have been placed in rows 8, 9, and 10. Notice that the constraints take the same form as the objective function, namely, a coefficient times W plus a coefficient times F. For example, the constraint for protein is

$$0.10W + 0.50F \leq 10$$

To have the spreadsheet compute the left side of this inequality, enter the coefficients in B8 and C8 as shown in Table 1.1. Then copy the objective function in D3 and D8 so that the ingredient weights will be multiplied by these coefficients. In fact, copy D3 into cells D5 through D10. To do this in Excel,

Click on cell D3.

Point to the tiny square in the lower right-hand corner of cell D3.

Drag this square to the lower right-hand corner of cell D10.

(For neatness, you may want to delete the formula from cell D4.)

6. *Types of constraints.* In the spreadsheet, column E is used to indicate the types of relationships in the constraints. In this problem, all constraints are of the $\geq$ type except the one in row 7. The spreadsheet uses the relationships in these cells. They are just entered as a reminder to the user.

7. *Limits on constraints.* The limits on the constraints are placed in cells F8 through F10 as shown in Table 1.1.

8. *Nonzero constraints.* The nonzero constraints can also be placed in the same form as the objective function by using the coefficients 1 and 0. For example, the constraint $F \geq 0$ can be expressed as

$$0W + 1F \geq 0$$

Nonnegativity constraints are entered in this form in rows 5 and 6.

9. *Total weight constraint.* If we use the same method, the constraint that

$$W + F \leq 100$$

is entered in row 7 as

$$1W + 1F \leq 100$$

10. *Slack.* Protein slack is computed with the following formula:

$$\text{Excel:} \quad \text{G8:} \quad = D8 - F8$$

This formula is copied into the cells above and below it.

11. *Solving the problem in a spreadsheet.* In a spreadsheet such as Microsoft Excel® 97, the problem is solved by the *solver* command found in the Tools menu. Three things must be specified: the cell that is to be minimized or maximized, the cells that contain values for the variables, and the cells that specify the constraints. These

are specified as follows:

  a. To access the optimizer menu, click on the **Tools** menu, then on **Solver**. The first time you do this, you must wait for the solver to load.

  b. To specify the cell to be optimized, click on the **Set Target Cell** box, then click on the cell with the objective function, in this case D3.

  c. To specify the type of optimization, click on the **Min** button.

  d. To specify the solution variables, click on the **By Changing Cells** box. Drag across the variable cells on the spreadsheet, in this case B2 through C2, so that they are surrounded by a moving marquee.

  e. To specify the first constraint, click on the [**Add**] button. Click on the cell with the first constraint, in this case D5. Click on the down arrow button in the middle of the dialog box and select ≥. Click on the **Constraint** box and click on the cell with the limit for the first constraint, in this case F5.

  f. Click on [**Add**] and repeat step e for each constraint. In this case, you will enter five constraints in this way. Remember that row 7 contain a ≤ constraint. After entering the last constraint, press [**OK**] instead of [Add]

  g. Once the above specifications are made, press [**Solve**] and wait for a solution. A dialog box will open indicating whether a solution could be found.

  h. Click on [**OK**] to remove the dialog box that appears when a solution is reached.

12. *The solution.* The solution, shown in Table 1.2, provides the following information:

  a. The values of the variables, 11.9 and 30.5 lb in this example, are in the variable cells, B2 and C2. This is followed by the total weight of 42.4 lb.

  b. The optimized value of the objective function, $1.21 in this example, is found in cell D3.

  c. Values of constrained quantities are in cells D8, D9, and D10.

  d. Slack values are in column G.

13. *Format.* You may want to use the Style|Number Format menu or the **Decimal** buttons in Excel to set decimal places for various cells.

Table 1.2 Spreadsheet solution to the dogfood problem

	A	B	C	D	E	F	G
1		W	F	Total			
2	Weight	11.86	30.51	42.37			
3	Cost	0.025	0.030	1.21	Min		
4						Limits	Slack
5		1	0	11.86	>=	0	11.86
6		0	1	30.51	>=	0	30.51
7	Weight	1	1	42.37	<=	100	−57.63
8	Protein	0.10	0.50	16.44	>=	10	6.44
9	Fat	0.12	0.15	6.00	>=	6	0.00
10	Fiber	0.75	0.20	15.00	>=	15	0.00

14. *Slack.* Notice that $>=$ constraints will have positive slack, indicating the amount that a constituent exceeds the lower limit. On the other hand, $<=$ constraints will have negative slack, indicating the amount that a constituent falls below the upper limit.

15. *Changes.* If any of your information changes, you can easily compute a new solution.

   a. To rerun the optimizer or solver:

   Excel: Click on **Tools**, **Solver**, and the [**Solve**] button.

   b. To change a cost or constraint limit, enter the new value in the appropriate cell and rerun the optimizer or solver.
   c. To add a new constraint:

   Excel: Click on **Tools**, **Solver**.

   Click on the [**Add**] button and specify a constraint.

   d. To remove a constraint, do not delete or modify the constraint on the spreadsheet. Instead, do the following:

   Excel: Click on **Tools**, **Solver**.

   Click on the constraint to delete.

   Click on the [**Delete**] button.

1.3   CHECK YOURSELF

If you have mastered this material, you should be able to do the following:

1. Define and use the following terms as they pertain to linear programming:

Objective function	Intersection of planes
Half plane	Slack
Extreme point	Nonnegative constraint
Constraint	Feasible region

2. State a linear programming problem in formal mathematical terms.
3. Convert ratio constraints to a form that can be used in linear programming.
3. Convert ratio constraints to a form that can be used in linear programming.
4. For a two-variable problem, show the feasible region and solve the problem graphically. From the solution, determine the value of the objective function, the values of the problem variables, and the actual values of constrained quantities and slack.
5. Set up, solve, modify, and resolve a problem on a spreadsheet.

1.4   LAB EXERCISES

1. Using the graph below, find the solution to the dogfood example with the following changes:

   a. The cost of Woof Meal is $0.040 per lb.

   b. The ratio of fat to lecithin must be less than or equal to 40. (See Sec. 1.2.4).

   Repeat this solution with Woof Meal costing $0.010 per lb. (Fido Bits costs $0.030 per lb in both cases.) What is the cost per 100 lb in each case?

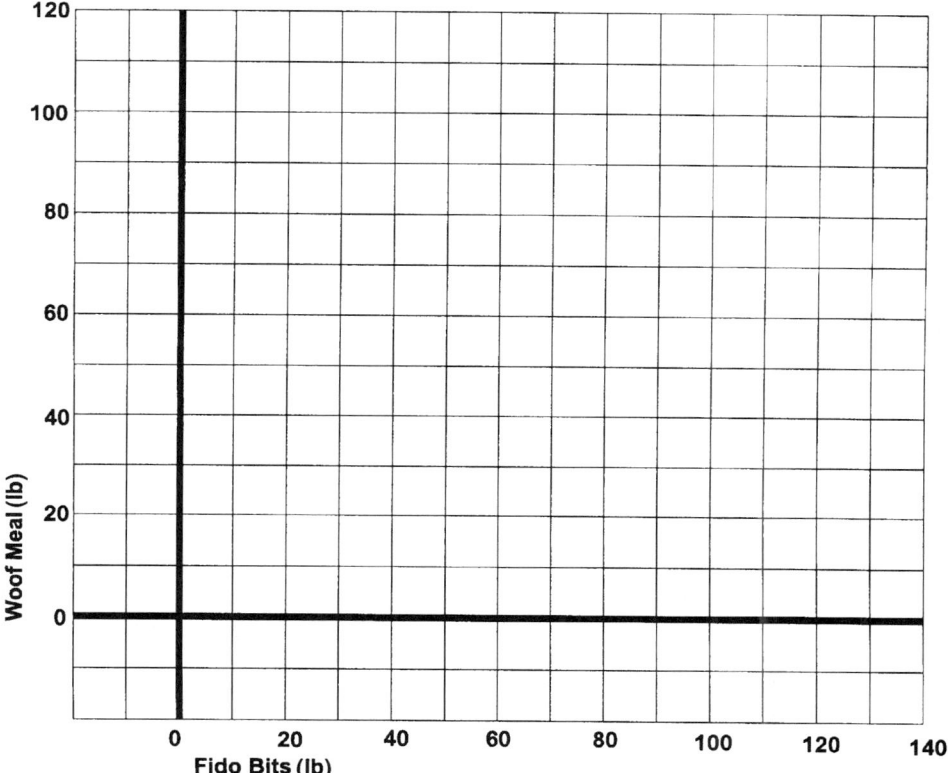

2. You are formulating a dessert that is to contain between 5 and 8% cocoa, at least 10% sugar, and between 4% and 7% raspberry essence. You have two ingredients available: C Powder containing 20% cocoa, 12% raspberry essence, and 22% sugar and R Powder containing 15% cocoa, 18% raspberry essence, and 66% sugar. A 10-lb bag of C powder sells for $2.00, and a 20-lb box R Powder sells for $9.00. The remaining ingredients have negligible cost.

   a. State the problem in formal mathematical form (see the 6th numbered entry in Sec. 1.2.1 and the 2nd numbered entry in Sec. 1.2.3).

   b. Using the graphical method, show the feasible region for a 100 lb batch. How many extreme points does this region have (see the 9th numbered entry in Sec.

1.2.2)? Can a dessert that meets all specifications be made without C Powder? Without R Powder?

c. Graphically determine the least-cost formulation (see items 10 through 12 in Sec. 1.2.2). Report the weights of each ingredient, the total cost, and the actual cocoa, raspberry essence, and sugar levels in the product. (Use a different color for the objective function so that it is easy to see.)

d. You decide to try a formulation high in vitamin C. C Powder contains 50 units of vitamin C per lb and R Powder 60 units per lb. Graphically determine a formulation that meets all constraints but maximizes vitamin C content instead of minimizing cost. Report the weights of each ingredient, the vitamin C content, the total cost, and the actual cocoa, raspberry essence, and sugar levels.

e. What formulation would maximize vitamin C if R Powder contained 100 units per lb.

f. Solve problems c, d, and e with a spreadsheet, making sure that it agrees with the graphical solution. Turn in a printout of each solution. (*Hint:* The spreadsheet will need to include two objective functions although only one will be used at a time.)

Use the graph below to find the solution.

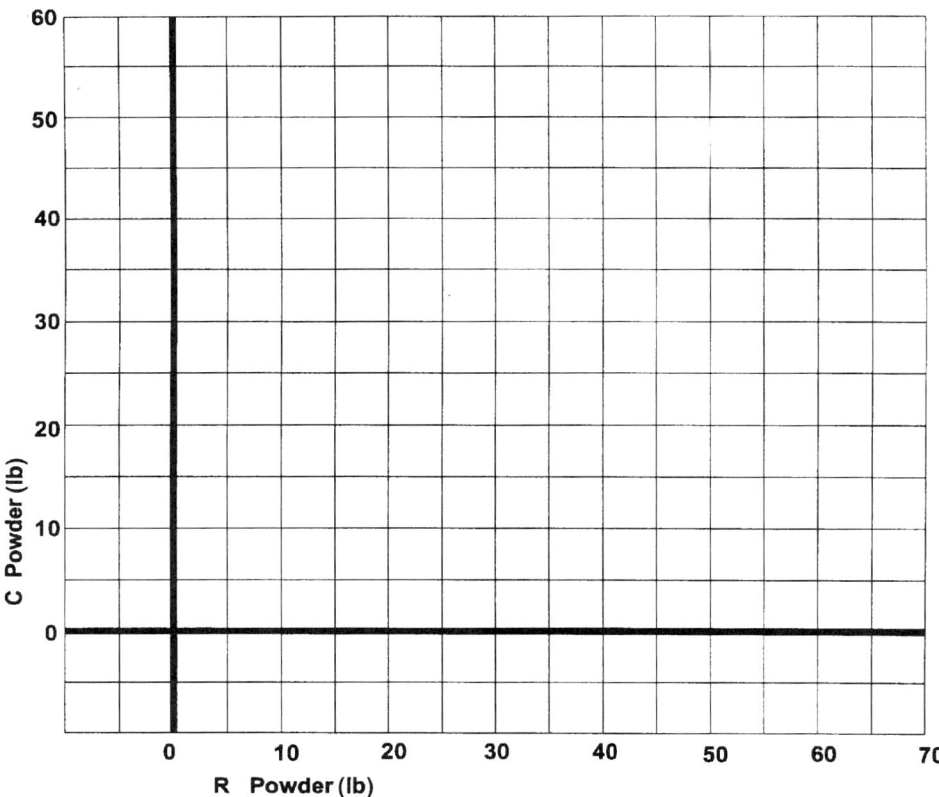

3. You are entrusted with the responsibility of formulating a mayonnaise. You have a number of ingredients available with the cost shown against them. In this exercise, you can practice formulating the product with various options:

Ingredients	Costs
• Oil (100% oil)	$0.58/lb
• Salted egg yolk (10% salt, 25% oil, 50% moisture)	0.93
• Salt	0.03
• Vinegar (10% acetic acid, 90% moisture)	0.26
• Mustard flour	0.71
• Water (100% moisture)	0.00

Initial Constraints
- Oil can vary from 70 to 80%.
- Salt must be less than 0.8%.
- Yolk can vary between 7 to 8%.
- Acid can vary between 0.2 to 0.5%.
- Moisture must be greater than 12%.
- Mustard can vary between 0.25 to 1.00%.
- Moisture content cannot exceed 50 times the acid content.
- The oil content cannot exceed 12 times the yolk content.
- Total weight of a batch is 100 lb.

a. Set up a spreadsheet and find a least-cost formulation. Print the spreadsheet.

b. After reviewing the formula, you decide to go for lower fat content and change the oil limits to 65 and 80%. In a sentence or two, describe the changes in the formulation.

c. Fat content can also be reduced by lowering the yolk content. If you change the yolk limits to 6.5 and 8% (retaining the change made in b), how does the formulation differ?

d. After reviewing this last formulation, you decide to limit moisture content to between 12 and 18% (replacing constraint 5). (Retain the changes made in b and c.) How does the formulation differ? Print the spreadsheet.

2

MATERIAL TESTING AND RHEOLOGY OF SOLID FOODS

2.1 INTRODUCTION

The complexity of modern food manufacturing processes and emphasis on quality require increased understanding of the role of material properties of solid and solidlike foods. Mechanical properties are generally defined as the stress-strain behavior of a material under static and dynamic loading, whereas rheology has been defined as a science devoted to the study of deformation and flow. In particular, the polymeric nature of starches and proteins suggests that the material properties should be determined in the context of basic polymer science principles. Generally, material testing procedures consist of small deformation nondestructive tests and larger strain destructive tests. The former are very useful for the characterization of various network structures common in many foods such as cheese. The latter are useful for determining the extensibility and ultimate strength of these structures. A combination of the two types of testing can be useful in understanding relationships between the micro (macro) structure and complex food properties such as texture.

2.2 INSTRUMENTS USED IN MATERIAL TESTING

2.2.1 Force Measurements

There are a number of force-measuring instruments available, such as Instron (the universal testing machine). This chapter refers to the TA-XT2 (Texture Technologies, Scarsdale, NY), a device that tests the strength and textural properties of food materials. It does this by deforming materials in various ways and measuring the force required to achieve that deformation. For example, fibers are usually tested by measuring the force needed to stretch and break them. Foods like cheese are usually tested by compressing or forcing them through a tiny orifice. The resulting force versus deformation data are a function of both the material properties and dimensions of the particular piece of material tested. To characterize the material independent of its dimensions, force-time data must be converted to stress-strain data. Other instruments are also available, but the TA-XT2 is widely used in the food industry.

2.2.1.1 Description The TA-XT2 is shown in Figure 2.1. It is linked to a computer, video monitor, and a control keyboard. The system is programmed through a Windows-based software. The sample is placed on the sample platform. The probe carrier can then be either lowered to compress the sample or raised to stretch the sample (for this, the sample must be attached at both ends). The texture analyzer can be programmed to hold a constant stress or strain, or a constant cross-head speed can be specified. Some of the possible tests that can be performed on the TA-XT2 are

1. Stress relaxation test
2. Penetration test
3. Creep test
4. 3-point bend test
5. Tensile test
6. Extensibility test

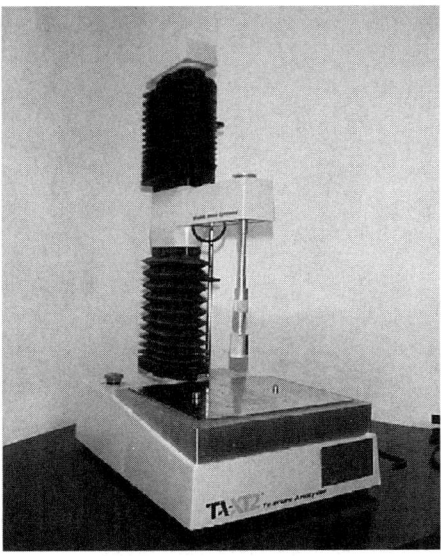

Figure 2.1 View of the TA-XT2 texture analyzer.

Measured torque

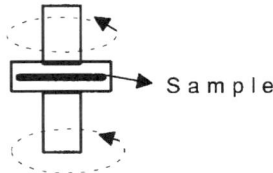

Applied torque

Figure 2.2 The operating principle of dynamic mechanical analysis.

2.2.2 Dynamic Mechanical Analysis (DMA)

Unlike the TA-XT2, a dynamic oscillatory rheometer is used mainly for small strain nondestructive tests either in oscillation mode, strain sweep, or stress relaxation mode. The sample is first placed onto the platform and the top plate is lowered to contact the sample (Figure 2.2). The bottom of the plate oscillates at a specified frequency and applies the torque to the sample. A transducer measures the torque transmitted through the sample. The results are given in terms of complex modulus (G^*), storage modulus (G'), loss modulus (G''), complex viscosity, and tan δ.

Tests that can be performed on the DMA include the following:

1. *Cure test.* Holding sample at constant temperature and frequency.
2. *Temperature sweep.* Increasing temperature at a fixed frequency.
3. *Strain sweep.* Increasing the strain amplitude at a fixed frequency.
4. *Frequency sweep.* Increasing the frequency of oscillation at a fixed strain.
5. *Stress relaxation.* Relaxation of stress at a constant strain.

Figure 2.3 View of an oscillatory rheometer.

One advantage of using the DMA is the ability to control the temperature over a wide range. Dynamic mechanical thermal analysis (DMTA) generally involves varying either temperature, strain amplitude, or frequency while the other two variables are held constant. Of course, for foods containing substantial water content the upper temperature limit is somewhat lower than 100°C. This chapter refers to a Bohlin (Bohlin Rheologi, Cranbury, NJ) VOR-M rheometer (Figure 2.3), but other instruments are also available.

2.3 BACKGROUND

In order to design and interpret material tests and their results, it is necessary to understand the basic concepts of stress and strain.

2.3.1 Stress

1. *Force.* The TA-XT2 measures the force needed to deform an object and records in grams, kilograms, pounds force (lb_f), or Newton (1 kg_f = 9.807 N). The force depends on the nature of the material, but since it also depends on the dimensions of the test sample, it is not a property of the material alone.

2. *Stress.* The force applied to an object is distributed throughout the entire object. If, at any point within the object, we draw a plane at right angles to this internal force, we can define the stress at that point as the magnitude of the force per unit cross-sectional area:

$$\text{Stress} = (\sigma) = \frac{F}{A} = \frac{\text{Applied force}}{\text{Cross-sectional area}} \qquad (2.1)$$

Clearly, the same force applied over a smaller cross-sectional area results in increased stress. Therefore, stress is defined as the intensity of force or a normalized force so to speak.

Example *If a rectangular bar that is 2 cm high, 1 cm thick, and 4 cm long is squeezed at the ends by a 4-N force, the stress at any point in the bar is*

$$\sigma = \frac{4 \text{ N}}{0.02 \text{ m} \times 0.01 \text{ m}} = 20,000 \frac{N}{m^2} = 20,000 \text{ Pa} = 20 \text{ kPa}$$

3. *Units of stress.* Stress is defined as the force per unit area, analogous to pressure. In fact, hydrostatic pressure is simply one form of stress and has the same units as stress.

4. *Compressive stress.* When an object is placed between a pair of opposing forces that are pointing toward each other as shown in Figure 2.4, the effect is to compress the object. The resulting stress is called the compressive stress. You apply a compressive stress when you squeeze a ball of dough between your hands.

5. *Tensile stress.* When an object is held by a pair of opposing forces that are pulling away from each other as shown in Figure 2.5, the effect is to stretch the object. The resulting stress is called the tensile stress. You apply a tensile stress when you stretch a rubber band.

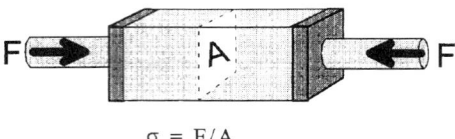

$$\sigma = F/A$$

Figure 2.4   Principle of compressive stress.

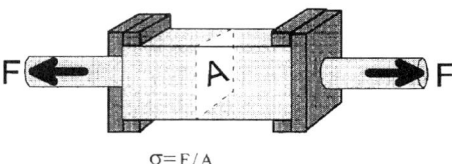

$$\sigma = F/A$$

Figure 2.5   Principle of tensile stress.

6. *Axial stress.* In both compressive and tensile stresses, the pair of applied forces exist along a common axis. These two stresses are, therefore, classified as axial stresses.

7. *Normal stress.* In computing either compressive or tensile stress on an object, you divide the applied force by the cross-sectional area of the object that is perpendicular to the axis of the force. Since this area is normal to the force, the stress is called the normal stress.

8. *Shear stress.* When a pair of forces are parallel but do not occur along a common axis, the effect is to skew the object. For example, if the top of a rectangular object is pulled to the right while the bottom is pulled to the left as shown in Figure 2.6, the object will become a parallelogram. This type of stress is called shear stress:

$$\text{Shear stress} = (\tau) = \frac{F}{A} \tag{2.2}$$

where τ usually denotes a shear stress.

9. *Tangential stress.* In computing shear stress, the magnitude of the forces is divided by the cross-sectional area of the object that is parallel to the forces. Since the forces are tangential, rather than perpendicular to the area, this is called a tangential stress.

10. *Isotropic stress.* A stress that comes equally from all directions, as with hydrostatic pressure, is called an isotropic stress. Isotropic stress is illustrated in Figure 2.7. It is identical to the pressure on the surface of the object:

$$\text{Isotropic stress} = (P) = \frac{F}{A} = \text{Hydrostatic pressure} \tag{2.3}$$

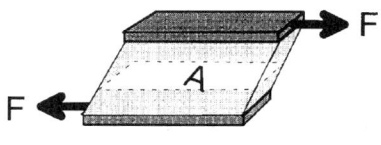

$$\tau = F/A$$

Figure 2.6   Principle of shear stress.

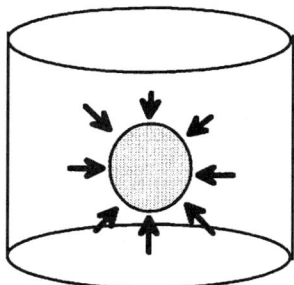

Figure 2.7 Principle of isotropic stress.

2.3.2 Strain

1. *Deformation.* When an object is stressed, one or more of its dimensions (say, L) usually changes. The magnitude of this dimensional change (ΔL) is referred to as deformation.

2. *Strain.* Under the same stress, a long object is expected to change more than a short object. Thus, deformation is a function of both the nature of the material and its dimensions. In order to have a property that is dependent only on the material, we define strain as the magnitude of the change divided by the initial dimension:

$$\text{Strain} = (\varepsilon) = \frac{\Delta L}{L_0} = \frac{\text{Change in length}}{\text{Original length}} \tag{2.4}$$

Both compressive and tensile strains are illustrated in Figure 2.8. This is usually referred to as engineering strain. True strain is given by

$$\varepsilon = \ln\left(\frac{L}{L_0}\right) \tag{2.5}$$

where $L =$ stressed length after elongation or compression.

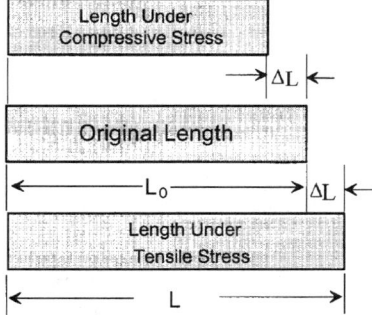

Figure 2.8 Principle of compressive or tensile strain.

Example   *A rubber band 3.00 in long is stretched to 4.50 in. The strain on this rubber band is*

$$\varepsilon = \frac{4.50 - 3.00}{3.00} = \frac{1.50}{3.00} = 0.50 \quad \text{(engineering strain)}$$

$$\varepsilon = \ln\left(\frac{4.5}{3.0}\right) = 0.41 \quad \text{(true strain)}$$

Thus, the engineering strain is only an approximation of the true strain for large deformation, but it is nontheless widely used.

3. *Units of strain.* Since strain is always the ratio of two lengths, it is always dimensionless.
4. *Axial strain.* When an object is subjected to a compressive or tensile strain, it will decrease or increase in length along the axis of the stress. Such a change is called the axial strain.
5. *Lateral (transverse) strain.* When an object is stretched, it usually gets thinner. When it is compressed, it usually becomes thicker. Thus, for any axial strain, there is usually a compensating strain at right angles to the force. This is called a lateral or transverse strain. In Figure 2.9, lateral strain is defined as

$$\varepsilon_{\text{Lateral}} = \frac{\Delta W}{W} \tag{2.6}$$

6. *Poisson's ratio.* The ratio of lateral strain to axial strain measures the extent to which changes in length are accompanied by changes in nonaxial dimensions, that is, the diameter of a cylinder. It commonly varies from 0 (no bulging) to a maximum of 0.5 (a bulge equal to half the compression). Referring to Figure 2.9, we see that Poisson's ratio is computed as

$$\text{Poisson's ratio } (\mu) = \frac{\varepsilon_L}{\varepsilon_A} = \frac{\Delta D/D}{\Delta L/L} \tag{2.7}$$

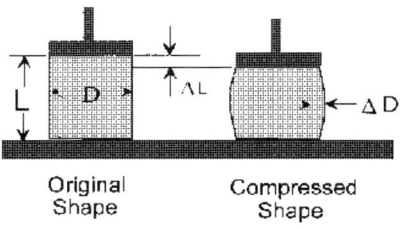

Original   Compressed
Shape   Shape

Figure 2.9   Principle of axial and lateral strain.

Example   *A cylinder of cheese 2.00 in long and 1.00 in in diameter is compressed along its axis to a length of 1.80 in. The diameter of the cylinder increases to 1.04 in:*

$$\text{Axial strain} = \frac{2.00 - 1.80}{2.00} = 0.10$$

$$\text{Lateral strain} = \frac{1.04 - 1.00}{1.00} = 0.04$$

$$\text{Poisson's ratio} = \frac{\text{Lateral strain}}{\text{Axial strain}} = \frac{0.04}{0.10} = 0.4$$

Poisson's ratio varies from 0 (no lateral contraction) to 0.5 for no volume change during deformation. Values near 0.5 are typical of elastomers (e.g., rubber), whereas values near 0 can be found for flexible foams, and intermediate values are associated with plastics and metals.

7. *Volumetric strain.* When an object is stressed in any way, such as under hydrostatic pressure, its volume changes. This is called volumetric strain and is computed as the change in volume divided by the initial volume at gauge or absolute pressure:

$$\text{Volumetric strain} = (\varepsilon_{\text{Vol}}) = \frac{\Delta V}{V_0} = \frac{\text{Change in volume}}{\text{Initial volume}} \tag{2.8}$$

Example   *A hard roll is found to occupy 4.00 in³. It is then placed in a tank of water and the pressrue is raised to 2 atm. Its volume changes to 3.92 in³. The volumetric strain at this pressure is*

$$\varepsilon_{\text{Vol}} = \frac{4.00 - 3.92}{4.00} = \frac{0.08}{4.00} = 0.02$$

8. *Shear strain.* Shear strain, as shown in Figure 2.10, is a distortion that takes place when the opposing forces are not in line with each other. Thus, the shear strain equals the tangent of the angle of deformation caused by the shear stress:

$$\text{Shear strain} = (\gamma) = \frac{\Delta L}{L} = \tan(\theta) \tag{2.9}$$

where γ usually denotes a shear strain.

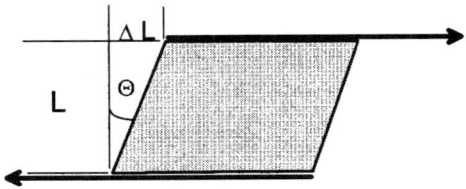

Figure 2.10   Principle of shear strain.

Example   *A piece of gelatin 1.5 in thick is held between two horizontal plates. The top plate is moved 0.1 in to the right, skewing the gelatin. The shear strain is*

$$\gamma = \frac{0.10}{1.5} = 0.067$$

and has units of radians.

The angle of deformation in this case is $\theta = \tan^{-1}(0.067) = 0.68°$, indicating that for small strains, $\gamma \sim \theta$.

2.3.3   Relating Stress to Strain

1. *Elastic material.* When a stress is applied to an object, it will usually deform. When the stress is removed, it may or may not return to its original dimensions. A material that returns to its original dimensions is said to be ideal elastic.

2. *Hooke's law.* Hooke's law states that the strain exhibited by an object is directly proportional to the applied stress, that is,

$$\sigma = E\varepsilon \tag{2.10}$$

where E is the proportionality constant. A material that obeys this law is said to be a "Hookean solid." In fact, most materials are Hookean in the limit of small strain.

3. *Young's modulus.* The constant E in Hooke's law is called Young's modulus:

$$\text{Young's modulus} = (E) = \frac{\sigma}{\varepsilon} = \frac{\text{Stress}}{\text{Strain}} \tag{2.11}$$

Young's modulus is a measure of a material's stiffness or resistance to deformation. A plot of stress versus strain for a Hookean material is shown in Figure 2.11. As the applied stress is increased, the strain increases in direct proportion. Young's modulus is the slope of this line. For Hookean materials, this modulus depends only on the material and not on its dimensions or the magnitude of the applied stress. It is, therefore, a useful measure of a material property. For example, a large stress applied to a breadstick results in only small deformation with a relatively large modulus,

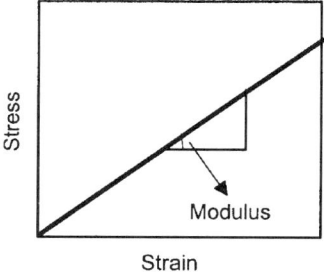

Figure 2.11   Principle of Hooke's law, that is, a linear response between stress and strain.

whereas an Angel Food cake that exhibits a smaller stress for the same deformation will have a smaller modulus. Of course, this concept of the modulus only applies in the linear region of a stress-strain curve.

4. *Units of moduli.* Since the denominator of a modulus is always dimensionless, the units of any modulus are the same as the units of stress, namely, force per unit area, for example, Pa[=] N/m^2.

5. *Bulk modulus.* We can similarly define the bulk modulus for volumetric changes:

$$\text{Bulk modulus} = (K) = \frac{P}{\varepsilon_{\text{Vol}}} = \frac{\text{Hydrostatic pressure}}{\text{Volumetric strain}} \qquad (2.12)$$

6. *Shear modulus.* The shear modulus is the ratio of shear stress to shear strain:

$$\text{Shear modulus} = (G) = \frac{\tau}{\gamma} = \frac{\text{Shear stress}}{\text{Shear strain}} \qquad (2.13)$$

3.4 Viscoelastic Properties

Some semisolid foods such as dough, gels, or cheese have both viscouslike and solidlike behavior. These materials are generally known as viscoelastic materials. In a dynamic mechanical test, a specimen is deformed by a strain that varies sinusoidally with time, resulting in a sinusoidally varying stress as shown in Figure 2.12. Dynamic testing allows for separation of stresses due to sinusoidally varying strain into its viscous and elastic components. The following analysis is based on Rosen (1993).

1. *Sinusoidal strain.* Sinusoidal strain is defined as

$$\gamma = \gamma' \sin \omega t \qquad (2.14)$$

where γ' = peak strain of a sinusoidally varying strain, ω = angular frequency (radian/s), t = time(s).

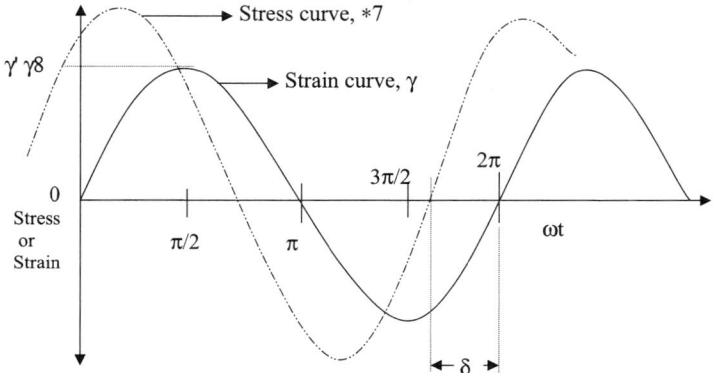

Figure 2.12 Schematic view of sinusoidally varying strain in a typical dynamic test.

2. *In-phase stress.* This component of the total stress ($\tau*$) is in phase with the strain and defines the storage modulus G':

$$\tau' = G' \cdot \gamma' \tag{2.15}$$

where τ' = in-phase stress and G' = storage modulus.

3. *Out-of-phase stress.* This component of the total stress ($\tau*$) is out of phase with the strain and defines the loss modulus G'':

$$\tau'' = G'' \cdot \gamma' \tag{2.16}$$

where τ'' = out-of-phase stress and G'' = loss modulus. Note that the phase angle (δ) will be zero for an ideal elastic material and 90° for a purely viscous material, whereas for a viscoelastic material phase angle varies between 0 and 90°.

4. *Complex modulus.* The complex modulus is the vector sum of the in-phase and out-of-phase moduli as shown by Eq. (2.17):

$$G* = \frac{\tau*}{\gamma*} = (G' + iG'') \text{ and the magnitude of } G*(|G*|) = (G'^2 + G''^2)^{1/2} \tag{2.17}$$

5. *Loss tangent (tan delta).* This is the ratio of the loss modulus to the storage modulus as shown by Eq. (2.18) and represents the ratio of the stress dissipated as heat to that stress stored elastically.

$$\tan \delta = \frac{G''}{G'} = \frac{\tau''/\gamma'}{\tau'/\gamma'} = \frac{\tau''}{\tau'} \tag{2.18}$$

6. *Work.* Work done during the first quarter cycle in a sinusoidally varying strain is given by Eq. (2.19):

$$W = \frac{(\gamma')^2}{2} G' + \frac{\pi}{4}(\gamma')^2 \, G'' \tag{2.19}$$

The first term above represents elastic or recoverable work, and the second term work dissipated or "lost" as heat. Work done in the second quarter cycle is the same except the sign of the stored energy term is negative, indicating that elastic energy is returned. However, there is always a dissipative loss. The total energy loss that is converted into heat within the material for a full cycle is given in Eq. (2.20) (Rosen, 1993):

$$W = 4.\left\{\frac{\pi}{4}(\gamma')^2 \, G''\right\} = \pi(\gamma')^2 G'' \tag{2.20}$$

An example of cyclic deformation, low G'' would be preferred in automobile tires, where low heat buildup is desired. On the other hand, in mixing bread dough, a combination of suitable G' and G'' values is needed to allow for some viscous flow, that is, mixing. Doughs are allowed to "relax" after mixing to dissipate the stored elastic energy of mixing.

2.4 TESTING METHODS

Thus, the material properties of a solid material are characterized by the above four constants: E, G, μ, and K. These four constants are related by

$$E = 3K(1 - 2\mu) = 2(1 + \mu)G \qquad (2.21)$$

This means that only two properties need to be determined experimentally. For elastomer ($\mu = 0.5$), the above equation reduces to the well-known equation:

$$E = 3G \qquad (2.22)$$

The above relationship can be used to convert stress-strain data obtained in shear or tensile mode and vice versa. These type of material constants are increasingly being used to characterize food materials. Some of the testing procedures used to determine them will be the focus of this chapter.

2.4.1 Compression Tests

In simple compression tests, a common objective is to determine the Young's modulus of the material. This is done as follows using cheese as an example:

1. Turn on the TA-XT2 and computer or another force-measuring instrument.
2. Enter the texture analyzer program.
3. Enter the cross-head speed (say, 2 mm/s), sample surface area (say, 506 mm²), compression time (60 s), compression distance (10 mm).
4. Cut a cheese sample 20 mm high, 25 mm in diameter using a cork borer.
5. Place the sample under the probe and start your test:
   a. This example refers to the test result shown in Figure 2.13. A cheese sample 20 mm high with a 25 mm diameter was compressed (50% strain) using a TA-XT2 as illustrated in Figure 2.14.
   b. At point B, the probe begins to compress the cheese sample. The force read by the instrument is shown on the Y axis.

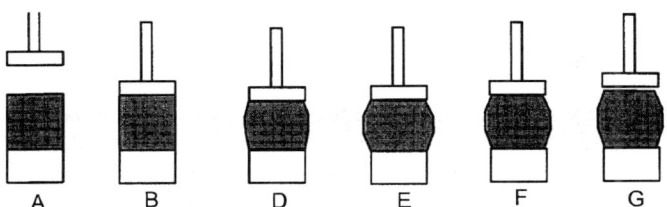

Figure 2.13 Schematic view of various stages in a compression test on a cheese sample.

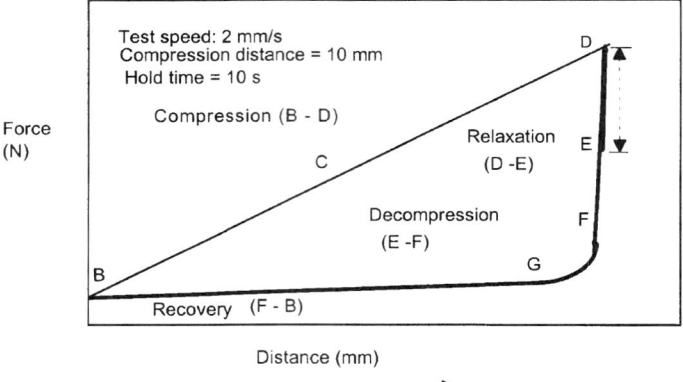

Figure 2.14 Typical force versus distance curve obtained for a cheese sample during a compression/relaxation/decompression sequence using the TA-XT2.

   c. At point D the sample has been compressed 10 mm and is then held for 10 s. The cheese is partly compressed and remains stationary until point E.

   d. With no further movement of the probe, the force no longer rises. However, between points D and E, the cheese is still experiencing a compressive strain, but the force decreases rather than remaining constant. This gradual reduction in force is called stress relaxation and is typical of viscoelastic materials.

   e. At stage E, the probe begins to rise and the force drops quickly.

   f. At point F, the probe is no longer in contact with the sample.

   g. At point G, the probe pulls away from the cheese and the force becomes 0 again.

The peak force at point D was 35 Newtons. The surface area $= 0.000506 \, \mathrm{m}^2$. Therefore, stress $= 35/0.000506 = 69{,}170 \, \mathrm{Pa}$ or $69.17 \, \mathrm{kPa}$. The force at point B is $0 \, \mathrm{N}$ and its stress is $0 \, \mathrm{kPa}$. Therefore, change in stress $\Delta\sigma = (69.17 - 0) = 69.17 \, \mathrm{kPa}$. In addition, the change in strain between points D and B is $\Delta\varepsilon = 0.5 - 0 = 0.5$.

$$\text{Elastic modulus } E = \left(\frac{\Delta\sigma}{\Delta\varepsilon}\right) = 138.34 \ \mathrm{kPa}$$

Note that the stress-strain curve is linear (or nearly so) in this region. Between points D and E, the modulus is a function of time unlike an ideal elastic material where E would be constant at constant strain.

2.4.2 Energy Considerations

1. A typical force versus distance curve for a compression test of some product such as cheese is shown in Figure 2.14. If the horizontal coordinate of this chart is converted to cross-head movement with time, we obtain a plot as shown in Figure 2.15. In Figure 2.14, the area under the curve from $B{-}D$ represents the integration of force and distance and, hence, the work consumed in compressing the cheese.

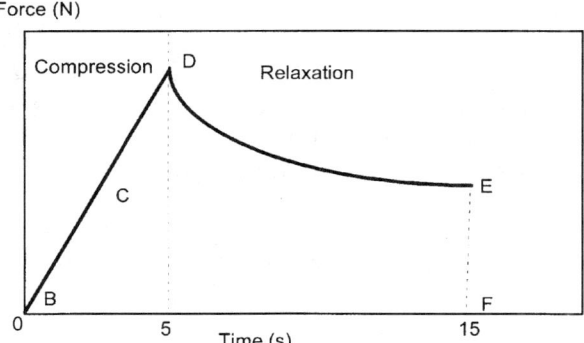

Figure 2.15 Typical force versus time curve obtained for a cheese sample during compression testing on a TA-XT2.

2. A curve representing the loading and unloading of the force versus cross-head movement is shown in Figure 2.16. The work of compression is given by the area *BCDEXFB* in Figure 2.16.

3. While the cross-head is stationary (10 s at *D*), internal rearrangements (at constant strain) at the molecular level within the cheese allow partial relaxation of the force exerted by the cheese against the cross-head and dissipate some of the stored energy. Since there is no deformation or movement during relaxation, none of this energy appears as work and so it must be lost as heat. This is shown as the vertical line (*DE*) in Figure 2.16.

4. When the cross-head moves up from *E* to *F* as shown earlier in Figure 2.13, it is pushed by the Cheese as it undergoes partial elastic recovery, so the cheese is doing work on the cross-head. In the process, the remainder of the stored energy is recovered. This energy is represented by the light shaded area *EXFE* in Figure 2.16.

5. The lost energy is the difference between the energy stored and the energy recovered and is represented by the dark shaded area *BCDEFB* between the compression and recovery curves. Although we stated above that this loss takes place just between *D* and *E*, in reality, it takes place throughout the entire process, including the initial compression.

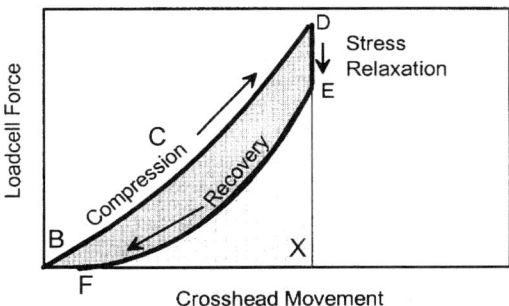

Figure 2.16 Typical force versus cross-head movement curve obtained for a typical cheese sample during compression testing on TA-XT2.

In an ideal elastic material such as rubber, there would be no viscous dissipation loss and the top portion of the curve in Figure 2.16 (*D–E*) would be horizontal. In addition, points *B* and *F* would exist at the same place. In a purely viscous material such as oil or water, there would be no elastic recovery portion of the curve. Materials that exhibit the curves shown in Figures 2.14, 2.15 and 2.16 are termed viscoelastic since they exhibit both viscous and elastic properties, the relative degree to which one predominates over the other depending on the timeframe of the experiment.

2.4.3 Tensile Test

In a tensile test, material is stretched rather than compressed. Such tests are a useful means of characterizing the material properties of fibrous or elastomeric materials. This type of test simulates the way you might sometimes pull a bread product, licorice, Mozzarella cheese, or jerk with the teeth rather than chewing it.

In this test, the test material is clamped between two jaws and these jaws are moved apart. Alternatively, the one end is fixed and the other end is pulled. The cross-head continues to move throughout the test. A typical curve that might be obtained with a tensile test is shown in Figure 2.17.

1. *A–B* represents the Hookean proportional limit.
2. *C* is the yield point and represents the yield stress. Although the cross-head continues to move and stretch the material, the stress may decrease as a neck is formed. This part of the curve will vary greatly from material to material. For brittle materials, fracture may occur at *C* and the force may drop down to 0. For some softer materials, plastic flow may continue for some time (*C–D*) and this portion of the curve may be quite long.
3. *E* is the ultimate strength of the material and it ruptures somewhere beyond point *E*. Several material properties can be determined from this curve:
    - *Young's modulus.* It measures the resistance of the material being stretched and is represented by the slope between *A* and *B*. This is referred to as the tensile

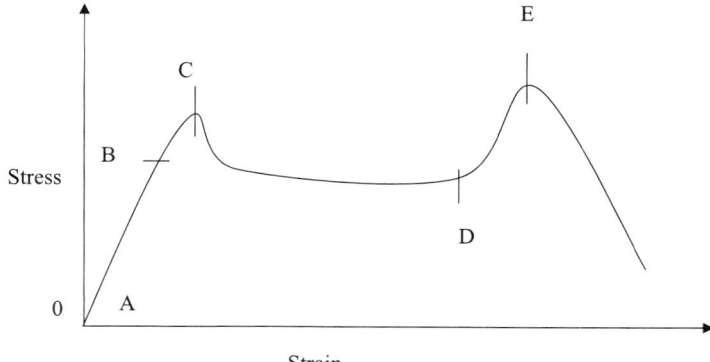

Figure 2.17 Schematic engineering stress-strain curve showing several critical points (not all points will be seen with all products).

modulus. A weak rubberband would have a low modulus, whereas a heavy string would have a high modulus. As with the compression test, this modulus is the ratio of stress to strain:

$$E = \frac{\Delta\sigma}{\Delta\varepsilon} \tag{2.23}$$

- *Critical strain.* A measure of the amount of deformation before internal yield either by fracture or necking. Elongation is expressed as a percent of the original length:

$$(\gamma_c)\% = 100\left(\frac{\Delta L}{L_0}\right) = 100\left\{\frac{L - L_0}{L_0}\right\} \tag{2.24}$$

where L_0 = the initial length of the material (at A) and L = the length of the material at yield (at C).
- *Yield stress.* The stress at γ_c.
- *Ultimate strength.* Stress at point E.
- *Tougness or modulus of toughness.* The area under a force-deformation curve (say, up to point C or E) represents the work (N · m or J) up to either yield or ultimate strength, respectively. The area under the corresponding stress-strain curve would have units of Pa, that is, a modulus of toughness. This is equivalent to dividing the toughness by the sample volume.

In order to perform a tensile test, the test material must be clamped between jaws. In many cases, the pressure exerted by the jaws weakens the material and causes premature failure. This can be prevented by notching the material to a "dogbone" shape as shown in Figure 2.18. The clamps are applied to the wide end sections and the narrower center section becomes a weak point, which promotes failure away from the jaws.

Example *A piece of bread is cut into a dogbone and stretched. The dimensions of the piece are as follows: Width at the center = 1 cm (0.01 m), thickness at the center = 0.2 cm (0.002 m). Initially, the clamps' separation = 8.6 cm (0.086 m). Cross-head speed = 0.2 mm/min. Determine the modulus ΔL = 50 cm (0.05 m).*

Solution

$$\Delta\varepsilon = \frac{0.05 \text{ m}}{0.086 \text{ m}} = 0.58 \quad (58\% \text{ change in length})$$

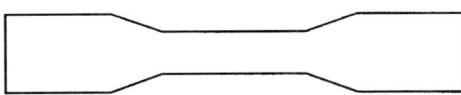

Figure 2.18 A typical notched test material cut into a "dogbone" shape for tensile testing.

The change in force over the measured interval is, say, 2 N. Therefore, the change in stress over this interval is:

$$\Delta\sigma = \frac{\Delta F}{\text{cross-sectional area}} = \frac{(2.0 \text{ N})}{(0.01 \text{ m})(0.002 \text{ m})} = 100 \times 10^3 \frac{\text{N}}{\text{m}^2} \quad (100 \text{ kPa})$$

The modulus is, therefore,

$$E = \frac{\Delta\sigma}{\Delta\varepsilon} = \frac{100 \text{ kPa}}{0.58} = 172.4 \text{ kPa} \quad \text{(referred to as the initial cross-sectional area)}$$

2.4.4 Flex Test

In a flex test, or 3-point bending test as it is sometimes called, a rectangular wafer of material such as a cracker is suspended across two parallel cylindrical rods as shown in Figure 2.19. A third rod, parallel to the first two, is fastened to the TA-XT2 machine and lowered onto the wafer midway between the supports Thus, the sample is flexed.

1. *Test dimensions*

   w = width of the wafer in the direction parallel to the support rods

   t = thickness of the wafer

   L = length of the span between the centers of the parallel rods

   D = deflection of the center of the wafer from its original position at any given moment in the test

   F = force being exerted to achieve the deflection D

2. *Flexural strain.* If a wafer is deflected D units from horizontal, the bottom surface is stretched as shown in Figure 2.20. The maximum strain in the outer fibers is computed by

$$\text{Maximum flexural strain} = (\varepsilon_f) = \frac{6Dt}{L^2} \tag{2.25}$$

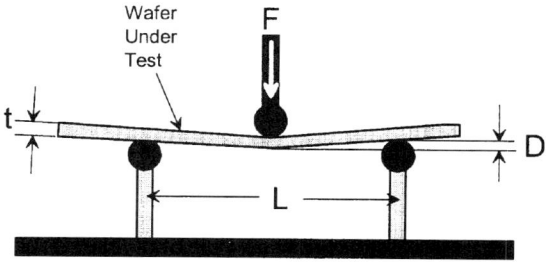

Figure 2.19   A typical flex test on an Instron or TA-XT2 machine.

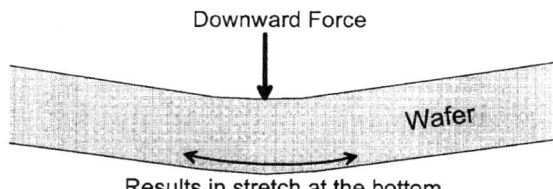

Downward Force

Wafer

Results in stretch at the bottom
with maximum stress and strain in the
outermost fibers occurring at midspan

Figure 2.20 A stretch at the bottom of a sample during a flex test.

Notice that this strain occurs in the horizontal direction, at right angles to the vertical deformation of the chip.

Example *If a chip 0.05 in. thick is suspended between rods 3.5 in. apart and is deformed 0.1 in. from the horizontal, the bottom surface of the chip will experience a strain of*

$$\text{Flexural strain} = \varepsilon_f = \frac{6(0.1 \text{ in.})\,(0.05 \text{ in.})}{(3.5 \text{ in.})^2} = 0.00245$$

In other words, the bottom surface at the bend will have increased in length by 0.245%.

3. *Flexural stress.* The maximum stress in the outside fibers at the midspan on this bottom surface is computed by

$$\text{Flexural stress} = (\sigma_f) = \frac{3FL}{2wt^2} \tag{2.26}$$

As with the flex strain, this stress is horizontal, at right angles to the applied force.

Example *If in the previous example, the chip was 1.5 in. wide and the deformation was brought about by a downward force of 0.05 lb, then the stress that is stretching the bottom surface has the following magnitude:*

$$\sigma_f = \frac{3(0.05 \text{ lb}_f)\,(3.5 \text{ in.})}{2(1.5 \text{ in.})\,(0.05 \text{ in.})^2} = 70.0\,\frac{\text{lb}_f}{\text{in.}^2} \quad \text{(psi)}$$

$$\sigma_f = \left(70.0\,\frac{\text{lb}_f}{\text{in.}^2}\right)\left(\frac{4.4482 \text{ N}}{1 \text{ lb}_f}\right)\left(\frac{39.37 \text{ in.}}{1 \text{ m}}\right)^2 = 4.83 \times 10^5\,\frac{\text{N}}{\text{m}^2} = 483 \text{ kPa}$$

4. *Flexural modulus.* It is the ratio of the change in maximum flex stress to the change in maximum flex strain:

$$\text{Flexural modulus} = (E_f) = \frac{\sigma_f}{\varepsilon_f} \tag{2.27}$$

Example In the previous example, a stress of 483 kPa produced a strain of 0.00245. The flex modulus of this wafer is

$$E_f = \frac{483 \text{ kPa}}{0.00245} = 1.97 \times 10^5 \text{ kPa} = 197 \text{ MPa}$$

5. *Critical stress.* If a material is stressed until it yields or fractures, the critical stress is the stress computed at the point of fracture. It is a measure of the strength of the material when one side is stretched relative to the other. In a fibrous material, for example, this is a measure of the strength of the fibers.
6. *Critical strain.* The strain computed at the point of fracture.
7. *Flex test.* A typical force distance curve on a flex test is shown in Figure 2.21.

   a. At *A*, the cross-head starts down.

   b. At *B*, the rod attached to the cross-head meets the test piece and begins to bend it.

   c. From *C* to *D*, there is a linear relationship between stress and strain. We will estimate the flex modulus from this part of the curve.

   d. At *E*, the test piece fractures but does not come apart. We measure maximum stress and maximum strain at this point.

   e. From *E* to *F*, the test piece continues to bend, but because of the fracture, gives much less resistance and the stress falls.

   f. At *F*, the piece falls apart and the stress drops to 0.

Example A tortilla chip (specially shaped for this test) is subjected to a flex test. Compute the flex modulus if the parameters of the test and the dimension of the sample are given as follows: width $w = 27$ mm, thickness $t = 7$ mm, and Span $L = 75$ mm. The distance traveled between C and D is

$$\Delta D = 2 \text{ mm} - 0.4 = 1.6 \text{ mm}$$

The change in strain between C and D is

$$\Delta \varepsilon_f = \frac{6 \Delta D t}{L^2} = \frac{6(1.6 \text{ mm})(7 \text{ mm})}{(75 \text{ mm})^2} = 0.012$$

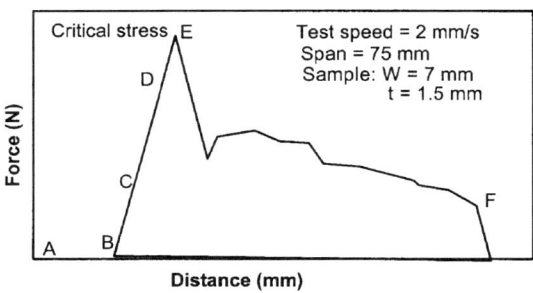

Figure 2.21 Typical force versus distance curve obtained during a flex test on a TA-XT2.

The change in force between C and D is

$$\Delta F = 70 \text{ N}$$

The change in stress between C and D is

$$\Delta \sigma_f = \frac{3\Delta FL}{2wt^2} = \frac{3(70 \text{ N}) (0.075 \text{ m})}{2(0.027 \text{ m})(0.007 \text{ m})^2} = 5,952 \text{ kPa}$$

The flex modulus is

$$E_f = \frac{\Delta \sigma_f}{\Delta \varepsilon_f} = \frac{(5.952 \text{ MPa})}{0.012} = 496 \text{ MPa}$$

The above represents an overview of typical material properties and testing procedures for solids. Be aware that "solids" originally meant load-bearing material (steel, copper, etc.). These tests were then adapted to polymers and plastics, and now are seeing their further adaptation to foods. Foods are not metals, but are referred to as soft solids. The major caveat to using these large strain tests is to be aware that the results will depend on the testing conditions, basically the cross-head speed. Of course, the material properties may also change with the temperature and composition of the sample. These tests are probably best used in a comparative mode, rather than an absolute mode. For example, the flexural modulus of a cracker probably should be interpreted in the context of moduli for other crackers of similar intended usage, or a target value known to represent a good product.

2.5 LAB EXERCISES

2.5.1 Objectives

1. To demonstrate testing for material properties on a typical force-measuring instrument.
2. To become familiar with the computations involved in compression, tensile, and flex tests.
3. To relate the material properties obtained with these instruments to the textural properties of some food materials.

2.5.2 Equipment

1. TA-XT2 or Instron-type instrument.
2. 1-inch cylindrical cutter to shape test pieces.
3. A micrometer to measure test pieces.

2.5.3 Materials

1. Approximately 1-in slabs of cheeses of three different hardnesses for compression testing: Whole milk Mozzarella, Part Skim Milk Mozzarella, and Provolone. Swiss or Cheddar can also be used for the hard cheese.

2. An appropriate material for a tensile test, for example, corn tortillas and wheat flour tortillas.

3. Rectangular tortilla chips, cracker, or a similar crisp material for flex tests.

2.5.3.1 Subjective Texture Measurement In order to interpret the measurement of stress, strain, and moduli, it is helpful to relate them to familiar materials. Measure these properties subjectively by squeezing various cheeses with your fingers. Since our fingers are not calibrated, we will simply rank the various cheeses, by assigning a rank of 1 to a cheese that is highest in a particular property, 2 to the cheese that is next highest, etc.

1. *Squeeze cheese.* Obtain a piece of each kind of cheese, apply a compressive stress by squeezing them between your fingers, and observe the strain (percent of compression).

2. *Judge relative stress.* To achieve about the same strain, which cheese requires the greatest stress (rank 1)? Which the next greatest (rank 2)?, etc.

3. *Judge relative strain.* For the same applied stress (finger pressure), which cheese shows the greatest strain (rank 1)? Which the next greatest (rank 2)?, etc.

4. *Judge relative moduli.* Based on your answers to 2 and 3, which cheese do you think has the greatest modulus (rank 1)? Which the next greatest (rank 2)? etc. Remember that the modulus is basically the firmness of the cheese. Report your data in Data Sheet 2.1.

2.5.3.2 Compression/Relaxation Test

1. Cut the cheese cylinder.
2. Trim it to the appropriate height.
3. Glue foil to the ends of the cheese cylinder.
4. Report test sample dimensions in Data Sheet 2.2.

2.5.4 Set Texture Analyzer or Instron Settings

1. Calibrate the instrument with appropriate standard load.
2. Set the safety knob at a safe level by moving down the probe near the platform. An approximate gap of 2 mm may be kept.
3. Attach the 1-in.-diameter cylindrical probe with the flat bottom onto the cross-head of the instrument.
4. Select the units of force, distance, and speed, etc.
5. Select the parameters as shown in Table 2.1.
6. Set deformation to about 25%, which should be within the elastic limit of the samples.
7. Run the test and save the raw data.

Table 2.1 Parameters used in compression/relaxation test

Parameters	Description	Values
Pre-test speed	Cross-head speed prior to trigger force	2 mm/s
Test speed	Cross-head speed during compression	1 mm/s
Post-test speed	Cross-head speed after test	5 mm/s
Distance	Compression distance (say, 25% strain)	2.5 mm
Relaxation time	Time for relaxation	180 s

2.5.5 Calculations

1. Separate the data into a compression test and relaxation test.

2. Graph stress versus strain for each compression test. How do the cheeses differ?

3. Locate the initial linear portion of the curve. Compute the elastic modulus (stress/strain) during the first stages of compression. Locate a linear portion of the curve immediately after the start of the test. Alternatively, you may arbitrarily draw a line connecting the origin to a point, say, 2% strain. This is called the secant method of determining the modulus.

4. Compute the compression modulus (E) of the material over this interval. Is this value a material property? How do the relative moduli of the three cheeses compare with your subjective judgement?

5. For each cheese, determine the degree of relaxation expressed as a percentage of the maximum stress. Which cheese shows the most relaxation? What does this tell you about the three cheeses?

6. Report the results in Data Sheet 2.2.

2.5.6 3-Point Bend Test

In this test, we will compare the fracture properties of different types of crackers such as saltines and graham crackers, etc.

1. Attach the 3-point bend rig to the TA-XT2 machine. Measure the gap between the parallel bars on the platform. Make sure that the gap between the two bottom bars is sufficient to support all samples and the top blade will lower in the middle of the bottom bars.

2. Calibrate the force and set the safety knob as done earlier in Section 2.5.4.

3. Select the parameters as shown in Table 2.2.

4. Measure carefully all the dimensions of each cracker.

5. Place the sample on the parallel bar support directly under the probe. Lower the probe to about 3 mm above the sample to make sure it is centered.

6. Run the test and save the raw data.

Table 2.2 Parameters for 3-point bend test

Parameters	Description	Values
Pre-test speed	Cross-head speed prior to trigger force	2 mm/s
Test speed	Cross-head speed during compression	1 mm/s
Post-test speed	Cross-head speed after test	5 mm/s
Distance	Vertical travel	4.0 mm

2.5.6.1 Calculations

1. *Fracture point.* Compute the critical flex stress and critical flex strain at fracture.
2. *Initial modulus.* Pick two points on the initial straight portion of the upward slope and use these points to compute the flex modulus.
3. Determine the modulus of toughness, that is, the work consumed per unit volume up to the critical strain.
4. Compare a force versus distance and stress versus strain graph. How and why do they differ?
5. Compare the different samples, and report the results in Data Sheet 2.3.

2.5.7 Extensibility Test

In this test, we will compare the difference in extensibility between corn tortillas and wheat flour tortillas. Fat-free or low-fat and regular tortillas may also be used.

2.5.7.1 Equipment and Materials

1. Attach the rounded 1-in. cylindrical probe onto the TA-XT2 instrument. (*Note:* This is more of a "stretch" test. A conventional tensile test can also be done.)
2. Calibrate the force and set the safety knob.
3. Set the test parameters as shown in Table 2.3.
4. Prepare both a wheat flour and corn flour tortilla.
5. Cut the tortilla into a square (4 in. × 4 in.).
6. Punch holes into the corners.
7. Place the tortilla in the rig and secure them appropriately with screws on the frame.
8. Place the sample on the platform directly under the probe. Lower the probe to about 3 mm above the sample to make sure it is centered.
9. Run the test and save the raw data.

Table 2.3 Parameters for extensibility test

Parameters	Description	Values
Pre-test speed	Cross-head speed prior to trigger force	2 mm/s
Test speed	Cross-head speed during compression	1 mm/s
Post-test speed	Cross-head speed after test	5 mm/s
Distance	Vertical travel (make sure there is at least 20 mm of clearance between the sample and platform)	20.0 mm

2.5.7.2 Calculations

1. Graph force versus deformation with both tortillas on the same axis.
2. What does this tell you about tortillas? Which is more extensible?
3. Where are the rupture points? What do you predict the graphs would look like if the samples were at refrigeration temperatures?
4. How would differences in sample thickness affect your results? Refer to the flex test for analysis.
5. Report the results in Data Sheet 2.4.

(*Note:* Due to the geometry of this particular test, it is not amenable to the determination of material properties. However, Morgenstein *et al.* [1996] does present an engineering analysis of such a test.)

2.5.8 Strain Sweep

2.5.8.1 Prepare Sample

1. Weight out the gluten and water.
2. Mix to uniformity.
3. Extrude the mixture through pastamaker 15 times.
4. Allow it to hydrate for 1 h.
5. Compress the mixture between plates (2.5 mm) for 1 h.
6. Cut a 1-in. dimeter sample.
7. Cut the cheese sample to the same dimension.

2.5.8.2 Set DMA Settings

1. Set to strain sweep.
2. Set the temperature control unit to 25°C.

1.5.8.3 Run Test

1. Place the sample on the lower plate.
2. Lower the top plate to 2.5 mm thickness.
3. Apply mineral oil to the sides of the sample to minimize moisture loss.
4. Run the test and save the raw data.

5.8.4 Calculations

1. Plot $G*$ (complex modulus) versus % strain.
2. Compare the cheese and gluten in terms of $G*$ and critical strain values. How does $G*$ compare to the modulus determined via TA-XT2. Why might they differ?
3. Report the results in Data Sheet 2.5.

2.5.9   Report

Your report should have a separate part for compression, 3-point bend, extensibility, and strain sweep tests. Each part should include the following:

1. The data tables for each test.

2. The calculations of the compression and flexural moduli and maximum stress and strain of each material.

3. *Compression test.* Compare the cheeses. Was the order of moduli what you predicted? What do the relative moduli tell you you about the three kinds of cheeses?

4. Compare the degree of relaxation. What do these say about the three cheeses? (Think carefully about what is going on in the cheese during the time relaxation is being measured.)

5. *Flex test.* Compare the chips or crackers. Do any of the test measurements appear to be related to subjective judgements? Which ones? How are they related? Can you suggest any ways of using these flex tests?

6. *Extensibility test.* Which sample is more extensible and why? How is this information useful? How do you define extensibility in terms of force and rupture point?

7. Discuss whether the test results appear to differentiate the two tortillas.

8. Discuss your results in terms of the composition of the sample, for example, moisture content, fat content, ratio of water to protein (cheeses), and gluten content for the tortillas.

2.5.10   Additional Exercise

1. Determine the effect of temperature on the material properties of cheese and tortillas (e.g., refrigerate the sample before testing).

2. Determine the effects of cross-head speed on the material properties. Discuss this in the context of viscoelasticity.

2.6   SUGGESTED READINGS AND REFERENCES

1. S. Timpshenko and G. H. MacCullough. *Elements of Strength of Materials*, 3rd ed. New York: Van Nostrand, 1949.

2. S. L. Rosen, "Linear viscoelasticity." In *Fundamental Principle of Polymeric Materials*, 2nd ed. New York: John Wiley & Sons, 1993, Chap. 18.

3. R. C. Progelhof and J. L. Throne, "Testing for design." In *Polymer Engineering Principles: Properties, Process Tests for Design*. New York: Hansen Publishers, 1993, Chap. 6.

4. M. P. Morgenstern, M. P. Newberry and S. E. Holst, "Extensional properties of dough sheets." *Cereal Chemistry*. 73:478 (1996).

5. ASTM D638. Standard test method for tensile propoerties of plastics. American Society for Testing Materials, Philadelphia, PA.

6. ASTM D790. Standard test method for flexural properties of unreinforced and reinforced plastics and electrical insulating materials. American Society for Testing Materials, Philadelphia, PA.

DATA SHEET 2.1

Subjective evaluation of texture rankings of cheese samples

Cheese	Relative stress required for the same strain	Relative strain obtained from the same stress	Relative modulus

DATA SHEET 2.2   Compression test

Test sample dimensions

Material	Diameter	Length	Cross-sectional area	Volume

Test results obtained on compression test using TA-XT2

Material	Initial stress (F/A)	Final stress (F/A)	% Relaxation	Ranking

DATA SHEET 2.3   3-Point bend test

TA-XT2 test parameters

Cross-head speed	
Tool diameter	
Support span	

Test sample dimensions

Material	Width	Thickness

Test sample results

Material	Flex modulus	Fracture stress	Fracture strain

DATA SHEET 2.4   Extensibility test

Test parameters on TA-XT2 or Instron

Cross-head speed	
Full-scale force	
Tool diameter	

Test sample dimensions

Material	Length	Thickness	Width

Test results

Material	Rupture distance	Rupture stress (kPa)

DATA SHEET 2.5   DMA strain sweep test

DMA testing parameters

Sample material			
Frequency			
Plate diameter			
Sample temperature			
Sample thickness			

Test Results

Material	Linear viscoelastic range	G^* in linear region
Gluten		
Cheese		

3

RHEOLOGY OF LIQUID AND SEMISOLID FOODS

3.1 BACKGROUND

Rheology is related to the study of deformation and flow. It plays an important role during the development, manufacture, and processing of foods and food products. Liquid foods such as milk, honey, fruit juices, beverages, and vegetable oils exhibit simple flow properties. Thicker materials such as creamy salad dressings, ketchup, and mayonnaise behave in a more complicated manner. Semisolid foods such as peanut butter and margarine also behave like both solids and liquids. Most of these food materials are transported by pumping at one stage during processing or packaging, and therefore, their flow behavior properties are important for determining the power requirements for pumping, for the sizing of pipes, and furthermore, how they relate to sensory characteristics such as food texture. The transport of liquid food by pumps is directly related to liquid properties, especially density and viscosity.

The flow behavior is also important in the design of processes and operations. For example, it is important to determine the type of flow whether turbulent or laminar in heat exchangers. An assumption of simple Newtonian flow can result in error in the estimation of holding time and other equipment design. Rheological properties also serve as a means of controlling or monitoring a process. For example, the apparent viscosity of a general food reduces during enzymatic hydrolysis, whereas apparent viscosity increases during protein denaturation.

3.1.1 Viscosity

Viscosity is a liquid property that describes the magnitude of the resistance due to shear forces within the liquid. When a fluid is confined between two parallel plates of infinite

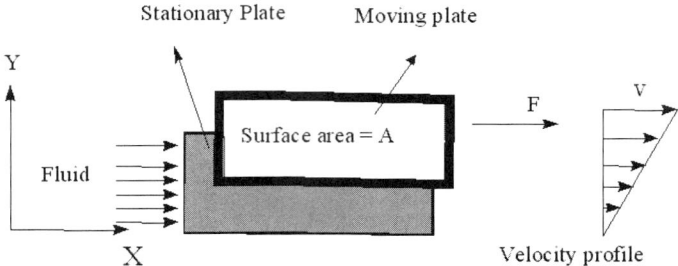

Figure 3.1   Imaginary representation of a Newtonian fluid flowing between a stationary plate and moving plate (F = force acting on the plate, v = velocity of moving fluid).

dimensions, the influence of shear force can be visualized as shown in Figure 3.1. In this scenario, the lower plate is held stationary and the force F is applied on the upper plate to produce a velocity v. This results in a velocity profile within the fluid. The velocity near the stationary plate is zero, whereas the liquid near the top plate will be moving at velocity v in m/s.

The shear force F on the plate area A will have shear stress:

$$\tau = F/A \quad (N/m^2)$$

As the distance between the plate is y, the velocity gradient can be described as dv/dy. This gradient is a measure of the rate of strain or shear rate being applied to the fluid.

3.1.2   Newtonian Fluids

For an ideal Newtonian fluid, the shear stress is a linear function of shear rate, and the proportionality constant for the relationship μ is called the dynamic viscosity:

$$\tau = -\mu \frac{dv}{dy} \tag{3.1}$$

Many food materials such as milk, apple juice, orange juice, wine, and beer exhibit Newtonian behavior. For Newtonian fluids, viscosity can be determined by applying a single shear rate and measuring corresponding shear stress. But to be accurate, it should always be estimated at several shear rates. The unit of viscosity is $N\cdot s/m^2$, which is Pa·s, whereas as in the cgs system, it is dyne·s/cm^2 which is also called poise.

$$1\ P = 100\ \text{centipoise (cP)}, \qquad 1\ cP = 10^{-3}\ \text{Pa·s or 1 mPa·s}$$

Humans can detect viscosity differences as low as 1 cP.

3.1.3   Non-Newtonian Fluid

For most agricultural materials, including food products such as cream, sugar, syrup, honey, and salad dressing, the relationship between shear stress and shear rate is not linear and these fluids are known as non-Newtonian. Some of these materials have a yield stress,

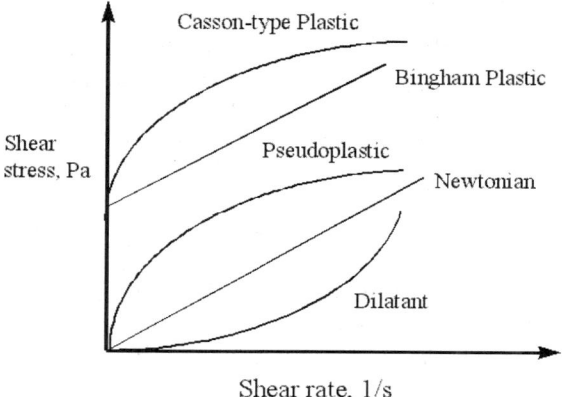

Figure 3.2 Shear stress versus shear rate for Newtonian, pseudoplastic (shear thinning), and dilatant (shear thickening), Bingham plastic, and Casson-type plastic fluids.

which must be attained before linear flow begins. These are called Bingham plastic-type fluids, examples are tomato ketchup, tomato paste, etc.

- In **non-Newtonian** fluids, the ratio of shear stress to shear rate will change with shear rate and this ratio at a given shear rate is called an **apparent viscosity**.
- The most common behavior is **pseudoplastic** or **shear thinning**, in which the shear stress versus shear rate curve is convex toward the shear stress axis as shown in Figure 3.2. These types of materials show a decrease in viscosity on shearing and examples include salad dressing, peanut butter, etc. The viscosity of some common foods is illustrated in Table 3.1.
- For **dilatant-fluids** or **shear thickening**, the shear stress versus shear rate curve is concave toward the shear stress axis. These fluids become thicker and viscosity increases on shearing. Some starch suspensions fall into this category.
- If the fluid has a yield stress and shear stress versus shear rate curve is convex toward the shear axis, then the fluid is called a **Casson-type plastic**.

Table 3.1   Viscosity values of some Newtonian food products

Product	Temperature °C	Viscosity (mPa·s)
Water	0	1.8
Water	20	1.0
Water	45	0.6
Milk, homogenized	20	2.0
Milk, homogenized	40	1.1
Corn syrup (48% solids)	27	5.3
Cream (10% fat)	40	1.5
Honey	27	4.8
Apple juice (brix 20°)	27	2.1
Corn oil	25	5.7
Peanut oil	25	6.6
Soybean oil	30	4.1

3.1.4 Rheological Models

3.1.4.1 Power Law Model   Many rheological models are used to describe the properties of the materials during flow and deformation. In most cases, the shear stress (τ) versus shear rate (dv/dy) curves for pseudoplastic and dilatant materials can be described using a simple power law model as shown in Eq. (3.2):

$$\tau = m\left(\frac{dv}{dy}\right)^{n} \tag{3.2}$$

In Eq. (3.2), m is usually called the **consistency coefficient** with units Pa·s^n and n is called the **flow behavior index**, which is unitless. **Newtonian fluid** is a special case of this model, where $n = 1$ and m is the dynamic viscosity. If $n < 1$, the fluid is pseudoplastic, if $n > 1$, it is dilatant.

3.1.4.2 Herschel–Bulkley Model   In the Herschel-Bulkley model (Eq. 3.3), the yield stress term (τ_0) has been added to describe plastic and Casson type plastic behavior:

$$\tau = m\left(\frac{dv}{dy}\right)^{n} + \tau_0 \tag{3.3}$$

Some flow behavior properties of food that follow the Herschel-Bulkley model are shown in Table 3.2.

3.1.4.3 Casson Model   In the Casson model (Eq. 3.4), the shear stress versus shear rate curve can be transformed into a straight line by plotting the square root of the shear stress versus square root of the shear rate. Chocolate is a notable example of this type of fluid:

$$\tau^{1/2} = m\left(\frac{dv}{dy}\right)^{1/2} + \tau_0^{1/2} \tag{3.4}$$

Table 3.2   Values of consistency coefficient (m), flow behavior index (n), and yield stress (τ_0) for selected foods

Product	Temperature (°C)	Shear rate, 1/s	m (Pa·s^n)	n	τ_0 (Pa)
Ketchup	25	10–560	18.7	0.27	32
Applesauce	20	3.3–530	16.7	0.30	0
Banana purée (17.7 brix)	23.8	28–200	6.08	0.43	0
Mayonnaise	25	30–1,300	6.4	0.55	0
Tomato juice concentrate (25% solid)	32.2	500–800	12.9	0.41	0
Blueberry pie filling	20	3.3–530	6.1	0.43	0
Chocolate, melted	46	—	0.57	0.57	1.16
Mustard	25	30–1,300	19.1	0.39	0
Peach purée (20% solid)	26.6	80–1,000	13.4	0.4	0
Comminuted batter meat (15% fat)	15	300–500	693.3	0.16	1.53
Orange juice concentrate (42.5° brix)	25	0–500	4.121	0.58	0

3.1.5 Temperature Dependency of Fluids on Viscosity

The viscosity of fluids decreases with an increase in temperature. For some fruit juices, the temperature effect can be described using an Arrhenius-type relationship as shown in Eq. (3.5):

$$\mu = \mu_0 \exp\left(\frac{E_a}{RT}\right) \qquad (3.5)$$

where μ is the viscosity, μ_0 the viscosity at reference temperature. E_a the activation energy, T the absolute temperature, and R the gas constant. A plot between $\log \mu$ versus $1/T$, the reciprocal absolute temperature, can be used to determine the values of μ_0 and activation energy E_a.

3.1.6 Time-dependent Viscosity

In some cases, the apparent viscosity of fluid changes with time, as the fluid is continuously sheared. If the apparent viscosity decreases with time, the fluid is called thixotropic, and if it increases with time, it is called rheopectic as illustrated in Figure 3.3. If the shear stress is measured as a function of shear rate, and if first shear rate is increased and then decreased, hysteresis will occur in the shear stress versus shear rate curves.

3.1.6.1 Thixotripic In the case of thixotropic foods, the material structure breaks down as shearing action continues. This type of food material includes gelatin, cream, shortening, and salad dressing, etc.

3.1.6.2 Rheopectic In the case of rheopectic fluids, the structure builds up as shearing continues. This type of behavior is not common in the food system, but can occur in a highly concentrated starch solution over long periods of time.

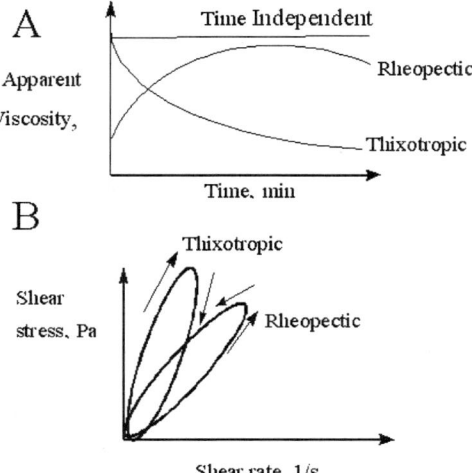

Figure 3.3 Behavior of time-dependent fluids (A, apparent viscosity as a function of time; B, shear stress as a function of shear rate).

3.2 VISCOSITY MEASUREMENT

A rheological measurement is taken by imposing a well-defined stress and measuring the resulting strain or shear or vice versa. The most commonly used experimental geometries for achieving steady shear flow are:

1. Capillary tube viscometer
2. Rotational viscometer
   - Searle-type
   - Couette-type

The use of narrow-gap rheometers such as a cone and plate is limited to relatively small shear rates. At high shear rates, end effects arising from the inertia of the sample make measurement invalid. The edge and end effects result mainly from the finite dimensions of the system, the shape of the free surface, related surface tension, and fracture of the samples.

3.2.1 Capillary Viscometer

In a capillary tube viscometer, viscosity measurement is based on the pressure force that is sufficient to overcome the shear force within the liquid and produces liquid flow at a given rate. Consider a small capillary viscometer of length L and internal radius r as shown in Figure 3.4 to measure the liquid viscosity. The shear forces are operating on all internal liquid surfaces for the entire length L and distance r from the tube center. Shear stress, τ force F per unit area, can be calculated as

$$\tau = \frac{F}{2\pi rL} \tag{3.6}$$

and the pressure drop ΔP across the capillary is given by

$$\Delta P = \frac{F}{\pi r^2} \tag{3.7}$$

Substituting the value F from Eq. (3.7) into Eq. (3.6) gives

$$\tau = \frac{\Delta P(\pi r^2)}{2\pi rL} = \frac{\Delta Pr}{2L} \tag{3.8}$$

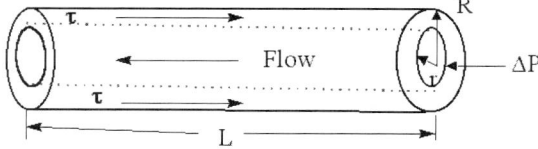

Figure 3.4 Schematic diagram showing the shear stress τ and pressure P balance for a section of capillary tube to measure viscosity.

According to Eq. (3.8), shear stress τ will increase from a value of 0 at the center of the tube to $\Delta PR/2L$ at the tube wall. By substituting the value of shear stress τ in shear stress and shear rate into Newtonian viscosity (Eq. 3.1), we get

$$\frac{\Delta Pr}{2L} = -\mu\frac{dv}{dr} \tag{3.9}$$

Rearranging Eq. (3.9) and integrating from the tube wall at radius R, where $v = 0$ to any location r within the velocity profile v, yield

$$\int_0^v dv = -\frac{\Delta P}{2\mu L}\int_R^r r\,dr \tag{3.10}$$

Therefore, velocity can be shown as

$$v = -\frac{\Delta P}{4\mu L}(R^2 - r^2) \tag{3.11}$$

The velocity profile of a liquid with viscosity μ can be shown by Eq. (3.11) at location r, when a pressure ΔP is applied across a length of capillary tube L.

By considering a cross-section area of a circular shell within the tube as

$$dA = 2\pi r\,dr \tag{3.12}$$

The volume of the flowing liquid in the shell can then be calculated by multiplying the cross-section area (Eq. 3.12) with the velocity:

$$d\dot{V} = (2\pi r\,dr)(v) \tag{3.13}$$

By integrating Eq. (3.13) from the tube center at $r = 0$ to the tube wall $r = R$ and rearranging, we get Eq. (3.14), which is also known as the classical Hagen–Poiseuille equation:

$$\mu = \frac{\pi\Delta PR^4}{8L\dot{V}} \tag{3.14}$$

where R = internal radius of the tube and $\dot{V}$ = volumetric flow rate.

Since the liquid is Newtonian, any flow rate-pressure combination will give the same viscosity:

$$\Delta P = \frac{\rho Vg}{A} = \rho hg = \frac{N}{m^2} = Pa \tag{3.15}$$

Substituting the value of ΔP in Eq. (3.14) and measuring the time t during the flow of fluid of a constant volume V, the viscosity can be determined from the following equation for a liquid of known density ρ as shown:

$$\mu = \left(\frac{\pi \rho g h R^4}{8LV}\right) t \tag{3.16}$$

Kinematic ($v = \mu/\rho$) can be calculated easily from capillary tube viscometer such as the Cannon–Fenske type (Fig. 3.5) by measuring the time t for draining the liquid between two etched marks in capillary tube bulbs. The ratio of the reservoir should be greater than 10 so that the pressure drop due to the flow in the reservoir can be neglected. Thus, all the terms in the parenthesis in Eq. (3.16) are constant for a capillary viscometer and, therefore, kinematic viscosity can be determined as

$$v = c \cdot t \tag{3.17}$$

Kinematic viscosity is measured in Stokes. 1 Stoke $= 100 \text{ cS} = \text{cm}^2/\text{s} = 10^{-4} \text{ m}^2/\text{s}$. 1 cS $= 10^{-2} \text{ cm}^2/\text{s}$. Kinematic viscosity has the same unit of measure as the diffusion coefficient. Therefore, it is also called momentum diffusivity and is a function of the fluid molecular properties in turbulent flow.

The capillary viscometer constant c can be determined easily by obtaining values needed or by measuring the efflux time of a fluid of known kinematic viscosity. Once the viscometer constant c is known, the kinematic viscosity of test fluid can be estimated easily.

Example   *A capillary tube viscometer is being used to measure the viscosity of honey at $30°C$. The tube radius is $2.5\,cm$ and the length $25\,cm$. A pressure of $10\,Pa$*

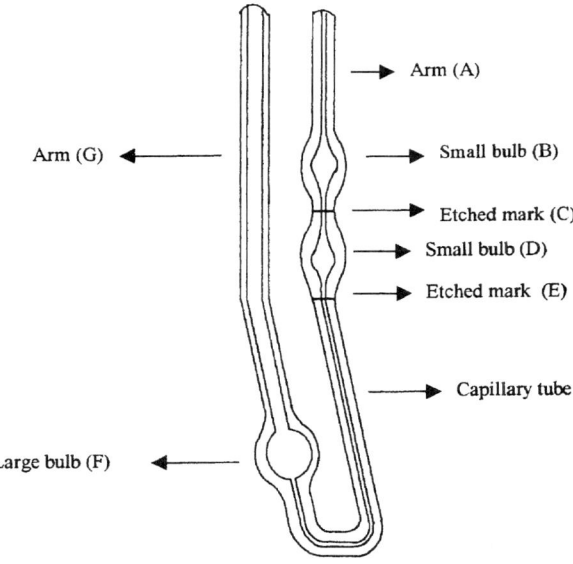

Figure 3.5   Typical Canon–Fenske-type capillary viscometer.

gives a flow rate of 1.25 cm³/s. Determine the viscosity of honey. You are given R = 2.5 cm or 0.025 m; L = 25 cm or 0.25 m; V = 1.25 cm³/s or 1.25 × 10⁻⁶ m³/s, and ΔP = 10 Pa.

Solution Substituting the given values in Eq. (3.14) gives viscosity as

$$\mu = \frac{3.1416 \times 10 \, (0.025)^4}{8 \times 0.25 \times 1.25 \times 10^{-6}} = 4.909 \text{ Pa·s}$$

3.2.2 Rotational Viscometer

The rheological parameters of non-Newtonian test fluid are estimated from the shear stress and shear rate relationship data generated with a co-axial cylinder viscometer, such as Brookfield LV, RV, or DV viscometers or the Haake Rotovisco RV series. In such visometer types, a spindle or sensing element rotates in a test fluid and measures the torque necessary to overcome the viscous resistance. The degree to which the spring is wound is detected by a rotational transducer, which is proportional to the viscosity of the test fluid.

In a coaxial cylinder rotational viscometer, liquid is placed in the space between the inner and outer cylinders. The measurement involves recording of the torque T required to turn the inner or outer cylinder at a given revolution per unit time. As torque $T = F*r$, $F = \tau \cdot A \, (A = 2\pi r L)$, where L is the length of the cylinder and r the radial location between the inner and outer cylinder. Therefore,

$$T = 2\pi r^2 L \tau \tag{3.17}$$

or shear stress is

$$\tau = \frac{T}{2\pi L r^2} \tag{3.18}$$

The shear rate γ for a rotational system is a function of angular velocity ω (2π N):

$$\gamma = r \cdot \frac{d\omega}{dr} \tag{3.19}$$

By substituting the values of shear stress and shear rate from Eqs. (3.18) and (3.19) in Eq. (3.1), we obtain the viscosity relationship

$$\frac{T}{2\pi \mu r^2 L} = r \frac{d\omega}{dr} \tag{3.20}$$

Further, to determine the angular velocity between the inner and outer cylinder, we can use the integration as

$$\int_0^{\omega_i} d\omega = -\frac{T}{2\pi \mu L} \int_{R_0}^{R_i} r^{-3} \, dr \tag{3.21}$$

By using the boundary conditions, at the outer cylinder radius R_0, the angular velocity $\omega = 0$, and at inner cylinder r_i, $\omega = \omega_i$ ($\omega = 2\pi$ N), integration leads to Eq. (3.22) to determine viscosity as shown:

$$\mu = \frac{T}{8\pi^2 NL}\left(\frac{1}{R_i^2} - \frac{1}{R_0^2}\right)$$ (3.22)

In a single-cylinder viscometer, the outer cylinder radius R_0 approaches infinity and, therefore, the last term in Eq. (3.22) can be omitted. Many single-cylinder rotational viscometers operate assuming that the wall of the vessel containing the fluid has no influence on the shear stress within the liquid. However, this assumption may not always be true for non-Newtonian fluids and, hence, should be carefully evaluated.

3.2.2.1 Co-axial Cylinder Searle-type System
In this type of rotational viscometer, the inner cylinder called a rotor rotates at a defined speed and the outer cylinder called a cup is held constant (Fig. 3.6). The rotating inner cylinder forces the liquid in the annular gap to flow, which offers it resistance depending on its viscosity characteristics. A torque-sensing element placed between the drive motor and shaft of the inner cylinder provides a direct measure of the sample viscosity. Most of the rotational viscometer is based on this working principle. However, these types of viscometer are limited when low viscous samples are to be measured, since the centrifugal force can turn the flow of liquid from a laminar region to turbulent flow, thus affecting the viscosity measurement.

3.2.2.2 Co-axial Cylinder Couette-type System
In this type of rotational viscometer, the outer cylinder rotates at a defined speed and forces the sample in the annular space to flow. The resistance of the liquid against being sheared transmits a velocity-related torque onto the inner cylinder that will be sensed by a torque sensor attached to it. It is measured by estimating just what counteracting torque is required to hold the inner cylinder stand still.

> ***Example*** *A single-cylinder rotational viscometer with a 2-cm radius and 5-cm length is being used to measure liquid viscosity. At 6, 9, and 12 rpm, the torque*

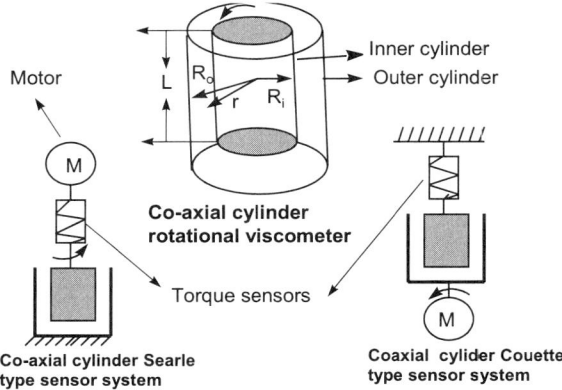

Figure 3.6 Schematic illustration of co-axial cylinder rotational viscometer (Searle- and Couette-types).

readings of 2.3, 3.7, and 5.0 10^{-3} N $\cdot$ cm were measured. Compute the viscosity of the liquid. You are given $R = 2\,cm$ or $0.02\,m$; $L = 5\,cm$ or $0.05\,cm$; $N_1 = 6\,rpm$ or $0.1\,rps$; $N_2 = 9\,rpm$ or $0.15\,rps$; $N_3 = 12\,rpm$ or $0.2\,rps$; $T_1 = 2.3 \times 10^{-3}\,N\cdot cm$ or $2.3 \times 10^{-5}\,N\cdot m$; $T_2 = 3.7 \times 10^{-3}\,N\cdot m$; $T_3 = 5.0 \times 10^{-3}\,N\cdot cm$ or $5.0 \times 10^{-5}\,N\cdot m$.

Solution Substitute the given values for the first set of data in Eq. (3.14):

$$\mu_1 = \frac{2.3 \times 10^{-5}}{8 \times (3.1416)^2 \times 0.1 \times 0.05} = 5.83 \times 10^{-5}\ \text{Pa·s}$$

Similarly for the second set of data, $\mu_2 = 6.25 \times 10^{-5}$ Pa·s and the third set of data, $\mu_3 = 6.33 \times 10^{-5}$ Pa·s.

$$\text{Average viscosity}\ (\mu_{avg}) = 6.14 \times 10^{-5}\ \text{Pa·s}$$

3.3 LAB EXERCISE

3.3.1 Objectives

The objectives of this lab exercise are to:

- Determine the Newtonian viscosity of a test fluid using a capillary viscometer.
- Determine the flow behavior curves of Newtonian, pseudoplastic, and dilatant types of food materials using co-axial cylinder viscometer.
- Determine the effect of temperature on the apparent viscosity of Newtonian food material.

3.3.2 Materials and Methods

3.3.2.1 Viscosity Measurement of a Newtonian Fluid

1. A capillary viscometer such as Cannon–Fenske, size 100 (kinematic viscosity range 2 to 10 cS) or 150 (kinematic viscosity range 6 to 30 cS).
2. Constant-temperature water bath.
3. Thermometer.
4. Suction rubber bulb.
5. Stopwatch.
6. 100-mL graduated cylinder.
7. 100-mL volumetric flask.
8. Pycnometer flask.
9. Distilled water, acetone, and trichloroethylene.
10. Viscosity standards: silicone oils, water.
11. Test fluids: apple juice, milk, water, noncarbonated drinks, etc.

3.3.2.1.1 Procedure

1. Fill up a 10-mL graduated cylinder with the test fluid.

2. Attach a suction bulb to arm G of the viscometer (Fig. 3.5). Invert the viscometer and dip arm A in the test fluid. Apply suction until the fluid level reaches the etched mark E. Return the viscometer to the upright position.

3. Place the viscometer in a desired controlled-temperature water bath and allow the temperature to equilibrate. Record the temperature.

4. Record the efflux time t for the test fluid to drain between the etched marks C and E by pressing the suction bulb.

5. Repeat the measurement by applying suction arm A to bring the test fluid level above the C mark.

6. Rinse the viscometer thoroughly, first with distilled water and then acetone. Dry the viscometer completely before each use.

7. Follow steps 1 to 5 above for each test fluid and viscosity standards.

8. Clean the viscometer after each standard solution, first with trichloroethylene and then acetone. Aspirate to dryness.

9. Determine the densities of each test fluid by filling them in tared 25-mL pychnometer flasks and weighing them accurately.

10. Determine the capillary viscometer constant C using viscosity standard data such as by measuring the efflux time for a fluid of known kinematic viscosity, as follows:

$$c = v_{known}/t_{known} \tag{3.23}$$

11. Estimate the kinematic viscosity of the test fluid using the viscometer constant and efflux time data.

12. Record the data in Data Sheet 3.1.

3.3.2.2 Viscosity Measurement of non-Newtonian Foods

1. Brookfield viscometer: model LV, RV, or DV.

2. Constant-temperature water bath.

3. Thermometer.

4. Four beakers, 600-mL capacity.

5. Suggested test fluids: pseudoplastic type such as banana purée, tomato paste, French dressing, and mayonnaise; dilatant type such as 50–55% cornstarch in water.

3.3.2.2.1 Procedure for Rotational Viscometer

1. Pour about 500 mL of the test fluid into a 600-mL beaker and place the beaker in a desired temperature-controlled water bath. Record the product temperature.

2. Press the auto zero button each time the power is turned on.

3. Carefully attach a suitable spindle to the viscometer shaft by avoiding any side thrust.

4. Enter the spindle number by pressing the spindle number entry access key.

5. Level the viscometer by adjusting screws on the mounting stand and the bubble level on the dial casing.

6. Insert the spindle in the test fluid up to the immersion groove cut in the spindle shaft.

7. Set the desired spindle speed by rotating the speed control knob.

8. Calculate the spindle multiplier constant (SMC) and shear rate constant (SRC) from the following equations and by using the viscometer torque constant (TK) data given for each specific viscometer model:

$$SMC = \frac{\text{Full-scale viscosity} \times \text{rpm}}{TK \times 10,000} \tag{3.24}$$

$$SRC = \frac{\text{Shear rate}}{\text{rpm}} \tag{3.25}$$

10. Enter the SRC and SMC values.

11. Run the viscometer spindle in the test fluid and record the shear stress data at various spindle speeds or shear rates.

12. Set the viscometer readout display to measure viscosity directly and record viscosity data at various shear rates.

13. Record the data for each product in Data Sheet 3.2.

3.2.2.2 Effect of Temperature on Product Viscosity In order to determine the effect of temperature on the test product viscosity, obtain shear stress and shear rate data at two more temperatures such as 40°C and 55°C by using rotational viscometer. Follow the same procedure as before to complete the experiment. Obtain experimental data at various shear rates in order to draw the curves for three different temperatures. Record the data in Data Sheet 3.3.

3.3.3 Results and Discussion

1. Report the capillary viscometer constant c by using the efflux time of standard solution.

2. Calculate the kinematic and dynamic viscosity of the Newtonian test fluid by using viscometer constant c and efflux time data.

3. Plot the shear stress and shear rate curves for Newtonian, pseudoplastic, and dilatant test fluid data obtained with the co-axial cylinder viscometer, that is, the Brookfield DV type. Do these curves indicate the desired characteristic behavior of the test fluids. If not, why did they deviate from their characteristic behavior?

4. Plot log shear sress versus log shear rate curves for the above test fluids and estimate their power law parameters such as the consistency index m and flow behavior index n from their intercept and slope values. Write the specific power law equation for each fluid.

5. Determine the viscosity of a Newtonian test fluid at three different temperatures and compute Arrhenius constant A and activation energy E_a by plotting $\log \mu$ versus $1/T$.

6. Compare the viscosity data, consistency index values m, and flow behavior index value n of the test fluids from the data available in the literature and discuss your results accordingly.

3.4 SUGGESTED READINGS AND REFERENCES

1. R. P. Singh and D. R. Heldman, 1981. "Transport of liquid foods." In *Introduction to Food Engineering*, Orlando, FL: Academic Press.

2. D. R. Heldman and R. P. Singh, 1981. "Rheology of processed foods." In *Food Process Engineering*. Westport, CT: AVI Publishing Co.

3. J. F. Kokini, 1992. "Rheological properties of foods." In *Handbook of Food Engineering* (D. R. Heldman and D. B. Lund, eds.) New York: Marcel Dekker.

4. J. R. Van Wazer, J. W. Lyons, K. Y. Kim and R. E. Colwell, 1963. *Viscosity and Flow Measurement*, New York: Interscience Publishers.

5. J. F. Steffe, I. O. Mohamed and E. W. Ford, 1986. "Rheological properties of fluid foods. In *Physical and Chemical Properties of Food* (M. R. Okos, ed.), St. Joseph, MI: ASAE.

6. G. Scchramm, 1981. *Introduction to Practical Viscometery*, Dieselstrasse, Germany: Haake Buchler Instruments Inc.

DATA SHEET 3.1

Date: _____

Capillary viscometer type: _____

Product specifications: _____

Capillary viscometer data for Newtonian test fluids

Product	Density (g/mL)	Viscosity (cP)	Efflux Time (s)
Standard			1.
			2.
			3.
Product 1			1.
			2.
			3.
Product 2			1.
			2.
			3.
Product 3			1.
			2.
			3.

DATA SHEET 3.2

Date: _____

Viscometer type: _____

Viscometer manufacturer: _____

Spindle no.: _____

Product specifications: _____

Rotational viscometer data for pseudoplastic and dilatant test fluids at 25°C

Product	Speed (rpm)	Shear Rate (1/s)	Shear stress (Pa)	Viscosity (cP)
Pseudoplastic				
Dilatant				

DATA SHEET 3.3

Date: _____

Viscometer type: _____

Viscometer manufacturer: _____

Spindle no.: _____

Product specifications: _____

Rotational viscometer data for Newtonian test fluids at different temperatures

Temperature ($^\circ$C)	Speed (rpm)	Shear Rate (1/s)	Shear Stress (Pa)	Viscosity (cP)
25				
25				
25				
25				
40				
40				
40				
40				
55				
55				
55				
55				

4

CONCEPTS OF HEAT
TRANSFER AND THERMAL
DEATH TIMES

4.1 BACKGROUND

Heat transfer is one of the most important unit operations in processing foods. Almost every process requires heat transfer either as heat input or heat removal to alter the physical, chemical, and biological characteristics of the product. During the storage of fruits, vegetables, meats and dairy products, heat is removed to cool the product for preservation for a longer period of time. Heating involves the destruction of pathogenic and other microorganisms responsible for food spoilage, thus making food safe and stable for longer period of storage. Heat transfer is governed by certain physical laws enabling us to predict heating phenomenon and determine optimum operating conditions.

This chapter reviews some basic ideas of heat transfer and the thermal destruction of microorganisms as a preparation for other thermal-processing-related chapters that follow. The following sections review some of the background needed to solve these problems.

4.1.1 Heat Transfer

4.1.1.1 Fourier's Law of Heat Conduction Fourier's law of heat conduction states that if a temperature gradient exists across a material, heat will be transferred in the direction of decreasing temperature at a rate that is proportional to the temperature gradient dT/dX and the area A through which the heat is moving. This is shown in Figure 4.1. The proportionality constant is characteristic of the particular material and is called the

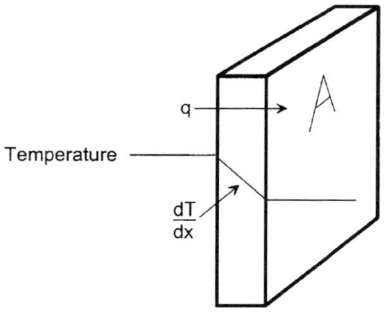

Figure 4.1 Fourier's law of heat transfer.

"thermal conductivity" of the material (Eq. 4.1):

$$q = -kA\frac{dT}{dx} \tag{4.1}$$

where q = rate of heat transfer in J/s (W) or Btu/h, k = the thermal conductivity of the material (W/m·K), A = the area of the conducting material perpendicular to the temperature gradient (m^2), and dT/dx = the rate of temperature change per unit distance (the thermal gradient). The minus sign in this equation indicates that heat moves from a high to low temperature or down the temperature gradient.

For a finite thickness and steady-state conditions, Eq. (4.1) becomes

$$q = -kA\frac{\Delta T}{\Delta x} \tag{4.2}$$

where ΔT = the temperature difference across the material and Δx = the thickness of the material. Fourier's law can be expressed in terms of the ratio of a driving force to thermal resistance as follows:

$$q = -\frac{\Delta T}{\Delta x/kA} = -\frac{\Delta T}{R} \tag{4.3}$$

where $R = \Delta x/kA$ = the thermal resistance of the material.

4.1.1.2 Newton's Law of Heat Convection When a fluid at one temperature comes into contact with a solid of a different temperature, a thermal boundary layer forms in the liquid as illustrated in Figure 4.2. Usually, both a velocity gradient and temperature gradient exist across this layer. Heat is transferred between the bulk of the liquid and the solid across this layer at a rate determined by the relationship (Eq. 4.4)

$$q = hA\Delta T \tag{4.4}$$

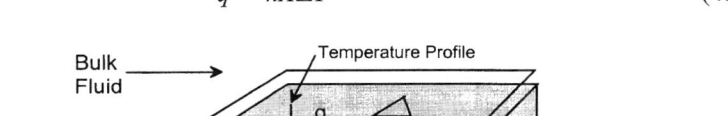
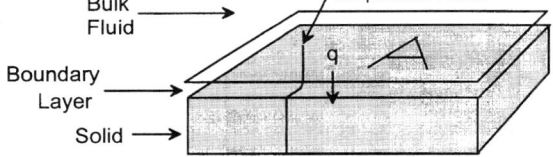

Figure 4.2 Newton's law of convective heat transfer.

where q = the rate of heat transfer (W or kcal/h or BTU/h), A = the area of contact between the liquid and solid (m^2), ΔT = the difference in temperature between the bulk of the liquid and the solid across the boundary layer, and h = a proportionality constant called the "convective heat transfer coefficient" that depends on the nature of the system.

As with Fourier's law, this equation can be expressed in terms of the ratio of a driving force to thermal resistance. Thus,

$$q = \frac{\Delta T}{1/hA} = \frac{\Delta T}{R} \tag{4.5}$$

where $R = 1/hA$ = the thermal resistance of the boundary layer.

4.1.1.3 Overall Heat Transfer Coefficient

When heat must pass through several layers in series, the thermal resistances of the various layers are summed. Any number of conductive and convective layers may be summed in this way:

$$\text{Total thermal resistance} = R = \sum_{i=1}^{n} R_i = R_1 + R_2 + R_3 + \dots \tag{4.6}$$

Using total resistance, we can express heat transfer through several layers in the form

$$q = \frac{\Delta T}{R} \tag{4.7}$$

The overall heat transfer coefficient is defined as

$$\text{Overall heat transfer coefficient } U = \frac{1}{RA} \tag{4.8}$$

Using this equation, we can express heat transfer through several layers in a form similar to the equations for Fourier's and Newton's laws:

$$q = UA\Delta T \tag{4.9}$$

Example   *A stainless steel plate 2 mm thick and 0.2 m^2 in area with conductivity 15 W/m·K is in contact with steam on one side and cold water on the other. The h for the steam is 3000 W/m^2 K and the h for the water 200 W/m^2 K. What is the total thermal resistance between the steam and cold water? What is the overall heat transfer coefficient of this system. What is the rate of heat transfer if the steam is at 100°C and the water at 20°C?*

Solution   This problem involves the heat transfer through series involving two convective layers and one conductive layer. The thermal resistances of each layer are

$$R_{\text{steam}} = \frac{1}{hA} = \frac{1}{\left(3000 \ \dfrac{W}{m^2 \ K}\right)(0.2 \ m^2)} = 0.0017 \ \frac{K}{W}$$

$$R_{\text{steel}} = \frac{\Delta x}{kA} = \frac{(0.02 \ m)}{\left(15 \ \dfrac{W}{m \ K}\right)(0.2 \ m^2)} = 0.0067 \ \frac{K}{W}$$

$$R_{\text{water}} = \frac{1}{hA} = \frac{1}{\left(200 \ \dfrac{W}{m^2 \ K}\right)(0.2 \ m^2)} = 0.025 \ \frac{K}{W}$$

The total thermal resistance is

$$R = R_{\text{steam}} + R_{\text{steel}} + R_{\text{water}} = 0.017 + 0.0067 + 0.025 = 0.033 \; \frac{K}{W}$$

The overall heat transfer coefficient is

$$U = \frac{1}{\left(0.033 \; \dfrac{K}{W}\right)(0.20 \text{ m}^2)} = 151 \; \frac{W}{\text{m}^2 \text{ K}}$$

The rate of heat transfer can be computed from either the resistance or overall heat transfer coefficient:

$$q = \frac{\Delta T}{R} = \frac{100 - 20 \text{ K}}{0.033 \; \dfrac{K}{W}} = 2.4 \times 10^3 \text{ W}$$

$$q = UA\Delta T = \left(151 \; \frac{W}{\text{m}^2 \text{ K}}\right)(0.2 \text{ m}^2)(100 - 20 \text{ K}) = 2.4 \times 10^3 \text{ W}$$

4.1.2 Unsteady-State Heat Transfer

The Schmidt plot and Gurney–Lurie-type charts are two ways of analyzing the unsteady-state diffusion of temperature changes. Before explaining the mechanics of these methods, let us review unsteady-state heat transfer. A slab of material is allowed to equilibrate to some initial temperature, say, T_0. As shown in Figure 4.3A, the temperature profile is horizontal, indicating that the temperature is uniform through the slab.

One side of the slab is then exposed to a new temperature that we will call T_1. For an instant, the temperature profiles look as shown in Figure 4.3B, with an abrupt temperature change at the surface. Soon, however, heat penetrates and the profile within the slab begins to assume the exponential shape as shown in Figure 4.3C. This shape arises because heat reaches the near side of the slab faster than the farther side. This profile occurs and constantly changes during unsteady-state heat transfer.

Eventually, if the external temperatures T_0 and T_1 remain unchanged, the slab will warm until the temperature profile becomes a straight line as shown in Figure 4.3D. This is the

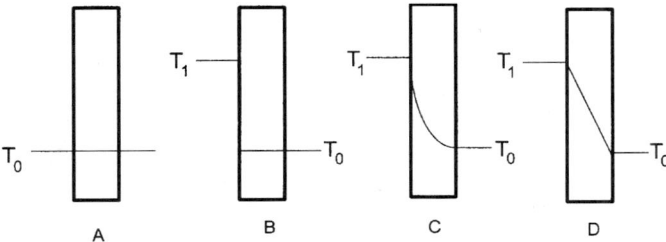

Figure 4.3 Temperature profile during unsteady-state heating of an object.

steady state and, although heat will continue to flow through the slab, no further change in the temperature will occur within the slab and the profile will remain constant.

4.1.2.1   Schmidt Plot Notation   The Schmidt plot is a method for approximating the exponential profile at various times between the start of heating and the development of steady-state conditions. It allows us to estimate the temperature at any point in the slab at any time after heating begins.

We will use the following notation as shown in Figure 4.4 to describe the Schmidt plot method:

1. *Distance variables.* (refer to Fig. 4.4). $x =$ any arbitrary distance into the slab in the direction of heat flow. At the heated surface, $x = 0$. $x =$ the total thickness of the slab. In the Schmidt method, X is to be divided into an aribtrary number of intervals. $n =$ the number of intervals into which X is divided. $\Delta x =$ the width of one interval. $\Delta x = X/n$.

2. *Time variables.* (refer to Fig. 4.5). $t =$ the time since the start of heating. $\Delta t =$ a specified time interval. Temperature profiles will be drawn Δt s.

3. *Temperature variables.* $T =$ the temperature at any distance (x) into the slab at any time. $T_0 =$ the initial temperature of the slab. $T_1 =$ the temperature to which the surface is exposed during heating (or cooling). $\alpha =$ the thermal diffusivity of the

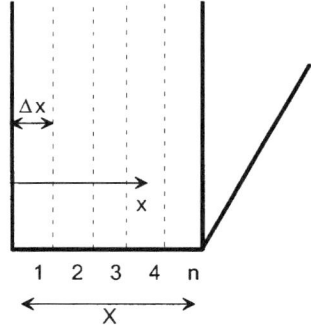

Figure 4.4   Slab divided into n intervals to indicate unsteady-state heat transfer in a Schmidt plot.

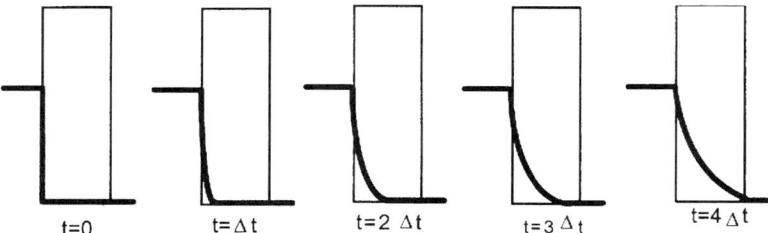

Figure 4.5   Temperature profiles at successive time intervals in a Schmidt plot.

slab, computed as

$$\alpha = \frac{k}{\rho C_p} \tag{4.10}$$

where k = the thermal conductivity of the slab, ρ = the density of the slab, and C_p = the heat capacity (specific) heat of the slab.

Thermal diffusivity can be interpreted as follows:

- k measures the rate at which heat passes through a material. The larger it is, the faster the material heats up.
- C_p measures the heat needed to raise a unit mass by $1°$.
- ρC_p measures the heat needed to raise a unit volume by $1°$. The larger it is, the slower the material heats up.
- α is the ratio of k to ρC_p and therefore indicates the relative rate at which a material heats up.

4.1.2.2 Schmidt Plot Procedure

1. *Select thickness interval.* Divide the thickness of the slab X into $n = 6$ to 12 intervals of thickness Δx. Draw vertical lines on a graph as shown in Figure 4.6 to represent these intervals.
2. *Compute time interval.* From the distance interval Δx that you selected and the thermal diffusivity α of the slab, compute the time interval Δt using the formula

$$\Delta t = \frac{(\Delta x)^2}{2\alpha} \tag{4.11}$$

3. *Compute thickness interval.* Alternatively, you can select a time interval and calculate the thickness interval with the inverse formula:

$$\Delta x = \sqrt{2\alpha\Delta t} \tag{4.12}$$

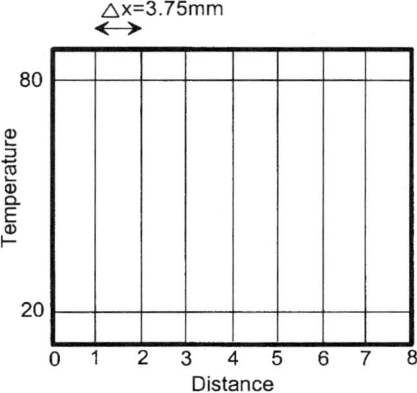

Figure 4.6 Plain layout of a Schmidt plot.

Example   *We have a slab of fish fillets at 20°C that is placed in contact with a plate at 80°C. The properties of the fish are thermal conductivity $= k = 0.5\,W/m\,K$, heat capacity $= C_p = 3.18\,kJ/kg\,K$, density $= \rho = 910\,kg/m^3$, area of one side $= A = 200\,cm^2$, and thickness $= x_1 = 3\,cm$. From Eq. (4.10), the thermal diffusivity of the slab is*

$$\alpha = \frac{k}{\rho C_p} = \frac{\left(0.5\ \dfrac{J/s}{m\,K}\right)}{\left(910\ \dfrac{kg}{m^3}\right)\left(3.18 \times 10^3\ \dfrac{J}{kg\,K}\right)} = 1.7 \times 10^{-7}\ \frac{m^2}{s}$$

If you divide the thickness of the slab into eight equal intervals, the length of each interval will be

$$\Delta x = \frac{3\ cm}{8} = \frac{0.03\ m}{8} = 0.00375\ m = 3.75\ mm$$

From these values, we calculate the time interval using Eq. (4.11):

$$\Delta t = \frac{(0.00375\ m)^2}{2\left(1.7 \times 10^{-7}\ \dfrac{m^2}{s}\right)} = 41.4\ s$$

4. *Set up graph.* On graph paper, label the horizontal axis to represent multiples of Δx. Label the vertical axis to represent temperatures that span the range from T_0 to T_1. Draw and label horizontal lines across the graph at T_0 and T_1 as shown in Figure 4.6.

Example   *If a slab is initially at 20°C and a temperature of 80°C is applied to one side of the slab, the resulting axes should look as illustrated in Figure 4.7.*

5. Draw first profile. To approximate the temperature profile after the first time interval Δt, draw a line connecting T_1 at distance $= 0$ with T_0 at distance $= 2$ as shown in Figure 4.7.

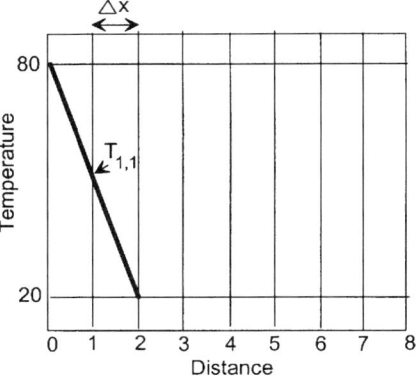

Figure 4.7   Temperature profile after Δt time interval in a Schmidt plot.

6. *Estimate first temperature.* The place where this line crosses distance 1 is an estimate of the temperature at distance Δx from the heated surface after time Δt. This point is labeled $T_{1,1}$, meaning the temperature at distance 1, time 1.

Example *In Figure 4.7, we estimate that after time $1\Delta t$ (41.4 s), the temperature at depth $1\Delta x$ (0.375 cm) into the slab is $50°C$.*

7. *Draw second profile.* The profile after the $2\Delta t$ s is approximated by drawing a line from $T_{1,1}$ to distance 3 at temperature T_0. Point $T_{2,2}$ estimates the temperature at distance $2\Delta x$ after time $2\Delta t$. [Notice that each line segment is drawn so that it connects to the previous estimation point(s) and crosses exactly two horizontal divisions.]

Example *From Figure 4.8, we estimate that after time $2\Delta t$ (82.8 s), the temperature at depth $2\Delta x$ (0.750 cm) into the slab has risen from $20°$ to $35°C$.*

8. *Draw third profile.* For the third time interval, you will find that it is possible to draw two line segments as illustrated in Figure 4.9 and use their midpoints to esimate $T_{1,3}$ (temperature at $1\Delta x$ after $3\Delta t$ s) and $T_{3,3}$ (temperature at $3\Delta x$ after $3\Delta t$ s).

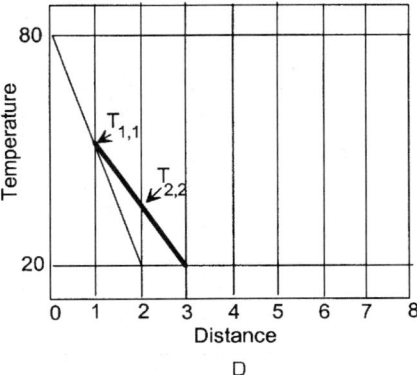

Figure 4.8 Temperature profile after $2\Delta t$ time intervals in a Schmidt plot.

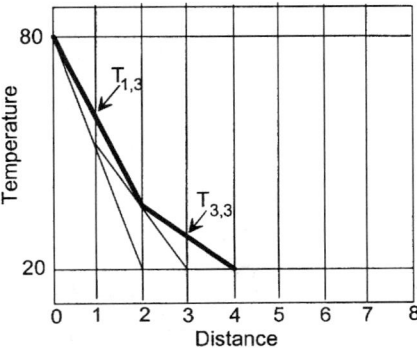

Figure 4.9 Temperature profile after $3\Delta t$ time interval in a Schmidt plot.

Example *From Figure 4.9, we see that after time $3\Delta t$ (124.2 s), the temperature $(T_{1,3})$ at a depth $1\Delta x$ (0.375 cm) is about 60°. At a depth of $3\Delta x$ (1.125 cm), the temperature $(T_{3,3})$ is about 30°C.*

9. *Draw additional profiles.* Figure 4.10 shows the lines for the next two time intervals. Notice that we are getting a progressively more detailed profile after every new interval. The number of intervals you plot depends on the accuracy of the estimates you wish to make.

Example *Figure 4.11 illustrates the temperature profile after time $8\Delta t$.*

Example *A temperature profile after eight time intervals is shown in Figure 4.12. It was constructed by tracing the outer lines in Figure 4.11. It indicates how this profile can give a good estimate at any distance into the slab. For example, at 3.5 distance intervals into the slab, the graph gives a temperature around 32°C. 3.5 intervals is*

$$3.5\Delta x = 3.5(0.375) = 1.3 \text{ cm}$$

so we conclude that after 331 s (8 × 41.4), a location 1.3 cm into the slab will have heated from 20 to 32°C.

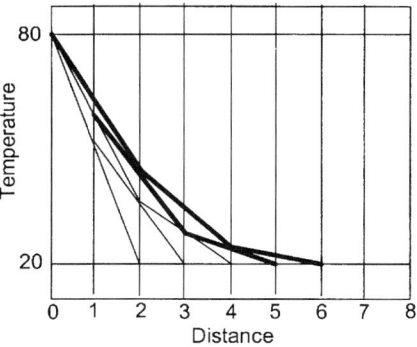

Figure 4.10   Temperature profile after $4\Delta t$ and $5\Delta t$ time interval in a Schmidt plot.

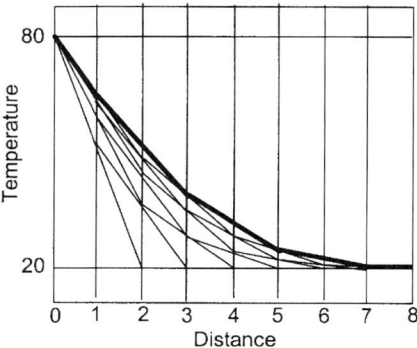

Figure 4.11   Temperature profile after $8\Delta t$ time interval in a Schmidt plot.

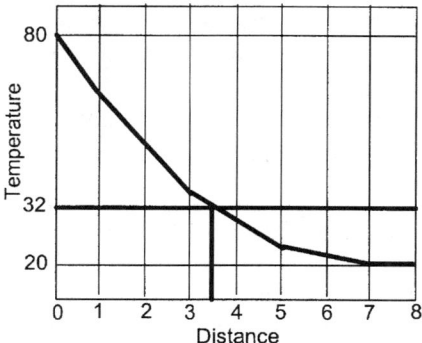

Figure 4.12 Temperature profile after $8\Delta t$ time interval in a Schmidt plot.

10. *Heating from both sides.* A temperature profile for a Schmidt plot when the slab is heated from both sides is shown in Figure 4.13. This figure covers the first four time intervals. At that time, the plot estimates the center temperature to be around 30°C.

4.1.2.3 Gurney–Lurie Charts Schmidt plots provide one way to track unsteady-state heat transfer. Gurney–Lurie charts are another way to estimate the time and temperature data under steady-state heat transfer.

1. *Biot number.* The rate of unsteady-state heating is affected by two factors:
   • The rate of heat transfer between the medium and object.
   • The rate of heat transfer within the object.

   In most cases, one or the other factor is limiting. You can determine which factor is limiting by computing the Biot number:

   $$Bi = \frac{hx_1}{k} \tag{4.13}$$

   where h = the convective heat transfer coefficient between the medium and object, k = the thermal conductivity of the object, and x_1 = the characteristic dimension of the object (volume/area).

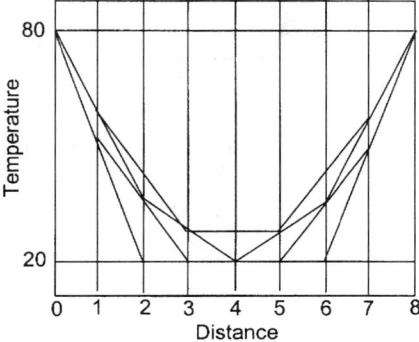

Figure 4.13 Temperature profile when heating from both sides of a slab after $4\Delta t$ interval in a Schmidt plot.

- When $Bi < 0.1$, we can assume negligible resistance within the object and that the rate of heating is limited by convective heat transfer.
- When $Bi > 40$, we can assume negligible resistance at the surface and that the rate of heating is limited by conductive heat transfer.
- When $0.1 < Bi < 40$, both factors are limiting.

4.1.2.4 Procedure for Infinite Plates and Cylinders

1. *Variables.* For situations where the Biot number is large, Gurney–Lurie-type charts are used to estimate the rate of heat penetration into an object. Figure 4.14 is such a chart for objects that resemble plates of large width compared to their thickness. This chart uses the following variables: x_1 = the half thickness of the plate, if heat is transferring through both surfaces, and the full thickness, if heat is transferring through only one surface. x = the distance between the center of the plate and the point under study. t = the time after the object is first immersed in the medium. T_1 = the temperature of the surrounding medium. T_0 = the initial temperature of the

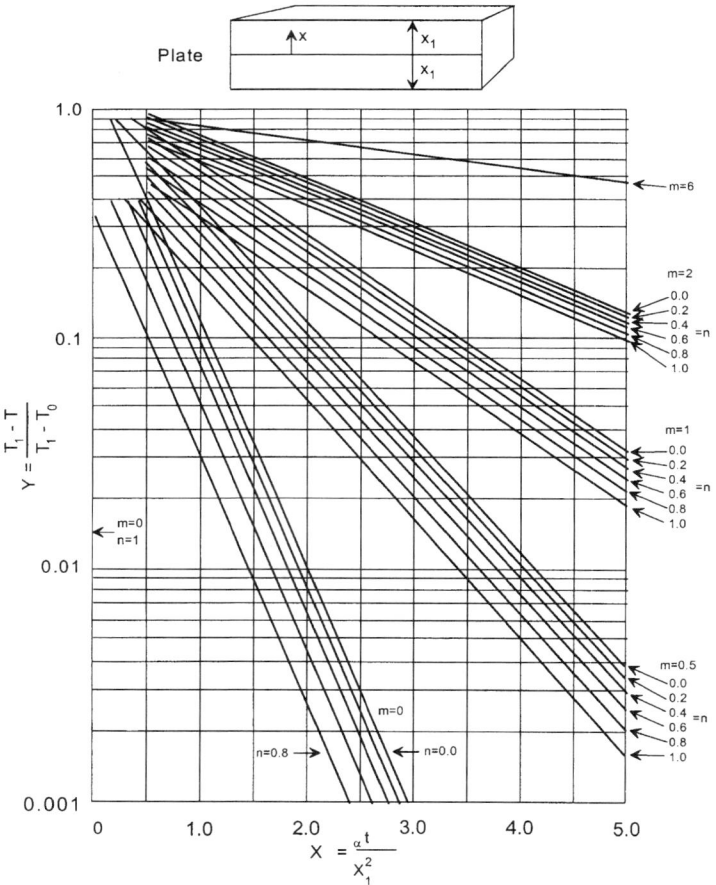

Figure 4.14 Gurney–Lurie-type chart for a plate of thickness $2X$.

object, assumed to be the same throughout the object. T = the temperature at the point under study at time t. α = the thermal diffusivity coefficient of the object = $k/\rho C_p$.

2. *Chart parameters.* From these variables, the following parameters are computed before using the charts:

$$n = \frac{x}{x_1} \tag{4.14}$$

$$m = \frac{k}{hx_1} \tag{4.15}$$

In some cases, the m value is close to the inverse of the Biot number. If time is specified and temperature is to be determined, compute

$$\text{Fourier number} = X = \frac{\alpha t}{x_1^2} \tag{4.16}$$

If a temperature has been specified and time is to be determined, compute

$$\text{Dimensionless temperature} = Y = \frac{T_1 - T}{T_1 - T_0} \tag{4.17}$$

3. *Read the chart.* If time has been specified, find X along the bottom of the chart (Figure 4.14 for plates, Figure 4.15 for cylinders). Go up to the group of lines that match your computed value of m and find the specific line for your computed value of n. Read Y off the vertical axis. If temperature has been specified, reverse the process, starting with Y on the vertical axis and reading X from the horizontal axis.

4. *Solve for the unknown.* Solve for time t using Eq. (4.16) or temperature T using Eq. (4.17)

Example *A slab of butter 34 mm thick is initially at 5°C. It is placed on an insulated surface and exposed on the top to air at 30°C. The convective heat transfer coefficient is 7.83 W/m² K. The properties of the butter are as follows: K = 0.192 W/m K, C_p = 2250 J/kg K, and ρ = 997 kg/m³. What is the temperature 10 mm below the surface after 4 h?*

The thermal diffusivity of the butter, computed by Eq. (4.10), is

$$\alpha = \frac{k}{\rho C_p} = \frac{\left(0.192 \ \frac{W}{m \ K}\right)}{\left(997 \ \frac{kg}{m^3}\right)\left(2250 \ \frac{J}{kg \ K}\right)} = 8.56 \times 10^{-8} \ \frac{m^2}{s}$$

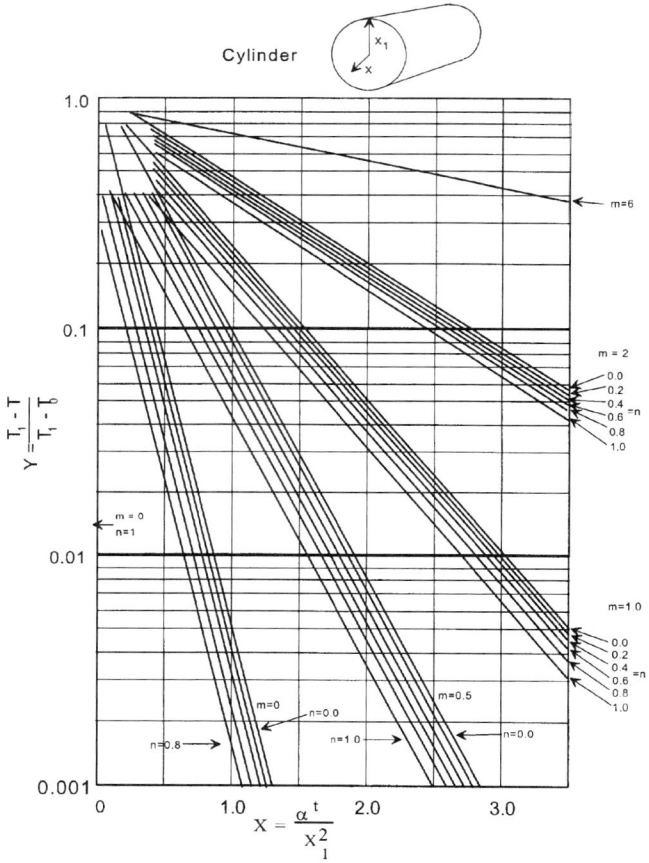

Figure 4.15   Gurney–Lurie-type chart for a cylinder of radius X_1.

Using Eqs. (4.14), (4.15), and (4.16), we determine that the needed parameters are

$$n = \frac{x}{x_1} = \frac{(34 - 10 \text{ m})}{(34 \text{ m})} = 0.71 \quad \text{(dimensionless)}$$

$$m = \frac{k}{hx_1} = \frac{\left(0.192 \ \dfrac{\text{W}}{\text{m K}} \right)}{\left(7.93 \ \dfrac{\text{W}}{\text{m}^2 \text{ K}} \right)(0.034 \text{ m})} = 0.71 \quad \text{(dimensionless)}$$

$$X = \frac{\alpha t}{x_1^2} = \frac{\left(8.56 \times 10^{-8} \ \dfrac{\text{m}^2}{\text{s}} \right)(4 \cdot 3600 \text{ s})}{(0.034 \text{ m})^2} = 1.07 \quad \text{(dimensionless)}$$

From the chart (Fig. 4.14), enter at X = 1.07 and go up to the group for m = 1, n = 0.8, and read Y = 0.37. Solving Eq. (4.17) for T, we get

$$Y = 0.37 = \frac{T_1 - T}{T_1 - T_0} = \frac{30 - T}{30 - 5}$$

$$T = 30 - 0.37(30 - 5) = 20.75 = \text{temperature 10 mm deep after 4 h}$$

4.1.2.5 Procedure for Finite Object

The procedure described above (Sect. 4.1.2.4) is useful when the characteristic dimension of an object is small compared to the other dimensions, that is, for plates that are wide compared to their thickness or cylinders that are long compared to their radius. Most objects are not like this. A can, for example, has a length that is not much greater than its diameter. For such finite objects, the following procedure can be used to estimate the temperature after a specified period of time:

1. For a cylinder of radius x_1 and length $2y_1$, compute the Fourier number separately for each of the two dimensions:

$$X_x = \frac{\alpha t}{x_1}, \qquad X_y = \frac{\alpha t}{y_1}$$

For a rectangular solid of dimensions $2x_1 \times 2y_1 \times 2z_1$, compute the Fourier number separately for each of the three dimensions:

$$X_x = \frac{\alpha t}{(x_1)^2}, \qquad X_y = \frac{\alpha t}{(y_1)^2}, \qquad X_z = \frac{\alpha t}{(z_1)^2}$$

2. Compute the appropriate m and n values for each Fourier number.
3. For each Fourier number, look up the corresponding Y value.
4. Compute the product of the Y values:

$$\text{For a finite cylinder:} \quad Y_{xy} = Y_x Y_y \tag{4.18a}$$
$$\text{For a rectangular solid:} \quad Y_{xyz} = Y_x Y_y Y_z \tag{4.18b}$$

5. Set the product equal to the dimensionless temperature and solve for T.

Example *A can of applesauce is placed in a steam chamber. The system has the following properties: conductivity of applesauce = k = 0.692 W/m K, density of applesauce = ρ = 1068 kg/m³, specific heat of applesauce = C_p = 3.95 kJ/kg K, radius of can = x_1 = 4.2 cm, length of can = 2y_1 = 12.8 cm, heat transfer coefficient of steam = h = 3000 W/m² K, initial temperature of sauce = T_0 = 5°C, and steam temperature T_1 = 105°C. What is the temperature after 1 h at a point along the cylindrical axis 2 cm from one end of the can?*

Solution   The Biot numbers are

$$Bi_x = \frac{\left(3000 \; \frac{W}{m^2 \, k}\right)(0.042 \text{ m})}{\left(0.692 \; \frac{W}{m \, K}\right)} = 182, \qquad Bi_y = \frac{\left(3000 \; \frac{W}{m^2 \, K}\right)(0.068 \text{ m})}{\left(0.692 \; \frac{W}{m \, K}\right)} = 295$$

Both are larger than 40 so the Gurney–Lurie approach is applicable. The thermal diffusivity of applesauce is

$$\alpha = \frac{k}{\rho C_p} = \frac{\left(0.692 \; \frac{W}{m \, K}\right)}{\left(1068 \; \frac{kg}{m^3}\right)\left(3.95 \times 10^3 \; \frac{J}{kg \, K}\right)} = 1.64 \times 10^{-7} \; \frac{m^2}{s}$$

Treating the can as a cylinder of radius 4.2 cm, we compute m, n, and X and read Y from the Gurney–Lurie-type chart for a cylinder:

$$m = \frac{1}{Bi} = \frac{1}{182} \approx 0, \qquad n = \frac{x}{x_1} = \frac{0}{4.2} = 0$$

$$X_x = \frac{\left(1.64 \times 10^{-7} \; \frac{m^2}{s}\right)(3600 \text{ s})}{(0.042 \text{ m})^2} = 0.33, \qquad Y_x = 0.2$$

Treating the can as a slab of half thickness 6.4 cm, we compute m, n, and X and read Y from the Gurney–Lurie-type chart for a plate:

$$m = \frac{1}{Bi} = \frac{1}{295} \approx 0, \qquad n = \frac{y}{y_1} = \frac{6.4 - 2.0}{6.4} = 0.68$$

$$X_y = \frac{\left(1.64 \times 10^{-7} \; \frac{m^2}{s}\right)(3600 \text{ s})}{(0.068 \text{ m})^2} = 0.13, \qquad Y_y = 0.3$$

Setting the product of the Y values equal to the dimensionless temperature, we obtain

$$Y_{xy} = Y_x Y_y = (0.2)(0.3) = 0.06 = \frac{105 - T}{105 - 5}$$

$$T = 105° - 0.06(105° - 5°) = 99°C = \text{the temperature at the desired location.}$$

4.1.3 Semilog Plots

In many natural processes, the rate at which some quantity changes is proportional to the quantity itself. This proportionality can be expressed with the differential equation

$$\frac{dX}{dt} = kX \tag{4.19}$$

where X = the quantity, dX/dt = the rate of change of the quantity with time (slope of the line), and k = a proportionality constant that may be positive or negative, depending on the phenomena.

> ***Example: Radioactive decay*** *Radioactive atoms disintegrate at a rate that is proportional to the number of atoms present. A plot of the number of surviving atoms over time is shown in Figure 4.16A. Notice that the decay rate (slope of the line) is proportional to the height of the line. Since the quantity is decreasing, k is negative.*

> ***Example: Bacterial growth*** *Until they become overcrowded, bacteria reproduce at a rate that is proportional to the number of bacteria present. The size of a bacterial population over time is depicted in Figure 4.16B. Again, the slope is proportional to the height of the line. Since the quantity is increasing, k is positive.*

> ***Example: Unsteady-state heating*** *When a cold object is dropped into a warm water bath, the rate at which the object's temperature increases is proportional to the difference in temperature between the object and water bath. Such a heating curve is shown in Figure 4.16C. In this case, the slope is proportional to the distance between the object temperature and water bath temperature. Since this difference is decreasing, k is negative.*

1. *X as a function of t.* To find the equation that relates X to time, rearrange Eq. (4.19) and integrate

$$\int \frac{dX}{X} = k \int dt \tag{4.20}$$

$$Ln(X) = kt + c$$

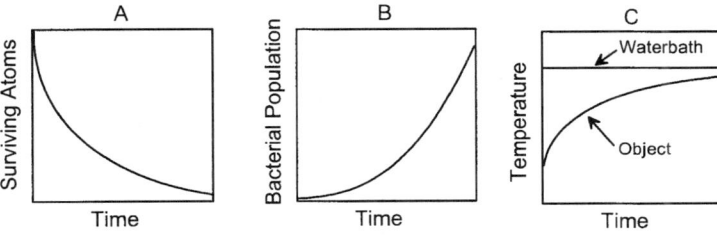

Figure 4.16 Different set of curves indicating exponential phenomena.

where c is the constant of integration. If you raise e (2.71828, base of natural logarithms) to the power of each side of this equation, you get

$$e^{Ln(X)} = e^{kt+c}$$
$$X = e^c e^{kt} \tag{4.21}$$
$$X = c' e^{kt}$$

where c' is a new constant equal to e^c. Equation (4.21) tells us that X is an exponential function of t. You will sometimes see Eq. (4.21) written in the following notation:

$$X = c' \exp(kt) \tag{4.22}$$

2. *Straight-line plot.* A straight-line equation with intercept c and slope k is shown as Eq. (4.20). This tells us that plotting the logarithm of X versus t will produce a straight line. Although this equation uses natural logarithms, it is easy to convert it to common (base 10) logarithms by dividing both sides of the equation by 2.303. The resulting equation is still that of a straight line, but will have a different slope and intercept:

$$\frac{Ln(X)}{2.303} = \frac{k}{2.303}t + \frac{c}{2.303}$$
$$Log(X) = k't + c'' \tag{4.23}$$

Example   *The data in the first two columns of Table 4.1 are plotted in Figure 4.17A and show an apparent exponential relationship. The third column of Table 4.1 shows the logarithms of the Y values in column 2. This column is plotted versus X in Figure 4.17B and fitted with a straight line. The intercept of this line is the value of the line where it crosses the Y axis (where X = 0) approximately −1.7. The slope is most easily determined by measuring the distance along the X axis required for the line to cross one unit on the log Y axis. In Figure 4.17B, this distance is 19 so the slope is*

$$\text{Slope} = b' = \frac{1}{19} = 0.053$$

Table 4.1   Data set showing an exponential relationship

X	Y	$Log(Y)$
5	0.02	−1.70
20	0.35	−0.45
35	0.95	−0.20
45	5.00	0.70

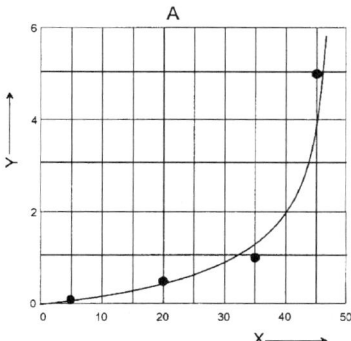 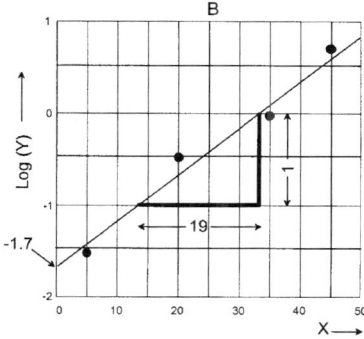

Figure 4.17 Plotting an exponential relationship using X versus values (A) and X versus Log(Y) values (B).

Thus, the equation of the line in Figure 4.17B is

$$Log(Y) = -1.7 + 0.053X$$

3. *Semilog paper.* We can avoid the computation of logarithms by using semilog paper. On semilog paper, one of the axes is logarithmically spaced as shown in Figure 4.18. This axis is labeled with the original units.

   Example *Data in columns 1 and 2 of Table 4.1 are plotted on a semilog paper in Figure 4.18. The Y axis is divided into three log cycles, each corresponding to a 10-fold increase in Y. For example, in the first log cycle, Y goes from 0.01 to 0.1; in the second log cycle, Y goes from 0.1 to 1, etc. The data from columns 1 and 2 are plotted directly on the graph and fitted with a straight line. The resulting plot is similar to the plot shown in Figure 4.17B.*

4. *Slope on semilog paper.* To determine the slope of a line from semilog paper, determine the distance along the linear axis required for the line traversing one log cycle. If the X axis is the log axis, this distance is the slope. If the Y axis is the log axis, the slope is the reciprocal of this distance.

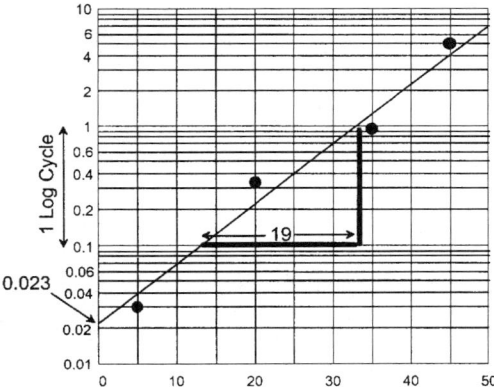

Figure 4.18 Semilog plot (Y axis in original units on a log scale).

Example   *In Figure 4.18, the line travels 19 units along the X axis while traversing one log cycle on the Y axis, so*

$$\text{Slope} = b' = \frac{1}{19} = 0.053$$

The intercept is the logarithm of the point where the line crosses the Y axis, that is,

$$\text{Intercept} = \text{Log}(0.023) = -1.64$$

So for this example, the equation of the straight line (Eq. 4.23) is

$$\text{Log}(Y) = 0.053X - 1.64$$

5. *Exponential form.* To convert an equation of the form

$$\text{Log}(Y) = a + bX$$

to exponential form, first convert to natural logarithms by multiplying each side of the equation by 2.303, then raise e(2.718282) to the power of each side:

$$2.303 \, \text{Log}(Y) = 2.303(a + bX)$$
$$Ln(Y) = 2.303a + 2.303bX$$
$$e^{Ln(Y)} = e^{2.303a + 2.303bX}$$
$$Y = e^{2.303a + 2.303bX}$$

To simplify the equation, remember that the log of a product is equal to the sum of the logs, that is, $\text{Log}(ab) = \text{Log}(a) + \text{Log}(b)$:

$$Y = e^{2.30a} e^{2.303bX} \qquad (4.24)$$

Example   *To convert the straight-line equation of Figure 4.18 to exponential form, first multiply each side by 2.303, then raise e to the power of each side:*

$$2.303[\text{Log}(Y)] = 2.303[(0.053) \, t - (1.64)]$$
$$Ln(Y) = 0.122t - 3.777$$
$$e^{Ln(Y)} = e^{0.122t - 3.777}$$
$$Y = e^{-3.777} e^{0.122t}$$
$$Y = 0.0229 e^{0.122t}$$

6. *Exponent of 10 form.* It is sometimes preferable to express the equation as an exponent of 10 rather than e. To do this, repeat the last procedrue but do not multiply by 2.303:

$$\text{Log}(Y) = a + bX$$
$$Y = 10^a 10^{bX} \qquad (4.25)$$

Example *Using the previous equation, we obtain*

$$\text{Log}(Y) = 0.053t - 1.64$$
$$Y = 10^{-1.64}10^{0.053t}$$
$$Y = 0.023(10)^{0.053t}$$

Notice that 0.023 is the original intercept as read from a semilog plot and 0.053 the slope determined from that plot.

4.1.4 Survivor Curves and *D* Values

When bacteria or bacterial spores are exposed to heat, they die at an exponential rate that can be determined from a semilog plot. The time required to cross one log cycle is called the *D* value ("decimal reduction") and its reciprocal is the slope. The smaller the *D* value, the faster it indicates the rate of destruction. We usually subscript the *D* to indicate the temperature at which it is measured.

Example *A culture containing 800 spores per mL is divided among several containers and subjected to a temperature of 245°C for different times up to 50 min. The number of survivors per mL is recorded in Table 4.2.*

The data are plotted on a four-cycle semilog plot as shown in Figure 4.19 and fitted with a straight line. The fitted line takes 14 min to cross each log cycle, indicating that there is a 10-fold reduction in spore count every 14 min. For example, it takes 14 min to go from 100 survivors to 10, another 14 min to go from 10 to 1, etc. Thus, D = 14 min and the slope of the line is

$$\text{Slope} = \frac{1}{14} = 0.0714$$

The equation for the line is

$$\text{Log}(N) = \text{Log}(800) - 0.0714t$$

or, in an exponent of 10 form:

$$N = 800(10)^{-0.0714t}$$

Table 4.2 Survival bacterial spores data at different time intervals

Time (min)	Spores/mL
0	800
10	190
20	27
30	6
40	1
50	0.2

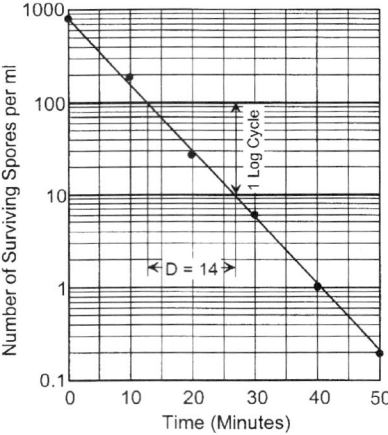

Figure 4.19   Semilog plot of survivor microorganisms versus time.

4.1.5   The Thermal Resistance Curve and *z* Value

D values are a function of temperature. As temperature increases, the rate of spore destruction increases and D decreases (Fig. 4.20A). It turns out that the change in the D value is an exponential function of temperature. So, when we again plot the values on a semilog plot, the resulting curve is called a thermal resistance curve as shown in Figure 4.20B. The increase in temperature in °F or °C required for D to decrease by one log cycle is called z.

> **Example**   The experiment described in the preceding section is repeated at 230, 245, and 260°F. The resulting data fit the curves shown in Figure 4.20A. From these curves, we get the following D values:

$$D_{230} = 40 \text{ min}, \quad D_{245} = 14 \text{ min}, \quad D_{260} = 5 \text{ min}$$

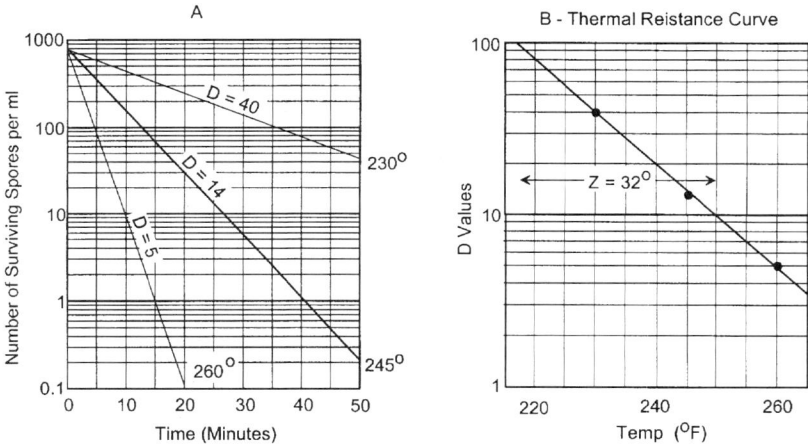

Figure 4.20   Plot of survivor curves to determine various D values (A) and a thermal resistance curve to determine the z value (B).

We plot D versus temperature on a semilog plot as shown in Figure 4.20B and fit the points with a straight line. This line crosses one log cycle every 32°F, indicating a 10-fold decrease in D every 32°F. This is z for this particular organism.

If we select a reference temperature (T_{ref}) and measure D at that temperature (D_{ref}), then the thermal resistance curve in Figure 4.20B has the equation

$$\text{Log}(D) = \text{Log}(D_{ref}) - \frac{1}{z}(T - T_{ref}) \tag{4.26}$$

Rearranging Eq. (4.26) gives

$$\text{Log}(D) - \text{Log}(D_{ref}) = -\frac{T - T_{ref}}{z}$$
$$\text{Log}\left(\frac{D}{D_{ref}}\right) = \frac{T_{ref} - T}{z} \tag{4.27}$$

Example   *250°F (121°C) is commonly selected as the reference temperature. C. botulinum has a z value of 18°F and a $D_{250} = 0.2$ min. What is the equation for the thermal resistance curve of C botulinum? What is the D value at 240°F?*

Solution   Since $T_{ref} = 250$ and $D_{ref} = 0.2$, the thermal resistance curve for *C. bolulinum* (Eq. 26) has the equation

$$\text{Log}(D) = \text{Log}(0.2) - \frac{T - 250}{18}$$
$$\text{Log}(D) = -0.70 - \frac{T - 250}{18}$$

From this, we compute

$$\text{Log}(D_{240}) = -0.70 - \frac{240 - 250}{18} = -0.14$$
$$D_{240} = 10^{-0.14} = 0.72 \text{ min}$$

In other words, at 250°F, t takes 0.2 min to reduce *C. botulinum* spores by a factor of 10. At 240°F, it takes 0.72 min to achieve the same reduction.

4.1.6   Thermal Death Time Curves

We are usually interested in reducing bacterial or spore counts by many factors of 10. A typical canning operation, for example, might reduce spore counts from an initial count of 10^3 to a final count of 10^{-9} or one spore per 10^9 cans. (When the spore count is less than 1 per can, it is called the "probability of a nonsterile unit or PNSU"). This is a reduction of

10^{-12} or 12 log cycles. We will define Y_n as the number of log cycle reductions to be achieved. Thus,

$$Y_n = \text{Log}(N_0) - \text{Log}(N) \tag{4.28}$$

where $N_0 =$ the initial bacterial or spore count and $N =$ the final bacterial or spore count or PNSU to be achieved. If we define F_T as the number of minutes required to achieve Y_n log cycles of reduction, then

$$F_T = Y_n D_T \tag{4.29}$$

Example *For a certain organism, $D_{245} = 12$ min. What is the time required to reduce a count of 10^4 to a PNSU of 10^{-6} (one can in a million) at $245°C$.*

$$Y_n = \text{Log}(10^4) - \text{Log}(10^{-6}) = 4 - (-6) = 10 \text{ log cycles}$$
$$F_{245} = Y_n D_{245} = 10(12 \text{ min}) = 120 \text{ min}$$

Example *For C. botulinum spores, $D_{250} = 0.2$ and $z = 18$. Graphically determine the time required to reduce the spore count from 10^2 to 10^{-6} at $240°F$. Verify your results algebraically.*

Solution In Figure 4.21, the D line is the thermal resistance curve for *C. botulinum*, passing through 0.2 min at 250°F and with a slope of 1/18. The number of log cycles in the required reduction is

$$Y_n = \text{Log}(10^2) - \text{Log}(10^{-6}) = 8$$

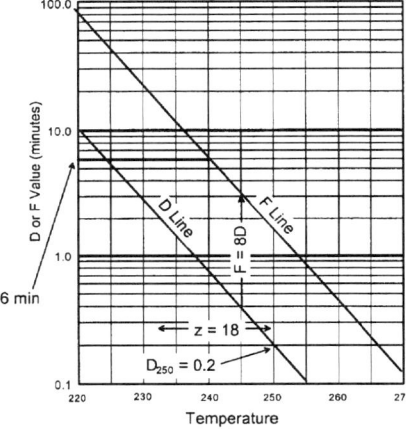

Figure 4.21 Plot showing thermal resistance (*D* values versus temperature) and thermal death (*F* values versus temperature) curves.

Pick any point on the D line and move it to Y_n times its value. Draw a line through this point parallel to the original line. For example, pick $D = 0.4$ at $245°F$ and move it to $8x\ 0.04 = 3.2$ at the same temperature. The resulting F line is the "thermal death time (TDT) curve" for an eight log cycle reduction. The required time at $240°F$ is read from this curve as $5.7\,min$. The D line in Figure 4.21 fits the equation

$$Log(D_T) = Log(0.2) - \frac{T - 250}{18}$$

The F line in the figure has the same slope and reference temperature, but is shifted up so its equation is

$$Log(F_T) = Log(8 * 0.2) - \frac{T - 250}{18} = 0.204 - \frac{T - 250}{18}$$

Substituting $240°F$ in this equation gives

$$Log(F_{240}) = -0.204 - \frac{240 - 250}{18} = 0.760$$

$$F_{240} = 5.75\ min$$

4.1.7 F_0

The destruction of *C. botulinum* is of such importance in canning that we have a special symbol for thermal death time at $250°F$ ($121°C$) when applied to this organism. We define $F_0 = F_{250}$ when $z = 18$. In order to compare two processes that may involve different times and different temperatures, we convert them to the equivalent F_0, either graphically or with the equations

$$Log(F_0) = Log(F_T) + \frac{T - 250}{18} \tag{4.30}$$

where T = the temperature at which a process is performed in minutes, F_T = the time the temperature is applied, and F_0 = the time required at $250°F$ for the same destruction of *C. botulinum*. Solving Eq. (4.30) for F_0, we have

$$F_0 = F_T 10^{\left(\frac{T-250}{18}\right)} \tag{4.31}$$

The number of log cycles for the destruction of *C. botulinum* can then be computed by

$$Y_n = \frac{F_0}{0.2} \tag{4.32}$$

Example *A culture containing a C. botulinum spore count of 10^2 is subjected to $242°F$ for 3 min. What is the probability of a nonsterile unit (PNSU) after this process.*

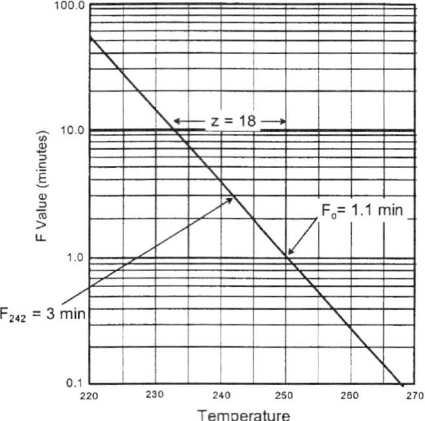

Figure 4.22   Plot showing thermal death time (*F* values versus temperature) curves to determine the F_0 value at 250°F.

Graphic Solution   In Figure 4.22 we draw a line through 3 min at 242°F with a *z* value of 18. The line crosses 250°F at 1.1 min, which is F_0 for this process.

Algebraic Solution

$$\text{Log}(F_0) = \text{Log}(3) + \frac{242 - 250}{18}$$

$$F_0 = 1.08$$

$$Y_n = \frac{1.08}{0.2} = 5.4 \text{ log cycles}$$

$$\text{Log}(N) = \text{Log}(10^2) - 5.4 = -3.4$$

$$N = 10^{-3.4} = 3.98 \times 10^{-4} = \text{PNSU}$$

4.2   LAB EXERCISES

1. The following exercise is based on the use of Fourier's law of conduction. A convection oven with dimensions $3 \times 3 \times 3$ is set at 375°F and is being used to roast poultry. The product needs to be roasted for 3 h in order to reach the desired temperature at the center. Five sides of the oven are well insulated with a 1-inch thick rockwool ($k = 0.030$ Btu/h ft °F). It can be assumed that the metal surfaces enclosing the insulation offer negligible resistance to heat flow. The front side of the oven is made of $\frac{1}{4}$-inch Pyrex glass ($k = 0.050$ Btu ft/h ft² °F).

   a. Calculate the heat loss (Btu/h) from the five insulated sides and the one glass side to a room at 70°F.

   b. If an additional 1 in. of glass wool ($k = 0.043$ Btu/h ft °F is added directly outside the rockwool, what is the total thermal resistance of the four layers (steel, glass, rockwool, and steel) if the steel layers are $\frac{1}{8}$ in. thick and have $k = 9.4$ Btu/h ft °F.

   c. Calculate the total heat loss with the added glass wool layer.

   d. Estimate the temperature at the boundary between the glass and rockwool layers.

   e. Calculate the heat lost if there was no insulation and the five walls were built of two $\frac{1}{8}$-inch stainless steel sheets attached back to back.

2. A batch of milk with a volume of 250 L and an initial temperature of 4°C is heated in a steam-jacketed, agitated vessel with a heating surface of 1.5 m².

   a. Calculate the time interval to heat the milk to 70°C if the overall heat transfer coefficient is 800 W/m² K. The specific heat of the milk is 3900 J/kg K and the density 1030 kg/m³. The steam temperature is 130°C. Use the equation

$$\frac{T - T_1}{T_0 - T_1} = \mathrm{Exp}\left(\frac{-UAt}{mC_p}\right)$$

   Note: $\mathrm{Exp}(x) = e^x$.

   b. Calculate three possible changes in the system that will each reduce the heating time of the same volume of the same material by exactly 25%.

3. A flat retort pouch (2.50 cm thick) containing a beef product (thermal conductivity $= 0.519$ W/m °C, specific heat $= 3.35$ J/g °C, density $= 1.12$ g/cm³) is at 76.7°C. It is suddenly introduced into a retort using steam at 118.5°C as the heating medium.

   a. Divide the thickness of the pouch into ten divisions (slices). On a regular graph paper with divisions in centimeters, use the Schmidt method to plot the temperature at each of the distance intervals for five time intervals.

   b. What is Δx and Δt for this plot?

   c. What is the temperature at the center after five time intervals? How long did it take to reach that temperature?

   d. What is the temperature at a depth of 0.75 cm after five time intervals?

   e. Using a Gurney–Lurie chart, determine the temperature at 0.75-cm depth after 10 min. Assume that the convective heat transfer from the steam is between 1000 and 5000 W/m² K.

4. A can with a diameter of 8 cm and length of 10 cm contains a paste with thermal conductivity of 0.245 W/m K, density of 992 kg/m³, specific heat of 3.8 kJ/kg K, and an initial temperature of 8°C. It is heated in an agitated hot water bath at 80°C with a convective heat transfer coefficient of 1500 W/m² K.

   a. What is the temperature at a point midway between the ends of the can and 3 cm from the side after 4 h?

   b. What is the temperature at the exact center after 4 h? (Do no more calculation than necessary.)

5. A spore suspension of *Bacillus stearothermophilus* contains 10^7 spores per mL. Samples of 0.01 mL of this suspension are heated in a phosphate buffer (pH 7.0) at

Table 4.3 Spore survival data

Time (min)	Number of Survivors/Sample	
	At 230°F	At 240°F
10	7.8×10^4	4.5×10^4
20	6.0×10^4	2.0×10^4
30	4.7×10^4	9.0×10^3
40	3.6×10^4	4.0×10^3
70	1.7×10^4	
100	8.0×10^3	

two temperatures for varying periods of time. The following counts/sample of surviving spores were obtained ($8.5E4 = 8.5 \times 10^4$):

a. Using two-cycle semilog paper, with time on the linear axis and survivor count on the log axis, plot the data shown in Table 4.3.

b. Draw separate straight lines to fit the data for each temperature. From these lines, determine the D value for each temperature.

c. To visualize the exponential nature of these curves, plot the same data on linear paper.

d. Using another piece of semilog paper with temperatures from 220 to 270°F on the linear axis and D values on the log axis, plot a thermal resistance curve and determine the z values for this organism.

e. Using the information from b and d, estimate the D value at 250°F.

f. Label 12-cycle semilog paper to cover spore counts from 10^3, down to 10^{-9}, and time from 0 to 100 min. Using the D value obtained in e, plot a survivor curve for 250°F. Determine the time required for a can with 10^3 spores to reach a PSNU of 10^{-9}.

g. Verify that your estimate of time in f agrees with that computed with the equation

$$\text{Log}_{10}(N) = \text{Log}_{10}(N_0) - \frac{t}{D}$$

h. On the same graph as your thermal resistance (D) curve (part d), plot a thermal death F curve for the reduction from 10^3 to 10^{-9}. Compare an estimate from this line with the estimates in f and g. From this curve, determine the time required to achieve this reduction at 260°C.

6. A canning factory produces 500,000 cans per day and operates for 40 days per year. Cans are processed to a PSNU of 10^{-9}.

   a. How frequently will you expect a can containing a spore to occur?

   b. What would the frequency be if you increased process time by D min?

7. By definition, F_0 is the processing time at 250°F with $z = 18$ (or 121°C with $z = 10$). How long must a can be processed at 225°F to obtain a F_0 of 8.0 min?

   a. Solve this problem graphically using three-cycle semilog graph paper. Scale the arithmetic axis so that it includes the range from 225 to 250. Scale the log axis so that it covers the D value range from 1 to 1000. Plot a line through 8 min at 250°F with a slope defined by $z = 18$ and read the time at 225°C.

b. Verify that you obtain the same answer if you solve this problem with the equation

$$\frac{F_0}{F_T} = 10^{(T-250/18)}$$

8. A container of food innoculated with *C. botulinum* spores is increased at 274°F for 12 s. (Disregard the time required to heat and cool the container.)

a. Graphically determine how long the can would have to be processed at 250°F to obtain the same pore destruction.

b. Verify the graphic solution algebraically.

c. Assuming an initial starting count of 2×10^3 per can and $D_{250} = 0.2$, what is the PNSU at the end of process?

4.3 SUGGESTED READING AND REFERENCES

1. M. Karel, O. R. Fennema, and D. B. Lund, 1975. *Physical Principles of Food Preservation*, New York: Marcer Dekker.
2. P. Fellows, 1988. *Food Process Technology; Principles and Practice*, Chichester, UK: Ellis Horwood Ltd.

5

CANNING, RETORT THERMAL PROCESSING, AND LETHALITY COMPUTATION BY GENERAL METHOD

5.1 INTRODUCTION

Food canning is the procedure for preserving food by sealing it in a hermetically sealed container and heating it to destroy pathogenic spoilage causing microorganisms and their spores, and to inactivate enzymes. Such products are said to be commercially sterilized. This process differs from pasteurization in which a lower level of heat treatment is used to destroy pathogenic organisms, leaving some spoilage organisms still viable.

Thermal processing may be carried out in batch retorts or continuous pressure cookers. A retort may be a still or agitating type, and may be designed to operate with saturated steam, or hot water. By processing under pressure, it is possible to use temperatures around 250°F (121°C), which greatly speed the destruction of microorganisms and spores.

In canning, heat is transferred through container walls to solid food by conduction and to liquid foods by convection, either natural or forced. The rate of heating of the food will depend on the nature of the heating medium, the conduction coefficient (thermal conductivity) of the can and food, and whether or not convection circulates the food within the can.

5.2 BACKGROUND

5.2.1 Retort Construction

A small retort is used in this chapter to test the processes used in the canning industry. A typical schematic diagram of a retort is shown in Figure 5.1. In general, the construction features of a retort are described below:

1. The retort chamber is designed and built according to the American Society of Mechanical Engineers Code and is made of heavy metal walls to withstand high pressures. In normal use, 15-psig pressure is used to attain temperatures around 250°F.

2. A strong door is provided with a locking system designed to withstand the pressure.

3. Support rods are attached within the chamber to an agitator disk. Cans are clamped to these rods for processing and the agitator disk can be rotated at speed selected by the user, if needed. The agitation circulates the contents of the cans and accelerates the rate of heat transfer within the can. In some retorts, special slip rings on the agitator shaft, such as those shown in Figure 5.2, provide connections between the rotating thermocouples and the stationary wires on the outside.

4. Thermocouples (TC) can be placed within the chamber and cans. These are attached to cables that carry thermocouple readings to the outside data logger for recording.

5. Steam, water, and air for heating/cooling are supplied through feed pipes. On/off and regulator valves control the flow through these feeds.

6. Vents allow steam and air to be removed from the chamber. A drain valve allows water to be removed.

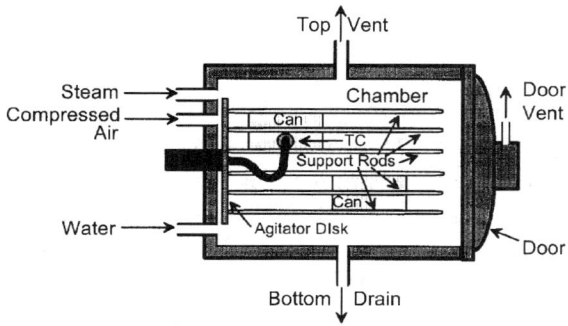

Figure 5.1 Schematic diagram of a typical horizontal still retort.

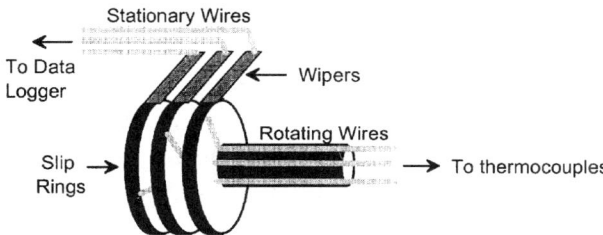

Figure 5.2 Typical slip rings connecting rotating thermocouple wires to the outside data logger.

5.2.2 Process Control

A control panel is used to operate the retort consisting of on/off valves to control steam, water, and air; a drain valve; vent valves to control temperature and pressure inside the retort.

5.2.2.1 On/Off Valves
Various on/off valves are used to control the feed, drain, and vent lines of the retort. The main valves of importance include:

1. *Upper vent.* Open to allow air to escape when steam is first introduced into the chamber. Open again after processing to allow steam and air to escape.
2. *Lower drain.* Open during turn-on and slightly open during processing to allow condensed steam to drain. Open again at the end to drain cooling water.
3. *Steam.* Open throughout the processing time. Closed during cooling.
4. *Water.* Introduce water into the chamber for cooling.
5. *Air.* Open during early cooling to maintain pressure. Open after cooling to force cooling water out of the chamber.

5.2.3 Running a Retort

A typical procedure for running a retort is as follows.

5.2.3.1 Procedure

1. *Loading*
   a. Load cans into the autoclave, secure the cans, and attach thermocouples.
   b. Plug in and turn on data logger. Verify thermocouple connections.
   c. If agitation is to be used, check the rpm and verify the proper stowage of cans.
   d. Close the retort's door tightly and lock the handle on top. Close all valves.
2. *Heating.* Steam under pressure is used to raise the temperature of the chamber to around 250°F and hold it there for enough time to destroy spores in the cans.
   a. Open steam, water, and air main valves. Open the steam line drain valve to let condensate out. Close the drain valve when steam starts coming out of the drain valve.
   b. Set the desired temperature on the temperature controller dial. Check that the pressure controller is set to 15 psig.
   c. Open the lower drain and upper vent valves.
   d. Open the control steam valve and vent valve on autoclave door.
   e. After the temperature reaches 212°F, vent the retort for 2 min.
   f. Close the upper vent valve, close the lower drain valve $\frac{3}{4}$, and close the vent valve on the door.
   g. Process as necessary.

3. *Pressure cooling.* Water is used to cool the cans. Compressed air is used to maintain pressure during cooling so that the contents of the cans do not boil and possibly explode.

   a. Close the steam and lower drain valves completely, set the temperature controller to 0.

   b. Open the autoclave air valve slightly.

   c. Open the cold water valve. Take care to maintain a constant pressure at 15 psig and keep water in the glass column below a specific level by adjusting the water control valve, upper vent, and air supply valve.

   d. Fill the autoclave with water to cover cans, close the cold water valve, and cool your product to approximately 212°F.

4. *Atmospheric final cooling.* Once the danger of boiling a product has passed, cans are cooled with water at atmospheric pressure. Then compressed air is used to drive the water out of the tank.

   a. Close the autoclave air and slowly open the upper vent.

   b. Open the cold water upper and lower drain valves, adjusting water flow to maintain a constant level in the autoclave.

   c. Cool the product to approximately 100°F.

   d. Close the cold water valve, upper vent.

   e. Open the lower drain wide and slightly open the autoclave air to push out water. Drain the autoclave until empty.

   f. Close all valves, shut off mains, turn off the data logger, remove your product, and clean the area.

   g. Open the steam, water, and air main valves. Open the steam line drain valve to let condensate out.

5.2.4 Temperature History of a Process

When food is processed, a typical retort chamber temperature profile such as that shown in Figure 5.3 results.

1. When food is initially loaded, the retort chamber is around room temperature.

2. When steam or hot water is introduced, the chamber temperature rises rapidly to the

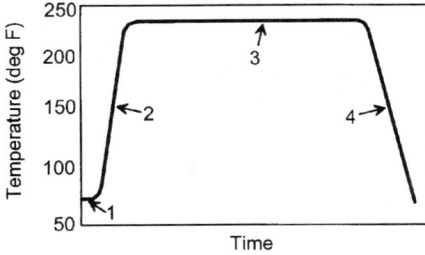

Figure 5.3 Typical temperature time profile in a retort.

value set on the controller. The time required to reach the processing temperature is called come-up time.

3. The chamber is held at this temperature for the required period of time. The time at this temperature is called heating time. The sum of come-up time and heating time is called processing time.

4. When cold water replaces the steam, the temperature comes down. The time required to reach approximately room temperature is called cool-down time.

Because of the thermal resistance and heat capacity of the food and container, the temperature of the food changes more slowly than the retort chamber. In particular, a point near the can center is slowest to change. A thick line shows a typical temperature history at the center of a can during the retort process as shown in Figure 5.4.

5.2.5 The Lethal Rate

As pointed out previously in Chapter 4, generally, *C. botulinum* spores are destroyed at a rate of one log cycle every 0.2 min in a phosphate buffer at 250°F. It has also been observed that at other temperatures, the processing time can be adjusted to an equivalent time at 250°F with the following equation:

$$F_0 = F_T 10^{[(T-250)/18]} \tag{5.1}$$

where F_T = minutes of processing at $T°$F, F_0 = minutes of processing at 250°F, and $10^{[(T-250)/18]}$ = is a conversion factor. Although strictly dimensionless, it is useful to think of the factor as having the units "minutes at 250° per minute at $T°$F." The 250 in this factor is the arbitrary but commonly used reference temperature. The 18 is the z value for *C. botulinum*.

1. In general, the conversion factor in Eq. (5.1) is called the lethal rate. In other words,

$$\text{Lethal rate} = L = 10^{[(T-250)/18]} \tag{5.2}$$

2. F_0 is called the lethality of a process.

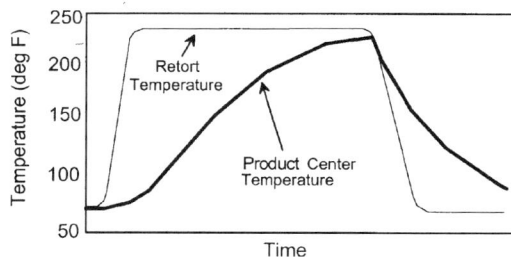

Figure 5.4   Typical product temperature monitored in the center of a can during retort processing.

3. Lethality F_0 of a process is computed by multiplying the lethal rate L by the time at temperature T (F_T), as shown in Eq. (5.1):

$$\text{Lethality} = F_0 = F_T \cdot L \tag{5.3}$$

The temperature in the can varies continuously as shown in Figure 5.5. We, therefore, cannot simply compute a single lethal rate for the process and multiply it by processing time. Instead, we compute lethal rates at various intervals along the process, and then integrate this rate over time. We will use the following example to show several ways that this can be done.

Example Typical time-temperature data taken from the center of a can during retort thermal processing are shown in Table 5.1. The lethal rate, computed using Eq. (5.2), is recorded in the third column. For example, the lethal rate at 30 min is computed as

$$L = 10^{[(241-250)/18]} = 10^{-0.5} = 0.32 \text{ min at 250 per min at } 241°$$

The lethal rates L versus time in Table 5.1 are plotted in Figure 5.5.

5.2.6 Computing Lethality

The lethality of the process is computed by integrating the lethal rate across time:

$$F_0 = \int_{10}^{55} L \, dT \tag{5.4}$$

Notice that, as indicated in Figure 5.5, lethality is a function of time, that is, $L = f(t)$. However, it is not a function that can be expressed as a simple equation and so the integration indicated in Eq. (5.4) cannot be readily performed. However, since integration

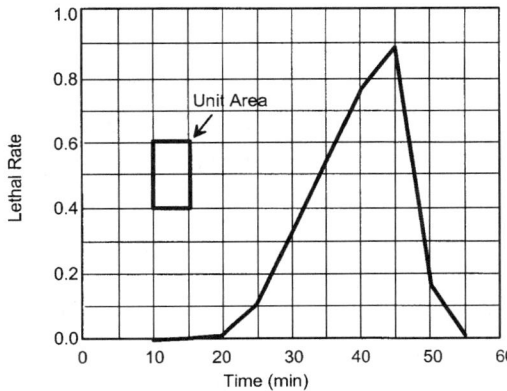

Figure 5.5 Typical plot of a lethal rate versus time obtained during retort processing.

Table 5.1   Typical thermal processing data obtained in a retort

Time (min)	Center Temperature (°F)	Lethal Rate
10	110	1.7×10^{-8}
15	180	1.3×10^{-4}
20	215	0.011
25	232	0.10
30	241	0.32
35	245	0.53
40	248	0.77
45	249	0.88
50	235	0.15
55	196	0.001

can be viewed as a summation of the area under a function, any method of computing areas will work. Several methods are described below to compute the area under a curve.

5.2.6.1   Counting Squares

1. Plot a lethality curve such as the one shown in Figure 5.5 on a graph paper that is ruled into squares. The smaller the squares, the more accurate the integration.
2. Count the number of graph squares that fall inside the curve. If a square falls on the curve, count it only if more than half the square falls inside the curve.
3. Draw a rectangle with an area of 1.0 min and count the squares in the rectangle.
4. Divide the number of squares under the curve by the number of squares in the unit area. The result is F_0.

Example   *In Figure 5.5, there are approximately 29 squares that are complete or more than half under the lethal rate curve. The unit area (5 min × 0.2 = 1.0) contains exactly two squares. The integral of the curve is*

$$F_0 = \frac{29}{2} = 14.5 \text{ min}$$

In other words, the process listed in Table 5.1 should destroy the same number of C. botulinum spores as if they were processed for 14.5 min at 250°F. Since the squares used in this example are large, the result is fairly inexact.

5.2.6.2   By Weight

1. Plot a lethality curve such as the one in Figure 5.5 on a graph paper.
2. Cut out the area between the curve and the horizontal axis and weigh it.
3. Cut out a rectangle of unit area and weigh it.
4. Divide the weight of the curve by the weight of the unit area to compute F_0.

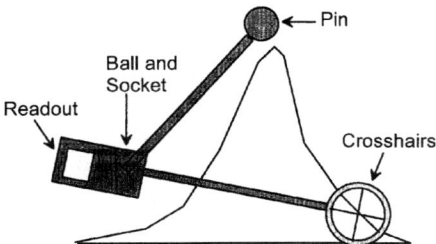

Figure 5.6 A polar planimeter.

5.2.6.3 Polar Planimeter A polar planimeter (Fig. 5.6) is a device that mechanically integrates a closed area. Use it as follows:

1. Draw a lethal rate curve on a graph paper and fasten it to a drawing board.
2. Place the arm with the cross-hair on the graph. Place the ball on the other arm in the socket of the first arm. Press the pin of the second arm into the drawing board.
3. Mark a starting point on the lethal rate curve and place the cross-hair over that point.
4. Read the planimeter.
5. Carefully trace along the curve. Do not backtrack. If you leave the line, trace an equal area on the other side of the line to compensate. When you reach the end, trace back along the horizontal axis. Continue until you have traced all the way around the area to be integrated and back to the starting point.
6. Read the planimeter and compute the difference between the two readings.
7. Repeat the process several times and average the readings.
8. In the same way, measure a unit area several times and average.
9. Divide the curve average area by the unit average area to obtain F_0.

5.2.6.4 Trapezoidal Rule The area under a curve can be approximated by a series of parallelograms as shown in Figure 5.7. The area of a parallelogram is computed as the

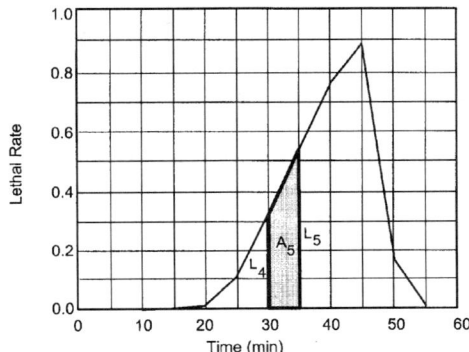

Figure 5.7 Integration of a curve area by using the trapezoidal method.

average height times the width:

$$A_i = \frac{L_{i-1} + L_i}{2} \Delta t \tag{5.5}$$

where A_i = the area of the ith parallelogram, L_i = the height (lethal rate) of the right-hand side of the ith parallelogram, L_{i-1} = the height of the left-hand side of the ith parallelogram, and Δt = the width of each parallelogram. The area under the curve is the sum of all parallelograms:

$$
\begin{aligned}
F_0 &= \sum_{i=1}^{n} = A_1 + A_2 + A_3 + \cdots + A_n \\
F_0 &= \sum_{i=1}^{n} = \Delta t \left(\frac{L_0 + L_1}{2} + \frac{L_1 + L_2}{2} + \frac{L_2 + L_3}{2} + \cdots + \frac{L_{n-1} + L_n}{2} \right) \tag{5.6} \\
F_0 &= \sum_{i=1}^{n} = \frac{\Delta t}{2} (L_0 + 2L_1 + 2L_2 + 2L_3 + \cdots + 2L_{n-1} + L_n)
\end{aligned}
$$

Equation (5.6) is used to compute the integral. Notice that because each side except the first and last is included in two parallelograms, they are included twice in the sum.

> **Example** *Using the lethal rates from Table 5.1, we determine that the integral by the trapezoidal rule is*
>
> $$
> \begin{aligned}
> F_0 &= \frac{5}{2}[1.70^{-8} + 2(1.30^{-4}) + 2(0.011) + 2(0.10) + 2(0.32) + 2(0.53) + \cdots \\
> &\quad + 2(0.15) + 1.0 \times 10^{-3}] \\
> F_0 &= \frac{5}{2} 5.25 = 13.08 \text{ min}
> \end{aligned}
> $$

5.2.6.5 Simpson's Rule   The trapezoidal rule approximates the lethal rate curve with a series of straight lines. Since the curve is continuous, this introduces a slight error into the integral. A more exact integral can be obtained by approximating the curve with a series of short parabolas. Simpson's rule requires that there be an even number of intervals or an odd number of lethal rates. It uses the equation:

$$F_0 = \frac{\Delta T}{3}[L_0 + 4(L_1 + L_3 + \cdots + L_{n-1}) + 2(L_2 + L_4 + \cdots + L_{n-2}) + L_n] \tag{5.7}$$

Notice that in this equation, the first and last L values are summed once. Of those remaining, the odd-numbered L values are summed four times each and the even-numbered values are summed two times each.

> **Example** *Table 5.1 has an even number of 10 values. However, the first two are so small that dropping the first one will have negligible effect. Using the rest, we obtain*
>
> $$
> \begin{aligned}
> F_0 &= \frac{5}{3}[1.3 \times 10^{-4} + 4(0.011 + 0.32 + 0.77 + 0.15) + 2(0.10 + 0.53 + 0.88) \\
> &\quad + 1.0 \times 10^{-3}] \\
> F_0 &= \frac{5}{3}(7.88) = 13.14
> \end{aligned}
> $$

5.2.6.6 Patashnik's Method Patashnik's method is a simple adaptation of the trapezoidal rule arranged so that it is easy to compute F_0 values while the retort is running. This makes it possible to stop the process when the desired F_0 is reached.

1. Temperature readings are taken at regular intervals and recorded every 5 min as shown in Table 5.2.
2. From each temperature reading, compute the lethal rate by the equation

$$L = 10^{[(T-250)/18]}$$

3. The first running total is equal to the first lethal rate. Subsequent running totals are equal to the previous running total plus the current lethal rate.
4. Lethality is computed by the equation

$$F_0 = \left(\text{Previous running total} + \frac{\text{Current lethal rate}}{2}\right) \times \text{Time interval}$$

Example *The calculations made at 30 min are*

$$L = 10^{[(241-250)/18]} = 0.3162$$
$$\text{Running total} = 0.1115 + 0.3162 = 0.4277$$
$$F_0 = \left(0.1115 + \frac{0.3162}{2}\right)5 = 1.348$$

5.2.6.7 Computer Spreadsheet A computer spreadsheet such as Excel that can be used to implement the trapezoidal rule is shown in Table 5.3.

1. Enter times into a column (column A). Times do not have to be evenly spaced.
2. In the next column, enter corresponding temperatures.

Table 5.2 Estimation of process lethality by Patashnik's method

Time (min)	Temperature (°F)	L	Running Total	F_0 (min)
10	110	1.69×10^{-8}	1.69×10^{-8}	
15	180	1.29×10^{-4}	1.29×10^{-4}	6.45×10^{-5}
20	215	0.0114	0.0115	0.039
25	232	0.1000	0.1115	0.308
30	241	0.3162	0.4277	1.348
35	245	0.5275	0.9552	3.457
40	248	0.7743	1.7295	6.712
45	249	0.8799	2.6094	10.847
50	235	0.1468	2.7562	13.414
55	196	0.0010	2.7572	13.783

Table 5.3 Integration of curve area (process lethality) on a computer spreadsheet

	A	B	C	D	E
	Time	Temperature	L	Area	F_0
1					
2	10	110	0.0000	—	0.000
3	15	180	0.0001	0.0003	0.000
4	20	215	0.0114	0.0287	0.029
5	25	232	0.1000	0.2784	0.307
6	30	241	0.3162	1.0406	1.348
7	35	245	0.5275	2.1093	3.457
8	40	248	0.7743	3.2544	6.712
9	45	249	0.8799	4.1355	10.847
10	50	235	0.1468	2.5668	13.414
11	55	196	0.0010	0.3694	13.783
12					
13			$F_0 =$	13.783	

3. In the third column, enter the formula to compute lethality and copy it down the column. For example, the formula in cell C2 of Table 5.3 is

$$\text{Excel:} = 10^{\wedge}((\text{B2} - 250)/18)$$

4. In the first cell of the fourth column, enter a formula to compute areas of parallelograms. For example, the formula in cell D3 of Table 5.3 is

$$\text{Excel:} = (\text{A3} - \text{A2})^*(\text{C2} + \text{C3})/2$$

5. To integrate the area for the entire process, enter a summation at the bottom of the area column. For example, the formula in cell D13 is

$$\text{Excel:} = \text{Sum}(\text{D3}:\text{D11})$$

6. To estimate the cumulative integrals throughout the process, enter equations to do this in the fifth column. In Table 5.3, start by placing the number 0 in cell E2. Then place the following formula in cell E3 and copy it down the rest of the column:

$$\text{Excel:} = \text{E2} + \text{D3}$$

Notice that the last sum in column E is the same as the sum in cell D13.

5.3 LAB EXERCISE

5.3.1 Prelab Questions

1. How are each of the following used in processing food in a retort: (1) steam, (2) water, (3) compressed air, (4) vents, and (5) drains?
2. Should you use air-to-open or air-to-close valves to control the flow of steam into a retort? Why? Suggest a use for the other kind of valve.
3. What will cause a flapper on a pneumatic controller to open? How does its opening affect the air pressure to a pneumatic controller valve?
4. Name seven methods for integrating a function that cannot be expressed as an equation. Which methods are based on trapezoids?

5.3.2 Objectives

1. Become familiar with thermocouples used for measuring temperatures of canned foods and how to install them in cans.
2. Learn to seal a can with or without vacuum.
3. Become familiar with the operation of the retort for thermal processing of cans.
4. Plot and analyze time-temperature data.
5. Determine the F_0 value using Patashnik's meethod during the process.
6. Calculate the F_0 value using graphical and numerical integration methods and compare the results.
7. Compare the F_0 of convective and conductive foods.

5.3.3 Materials

1. Retort.
2. Data logger.
3. Thermocouples and fittings to fit 211×300 cans.
4. Pan punch for installing thermocouples.
5. Vacuum can sealer.
6. Can opener.
7. Chart paper for data logger.
8. 211×300 cans and lids.
9. Tomato paste or other high-viscosity food for canning.
10. Tomato juice or other low-viscosity food for canning.

5.3.4 Procedure

5.3.4.1 Set Up the Data Logger

1. Turn on the data logger.

2. Set the logger to record temperature at the desired number of channel locations.

3. Set time intervals to record data every 1 to 2 min.

4. Test that all thermocouples are recording the temperature correctly.

5.3.5 Cans' Preparation and Retort Processing

1. Each student should install a thermocouple in a 211×300 can. Make sure that the tip of the thermocouple is at the center of the can. Check that the thermocouple connector seal is seated well.

2. Half the cans should be filled with a high-viscosity food that will not circulate due to convection. Tomato paste is a good choice.

3. Half the cans should be filled with a low-viscosity food that will circulate due to convection. Tomato juice is a good choice.

4. Vacuum seal your can.

5. Check that there are no visible leaks. Weigh the can so that you can verify its weight does not change during processing.

6. Attach the thermocouple in each can to a cable inside the retort. Record the thermocouple number and contents of each can.

7. Check that each thermocouple is working properly.

8. Start the data logger printing.

9. Run the retort according to the instructions given in Section 5.2.3.1.

10. Compute F_0 by Patashnik's method after each printout on the data logger.

5.3.6 Report

1. On one graph, plot the following data:
   a. Retort temperature versus time.
   b. Temperature of a high-viscosity food versus time.
   c. Temperature of a low-viscosity food versus time.

2. Determine the following from your graphs:
   a. The approximate come-up time of the retort (time required for the retort to go from the initial temperature to the maximum temperature).
   b. The rate of heating of the retort during come-up time (temperature rise divided by come-up time).
   c. The maximum retort temperature.
   d. The time at maximum retort temperature.
   e. The product temperature at steam off.
   f. The maximum product temperature reached.
   g. The cooling medium temperature.
   h. The rate of cooling of product.

3. Compute lethality and plot lethality versus time for a high- and low-viscosity food.

4. Compute the total lethality for a high-viscosity can be each of the six remaining methods described above, if Patahnik's method is already done.

5. Compute the total lethality for the low-viscosity can by one of the methods.

6. Calculate the process time and total heating times necessary to obtain $F_0 = 2$ min for the low-viscosity food. By how many log cycles will *C. botulinum* spores be reduced in this time?

7. Compare the F_0 values obtained by the different integration methods.

8. Compare the F_0 values of the high- and low-viscosity foods and suggest explanations for the difference.

5.4 SUGGESTED READING AND REFERENCES

1. D. B. Lund, 1975. "Heat processing." In (M. Karel, O. R. Fennema and D. B. Lund (eds.), *Principle of Food Science, Vol. 2: Physical Principles of Food Preservation*. New York: Marcer Dekker.

2. C. R. Stumbo, 1973. *Thermobacteriology in Food Processing*, 2nd ed., New York: Academic Press.

6

HEAT PENETRATION TEST AND THERMAL PROCESS DESIGN USING BALL'S FORMULA METHOD

6.1 INTRODUCTION

Canning is a thermal process in which food is brought to commercial sterility by sealing it hermetically in a container and then heating. When determining the merits of a process or developing a new one, you must examine the following factors:

- Product sterility
- Economy
- Product quality
- Product uniformity

Product sterility requires adequate thermal treatment; the more, the better. On the other hand, the other three factors are usually maximized by reducing the time and temperature of heat treatment. In order to balance these factors, it is important to know the rate of heating of the food so that adequate heat treatment can be achieved without over-processing.

Evaluation of thermal processes involves finding two separate sets of parameters:

1. The first set of parameters describe the kinetics of microbial destruction, including the following parameters discussed in the previous chapters:

    a. *D value*. Time required for a 10-fold reduction of organisms at a given temperature.

    b. *z value*. Number of temperature change, in degrees, required for a 10-fold change in the *D* value.

    c. *Lethal rate L*. It converts the actual heating time of a process at a specified temperature to the time required at 250°F that will achieve the same destruction of *C. botulinum* species.

    d. *Lethality F_0*. Time at 250°F that achieves the same destruction of *C. botulinum* as the process under study.

2. The second set of parameters describes the heat penetration characteristics of the system and includes:

    a. *Temperature response parameters f_h and f_c*. These describe the rate of heat penetration into a container and its contents during heating and cooling, respectively.

    b. *Lag factors j_h and j_c*. These describe the lag time before the heat penetration rate reaches f_h and f_c.

The lethal effect of heat on bacteria is a function of time, temperature, and the initial bacterial population in the product. To design or evaluate a heating process, it is necessary to know the heating characteristics of the slowest heating portion of the container, called the cold zone, the number of microorganisms of concern that are present, and their thermal resistance characteristic. Heat penetration tests are commonly used in the food industry to determine the appropriate process time for a food product to achieve commercial sterility.

Foods such as beef stew and cream-style corn require long processing times in still retorts. This may result in undesirable changes in color, consistency, and vitamin retention. Thermal processing in agitated retorts results in shorter processing times. There are several agitated retorts in commercial use. For example, the FMC Sterilmatic retort is used extensively in the vegetable processing industry. Other brand names include the Stock Rotomat and Orbitort. Some of them rotate the containers around their central axis, whereas others rotate them end over end.

Head space is defined as that portion of the container not occupied by the product. Both the head space and speed of agitation (rpm) can have a significant influence on the rate of heating and cooling, and hence, on the total processing time.

6.2 BACKGROUND

6.2.1 Heat Penetration Test

In a heat penetration test, a thermocouple is placed inside a container so that it measures the temperature of food at the slowest heating point, the so-called cool spot. For conductive heating foods or foods being agitated, this is usually the geometric center of the container. For convective foods that are not agitated, this spot may be a little below the geometric center. Two temperatures are collected over time:

1. The retort chamber temperature T_R.
2. The temperature at the cool spot in the food T.

The difference between these two temperatures provides the driving force that heats the food. This means that as the food temperature approaches the retort temperature, the rate of heating decreases in an exponential manner as shown in Figure 6.1.

To study this process, we define the following variables: t = the time from the start of processing (min), T = the temperature of the cool spot in the food at any time t, T_0 = the temperature of the cool spot in the food at the starting time ($t = 0$), and T_1 = the processing temperature of the retort.

The data shown in Table 6.1 are plotted in Figure 6.1. Since the temperature of the food T approaches the temperature of the retort T_1 exponentially, we compute the difference, $T - T_1$ in column 4 of Table 6.1, and plot it versus time on semilog paper in Figure 6.2. The resulting plot can be divided into two parts:

- The lag phase where the slope of the curve increases.
- The linear phase where the data fit a straight line.

Our objective is to describe both parts of this curve with a single linear equation. Fitting the linear phase is easy. Just draw a line through that part of the data as was done in Figure 6.2, observe that this line crosses the Y axis at 580 and takes 17.5 min to cross one log cycle. Since $T_1 = 240$ in this example, this line has the equation:

$$\log(240 - T) = \log(580) - \frac{1}{17.5}t$$

To put this equation in general terms, we write

$$\log(T_1 - T) = \log(T_1 - T_A) - \frac{t}{f_h} \tag{6.1}$$

where t = the processing time (min), T = the temperature at the cool spot in the food at time t, T_1 = the processing retort temperature, T_A = the apparent initial temperature needed to yield a straight line, and f_h = the time needed for the heat penetration curve to cross one log cycle (note the similarity between f_h and D and z).

Since the graph says that $(T_1 - T_A) = 580$ and $T_1 = 240$, we conclude that the apparent initial temperature is $540 - 580 = -340°F$. Now we know that the initial temperature was

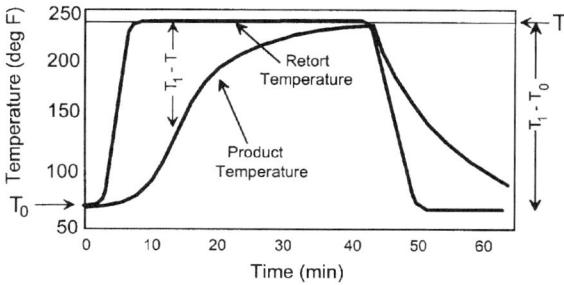

Figure 6.1 Temperature profiles of retort temperature and product temperature at the cool spot in a typical heat penetration test.

Table 6.1 Typical data for a heat penetration test

Time (min) t	Retort temperature T_R	Food temperature T	Difference $T_1 - T$
0	71	$70 = T_0$	170
5	152	75	165
10	$240 = T_1$	94	146
15	240	154	86
20	240	194	46
25	240	215	25
30	240	229	11
35	240	234	6
40	240	$237 = T_B$	3
45	158	195	
50	70	145	
55	$68 = T_2$	118	
60	68	100	

T_1 = Heating medium temperature.
T_2 = Cooling medium temperature.
T_0 = Initial product temperature.
T_B = Product temperature at shutoff.

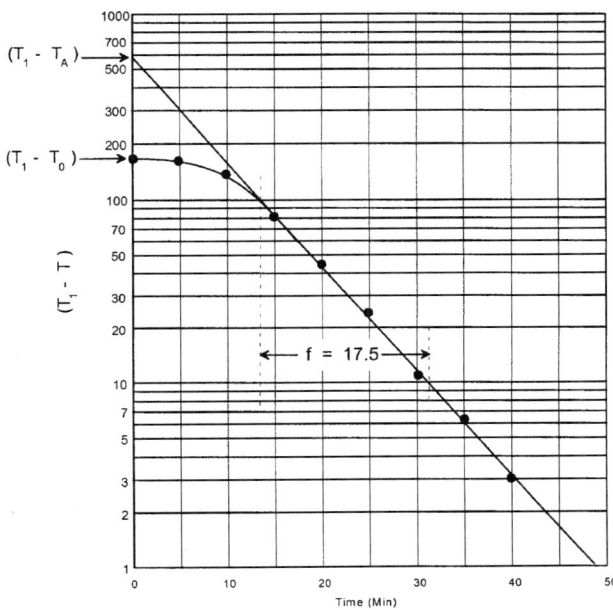

Figure 6.2 Typical plot of a heat penetration curve on semilog paper.

70°F, not -340, so what does the -340 represent? It is a fictitious temperature that we invented to make data fit a straight line which is not completely straight. But it would be more convenient if the intercept was determined by the real initial temperature. How can we do this and still have a linear equation?

To solve this problem, note that in Figure 6.2, the actual curve has an intercept at $\log(T_1 - T_0)$. The difference between this true intercept and the apparent intercept is

$$\text{Difference} = \log(T_1 - T_A) - \log(T_1 - T_0) \tag{6.2}$$

Thus, if we know the true initial value, we can find the apparent intercept by adding this difference. If we call this difference $\log(j_h)$, Eq. (6.2) becomes

$$\log(j_h) = \log(T_1 - T_A) - \log(T_1 - T_0) \tag{6.3}$$

Solve Eq. (6.3) for the apparent intercept, $\log(T_1 - T_A)$:

$$\log(T_1 - T_A) = \log(j_h) + \log(T_1 - T_0) \tag{6.3}$$

Substitute this into Eq. (6.1) and simplify:

$$\log(T_1 - T) = \log(j_h) + \log(T_1 - T_0) - \frac{t}{f_h} \tag{6.4}$$

$$\log(T_1 - T) = \log[j_h(T_1 - T_0)] - \frac{t}{f_h} \tag{6.5}$$

Equation (6.5) is the linear equation of choice, whose intercept is computed from the true initial temperature. The only requirement is that we have some way of determining j. To do this, simplify Eq. (6.3) as follows:

$$\log(j_h) = \log\left(\frac{T_1 - T_A}{T_1 - T_0}\right) \tag{6.6}$$

$$j_h = \frac{T_1 - T_A}{T_1 - T_0} \tag{6.7}$$

Thus, j_h is the ratio of two differences, one apparent and one real. Both differences can be read directly from the semilog plot. We can now determine f_h and j_h for a particular food and container from one set of data, then use them to predict the heating rate for the same product and container in situations with different values for T_0 and T_1.

Example 1   *From Table 6.1, we read $T_1 = 240°$, $T_0 = 70°$. From Figure 6.2, we read $T_1 - T_A = 580°$ and $f_h = 17.5\,min$. We compute j_h from Eq. (6.7):*

$$j_h = \frac{580}{240 - 70} = 3.4$$

Substituting these values into Eq. (6.5), we have

$$\log(240 - T) = \log[3.4(240 - 70)] - \frac{t}{17.5}$$

$$\log(240 - T) = 2.76 - \frac{t}{17.5}$$

or it can be written as

$$T = 240 - 10^{\left(2.76 - \frac{t}{17.5}\right)}$$

Using this equation, you can now predict the temperature at any given time. For example, at 37 min, the temperature is

$$T = 240 - 10^{\left(2.76 - \frac{37}{17.5}\right)} = 240 - 4.42 = 235.6°\text{F}$$

Example 2 *Assuming that f_h and j_h in Example 1 hold the same values, regardless of the initial food temperature and retort temperature, what is the temperature in a can after 28 min if the processing temperature in the retort is 246° and the food is initially at 87°?*

Solution Substituting the values in Eq. (6.5), we obtain

$$\log(246 - T) = \log[3.4(246 - 87)] - \frac{28}{17.5}$$

$$\log(246 - T) = 2.7 - \frac{28}{17.5} = 1.1$$

$$T = 246 - 10^{1.1} = 233.4°\text{F}$$

Example 3 *How long will it take for the same system to reach 250° if the initial temperature is 65° and the retort processing temperature 254°?*

Solution Substituting in Eq. (6.5), we get

$$\log(254 - 250) = \log[3.4(254 - 65)] - \frac{t}{17.5}$$

$$0.602 = 2.808 - \frac{t}{17.5}$$

$$t = 17.5(2.808 - 0.602) = 38.6 \text{ min}$$

6.2.2 Cooling Curves

In this example, Figure 6.1 has been redrawn in Figure 6.3A, with emphasis on the cooling curve. From this, it is possible to compute f_c and j_c values. To do this:

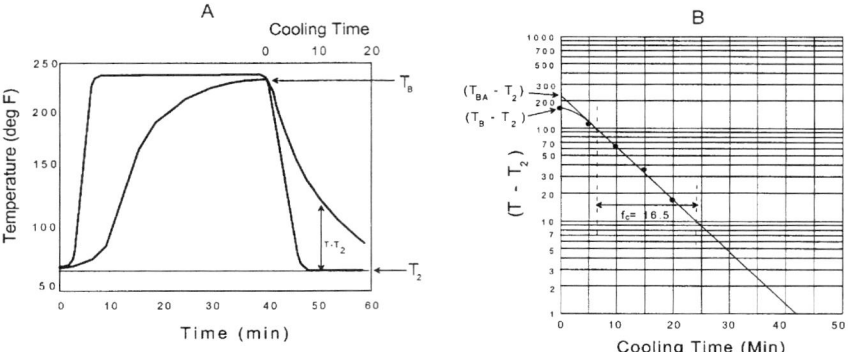

Figure 6.3   Highlight of a temperature profile during the cooling portion of a heat penetration test (A) and a temperature difference versus cooling time on semilog paper (B).

1. Treat the steam turnoff time as time 0 as indicated at the top of Figure 6.3A.
2. Let T_2 = the temperature of the cooling medium.
3. Let T_B = the temperature of the food at turnoff time.
4. Plot $(T - T_2)$ versus time on semilog paper as in Figure 6.3B.
5. Fit a straight line and determine f_c as the time required for the cooling curve to traverse one log cycle.
6. Let T_{BA} = the apparent temperature of the food at shutoff as read from this line.
7. Determine j_c and write out the equation for cooling.
8. Use the resulting equation to predict cooling time and temperature. Let us look at example 4 to determine j_c and cooling time.

Example 4   *Figure 6.3B shows a plot of food temperature minus cooling medium temperature $(T - T_2)$ versus time from the cooling portion of Figure 6.3A. From this plot, we read $f_c = 16.5\,min$ and $T_{BA} - T_2 = 240°$. From Table 6.1, we read the food temperature at shutoff as $T_B = 237°$ and the cooling medium temperature as $T_2 = 68°$. We compute j as*

$$j_c = \frac{T_{BA} - T_2}{T_b - T_2} = \frac{240}{237 - 68} = \frac{240}{169} = 1.42$$

The equation for the cooling curve is, therefore,

$$\log(T - T_2) = \log[1.42(T_B - T_2)] - \frac{t}{16.5}$$

with $t = $ cooling time. This equation can be used to predict cooling times and temperature in the same way as the heating curve equation. If we use the particular *values of T_2 and T_B in this data, the equation becomes*

$$\log(T - 69) = \log[1.42(237 - 68)] - \frac{t}{16.5}$$

$$\log(T - 68) = 2.38 - \frac{t}{16.5}$$

6.2.3   Ball's Formula Method

The next step is to use these heat penetration parameters to design or evaluate a process. Designing involves determining the time required to achieve a certain lethality. Evaluation involves determining the lethality achieved by the process. In the previous chapter, we used the general method to determine the lethality of a process. The problem with that method is that it requires experimental data for each new situation. If we use a new retort, change the initial product temperature or retort temperatures, or change the can size, we need a new set of experimental data.

Ball has proposed a formula for calculating lethality for a new situation using f and j values obtained experimentally for a particular product. This method involves using the same set of f and j values, which can be used with different initial temperatures and heating medium temperatures without further experimentation. Furthermore, formulas are available for converting f values to fit different can sizes.

Ball proposed the following simplification: The temperature curve of the retort slopes upward from time 0 to the time where the processing temperature is reached as shown in Figure 6.4A. During this "come-up" t_c time, the lethal rate is constantly changing. Ball proposed replacing this with a curve that remains at the starting temperature for 58% of the come-up time, then instantly changes to the full processing temperature as shown in Figure 6.4B. Experience has shown that this simplification gives reliable results.

6.2.3.1   Terms   First let us define the following terms shown in Figure 6.5. t_c = come-up-time = the time required for the retort chamber to reach the processing temperature. t_p = processing time = the time during which the retort maintains the processing temperature. t_h = total heating time = $t_c + t_p$. t_B = Ball *processing* time = $0.42t_c + t_p$. In the Ball formula method, the retort is assumed to be at processing temperature throughout Ball processing time, but no heat treatment occurs before the start of Ball processing time.

To use Ball's approach, we must move the apparent intercept to the start of the Ball processing time as shown in Figure 6.6.

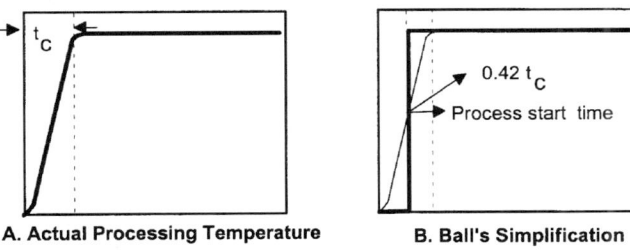

| A. Actual Processing Temperature | B. Ball's Simplification |

Figure 6.4   The process come-up time and start of Ball's processing time.

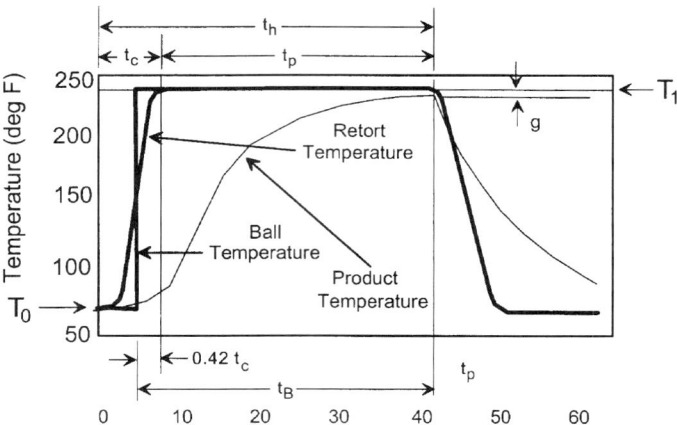

Figure 6.5   Typical curve showing various terms in the computation of Ball's processing time.

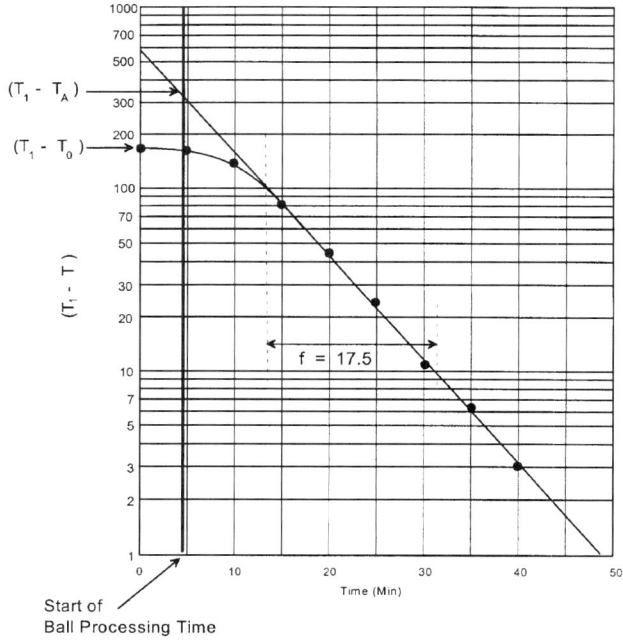

Figure 6.6   Typical temperature difference versus time curve on a semilog plot to show the start of Ball's processing time and Ball's apparent intercept point.

If we use Ball processing time, the heating curve equation becomes

$$\log(T_1 - T_b) = \log[\, j_h(T_1 - T_0)] - \frac{t_B}{f_h} \tag{6.8}$$

where T_1 = heating medium temperature, T_0 = initial temperature of the food, T_b = the maximum temperature of the food at the end of processing, t_B = Ball processing time, j_c is calculated using the intercept $(T_1 - T_A)$ with the start of the Ball heating time as shown in Figure 6.6.

If we define $g = T_1 - T_b$ = the difference between the maximum food temperature and heating medium temperature (see Fig. 6.5), Eq. (6.8) becomes

$$\log(g) = \log[j_h(T_1 - T_0)] - \frac{t_B}{f_h} \tag{6.9}$$

Using this equation, we can calculate g for any Ball processing time. Conversely, we can compute the required Ball processing time for any desired g with the equation:

$$t_B = f_h \log\left[\frac{j_h(T_1 - T_0)}{g}\right] \tag{6.10}$$

Example 5 *From Figure 6.5, we read the come-up time as $t_c = 8\,min$ and processing time as $t_p = 34\,min$. This makes the Ball processing time:*

$$t_B = 34 + 0.42(8) = 37.4 \text{ min}$$

The start of Ball processing time occurs at $0.58(8) = 4.6\,min$ and we have marked this on Figure 6.6. The fitted line crosses the time at 330, rather than 580, and becomes the intercept $T_A - T_1$. Using this, we compute j_h as

$$j_h = \frac{330}{240 - 70} = 1.94$$

and the equation for predicting temperature becomes

$$\log(240 - T) = \log[1.94(240 - 70)] - \frac{t}{17.5}$$
$$\log(240 - T) = 2.52 - \frac{t}{17.5}$$

6.2.3.2 Determining the Time of a Heat Sterilization Process The Ball formula method makes the following assumptions:

- That $f_h = f_c$, that is, the heating and cooling curves have the same slope.
- That $j_c = 1.41$.
- That the transition from heating to cooling is a segment of a parabola on a semilog plot.
- That the cooling medium temperature is 180°F below the heating medium temperature.

We can use Ball's method to calculate the Ball's processing time required for a process, if we have the following information:

- T_0 = initial product temperature.
- T_1 = the retort temperature.
- F_0 = the lethality to be achieved.

The following example shows how it is done.

Example 6   *A sterilization process is being designed to achieve $F_0 = 7\,min$. In other words, we want a process that is the equivalent in its destruction of C. botulinum to 7 min at 250°F. The parameters obtained in Example 5 were $f_h = 17.5\,min$ and $j_h = 2.0$. The initial temperature of the food is 170°F and the processing temperature 240°F. How much Ball processing time is required?*

Solution   We make the standard assumptions that $f_c = f_h = 17.5\,min$, $j_c = 1.41$, and $T_2 = T_1 - 180°$. The calculations are done in Table 6.2 and show that 57 min are required.

Example 7   *We are evaluating a process and need to determine its F_0 value. The processing temperature is 255°F. The initial temperature of the food is 95°F, and the f and j values are those determined in Example 5. The come-up time is 12 min and the processing time 23 min.*

Solution   The computation is shown in Table 6.3. This process has the same effect on *C. botulinum* as an exposure of 4.21 min at 250°F. If we assume that for *C. botulinum*, $D = 0.2\,min$, this should produce 21 (= 4.21/0.2) log cycles of spore reduction. If each can start with 10^4 spores, the PNSU of the finished cans will be $10^{4-21} = 10^{-17}$. If we use Eq. (6.9), the temperature in the center of the can at steam

Table 6.2   Computing sterilization time for a desired F_0

Line no.	Variable	Tabulation
1	F_0 (min)	7.0 min
2	f_h (min)	17.5 min
3	j_h	1.94
4	T_0 (°F)	70°F
5	T_1 (°F)	240°F
6	$L = 10^{[(T_1-250)/18]}$	$L = 10^{[(240-250)/18]} = 0.278$
7	$T_1 - T_0$	$240 - 70 = 170$
8	$j_h(T_1 - T_0)$	$1.94(170) = 330$
9	$\log[j_h(T_1 - T_0)]$	$\log(33) = 2.519$
10	$R = \dfrac{f_h \times L}{F_0}$	$R = \dfrac{17.5 \times 0.278}{7} = 0.695$
	Use a table or graph to obtain log(g) for this $R(f_h/U)$:	
11	$\log(g)$	-0.75
12	$\log[j_h(T_1 - T_0)] - \log(g)$	$2.519 - (-0.756) = 13.265$
13	$t_B = f_h\{\log[j_h(T_1 - T_0)] - \log(g)\}$	$17.5(3.265) = 57.1\,min$

shutoff is

$$\log(g) = \log[1.94(255 - 95)] - \frac{28}{17.5} = 0.89$$

$$g = 7.8°F \qquad T_{max} = 255 - 7.8 = 247°F$$

6.2.4 Changing Can Size

Larger cans take longer to heat than smaller ones. This is reflected in a larger f_h value for larger cans.

6.2.4.1 Conductive Foods

For conductive foods, the relationship between f_h and the can is dimensions and can be expressed by the Eq. (6.11):

$$\text{Can factor} = 4\alpha f_h = \frac{0.933d^2}{2.34 + (d/L)^2} \qquad (6.11)$$

where d = can diameter ($\frac{1}{8}$ in.), L = can length ($\frac{1}{4}$ in.), f_h = the heat penetration factor, and α = thermal diffusivity of the food.

The ratio of f values for two cans equals the ratio of their can factors. Thus,

$$\frac{f_a}{f_b} = \frac{(\text{Can factor})_a}{(\text{Can factor})_b} \qquad (6.12)$$

Example 8 *A heat penetration test determines that for a particular product, $f_h = 27$ min in a 307 × 509 can. What f_h value would you use for a 202 × 308 can?*

Table 6.3 Computing F_0 for a given processing time

Line no.	Variable	Tabulation
1	t_B (min)	$23 + 0.42(12) = 28.0$ min
2	f_h (min)	17.5 min
3	j_h	1.94
4	T_0 (°F)	95
5	T_1 (°F)	255
6	$L = 10^{[(T_1-250)/18]}$	$L = 10^{[(255-250)/18]} = 1.896$
7	$T_1 - T_0$	$255 - 95 = 160$
8	$j_h(T_1 - T_0)$	$1.94(160) = 310$
9	$\log[j_h(T_1 - T_0)]$	$\log(310) = 2.49$
10	t_B/f_h	$28.0/17.5 = 1.60$
11	$\log(g) = \log[j_h(T_1 - T_0)] - t_B/f_h$	$2.49 - 1.60 = 0.89$
	Use a table or graph to obtain $R(f_h/U)$ for given $\log(g)$:	
12	R	7.8886
13	$F_0 = \dfrac{f_h \times L}{R}$ (min)	$\dfrac{17.5 \times 1.896}{7.8886} = 4.21$ min

Solution   A 307×509 can has a diameter of $3\frac{7}{16}$ in. and length of $5\frac{9}{16}$ in., so d and L are

$$d = 3\tfrac{7}{16} - \tfrac{1}{8} = 3.4375 - 0.125 = 3.3125 \text{ in.}$$

$$L = 5\tfrac{9}{16} - \tfrac{1}{4} = 5.5625 - 0.25 = 5.3125 \text{ in.}$$

For a conductive food, the can factor for this can is

$$\text{Can factor} = \frac{0.933(3.3125)^2}{2.34 + (3.3125/5.3125)^2} = 3.751$$

Similarly, the can factor for a 202×308 can is 0.765. Using Eq. (6.12), we get

$$\frac{f_a}{f_b} = \frac{f_a}{27} = \frac{0.765}{3.751}$$

$$f_a = 27\left(\frac{0.765}{3.751}\right) = 5.5 \text{ min}$$

6.2.4.2 *Convective Foods*   For convective foods, the can factor is

$$\text{Can factor} = \left(\frac{rL}{r+L}\right)$$

where $r =$ can radius $- \frac{1}{16}$ in. $= d/2$ and $L =$ can length $= \frac{1}{4}$ in.

Example 9   *What would the f_h value have been in the last example if the food is heated convectively?*

Solution   For the 307×509 can, $r = d/2 - \frac{1}{16} = 3.4375/2 - \frac{1}{16} = 1.6563$, $L = 5.5265 - \frac{1}{4} = 5.3125$. The can factor is

$$\text{Can factor} = \frac{1.6563 \cdot 5.3125}{1.6563 + 5.3125} = 1.2626$$

Similarly, the can factor for the 202×308 can is 0.764? The f_h value for the 202×308 can is

$$f = 17.5\left(\frac{0.7647}{1.2626}\right) = 10.6 \text{ min}$$

6.3 LAB EXERCISE

6.3.1 Objectives

The objectives for this lab exercise are to:

1. Learn how to conduct a heat penetration test in can retorting process.
2. Calculate the parameters: f_h, j_h, f_c, and j_c.
3. Use Ball's formula method to design a process.
4. Modify the process for different can sizes.

6.3.2 Calculations

In the previous chapter, we had obtained two sets of data: one for a conductive heating food (tomato paste) and another for a convective heating food (tomato juice). Carry out the following calculations:

1. Determine T_0 and T_1, T_2, t_c, t_p, and t_B.
2. Plot the data $(T_1 - T)$ versus time from the starting time to shutoff time for each product on semilog paper.
3. Fit a straight line to the linear portion of the curve for each product.
4. Determine f_h and T_A (based on Ball's processing time) from the graph and compute j_h for each product.
5. What is the equation for the heating curve for each product? (To make this equation generally useful, represent all temperatures with variables.)
6. Based on these equations, how long would it have taken for each product to reach 235°F if the food had started at 88°F and the retort had been operated at 240°F?
7. Discuss and compare the specific f and j values for heating and cooling for each product obtained from this experiment. [What is the effect of the mode of heating (convection versus conduction) on these values? What is the effect of using Ball processing time rather than total heating time on these values? etc.?]
8. From the data, determine the cooling medium temperature T_2. Determine f_c and j_c for the cooling portion of the curve for each product.
9. You plan to modify this procedure for a retort that takes 7.5 min to reach an operating temperature of 258°F. Food will enter the process at 110°F. Using Ball's formula method, design a process that has the same effect on *C. botulinum* at 4.5 min at 250°F. (Disregard your experimentally determined f_c and j_c.) What is the Ball heating time and total heating time for your process?
10. Suppose that in a test run of the conductive heating food, you stopped the process 2 min sooner than your design specified. Assuming $D_{250} = 0.2$ and a starting count of 4.5×10^3 spores per can, what PNSU would you have oobtained?
11. In the previous chapter, we used 211×300 cans. We also plan to sell this product in 300×404 cans. We will use a retort with a come-up time of 8 min and an operating temperature of 254°F. The food will enter the process at 135°F. What should be the shutoff time for each product, for this can size?

6.4 NOMENCLATURE

Symbols Used in this Lab

Symbol	Meaning	Units
t	Time	min
t_c	Come-up time (from the start of processing until the retort reaches the processing temperature)	min
t_p	Processing time (from the end of the come-up until shutoff)	min
t_h	Total heating time $(t_c + t_p)$	min
t_B	Ball heating time $(0.42t_c + t_p)$	min
T	Temperature of product at time t	°F or °C
T_0	Initial temperature of product	°F or °C
T_1	Temperature of heating medium	°F or °C
T_2	Temperature of cooling medium	°F or °C
T_B	Temperature of product at shutoff (start of cooling)	°F or °C
T_A	Apparent initial temperature of product (used to force the log heating curve to fit a linear equation)	°F or °C
T_{AB}	Apparent shutoff temperature of product (used to force the log cooling curve to fit a linear equation)	°F or °C
f_h	Time required for the difference between product and heating medium temperature to be reduced by a factor of 10	min
j_h	Lag factor used to convert the initial temperature to the apparent initial temperature on log heating curve	
f_c	Time required for the difference between product and cooling medium temperature to be reduced by a factor of 10	min
j_c	Lag factor used to convert the initial temperature to the apparent initial temperature on log cooling curve	
g	$T_1 - T_B =$ difference between product and heating medium temperature at shutoff	°F or °C
R	f_h/U	
L	Lethal rate	
F_0	Lethality	min

6.4.1 Additional Computation Problems

1. Cans filled at ambient temperature in a factory in South Texas, where the average temperature in summer is 120°F, reach 245°F in 40 min, when the retort temperature is 250°F. How long would it take similar cans filled with the same product to reach 240°F if they are filled in a factory in North Dakota in winter, when the average temperature in the factory is 40°F? The retort in North Dakota also operates at 250°F. Assume there is no heating lag or jump.

2. One retort operates at 250°F and heats a food product from 90 to 210°F in 35 min. What must the retort temperature be if the same food product is to be heated from 60 to 230°F in the same amount of time? Assume $j_h = 1.0$.

3. The following time–temperature data are taken for salmon in a 202×214 can in a retort operating at $250°F$:

Time (min)	Temperature (°F)
0	90
5	115
10	165
15	196
20	216
25	228
30	236
35	241
40	244.5
45	246.6

How long will it take to heat salmon from 80 to $249°F$ in a 401×411 can? Assume that j_h remains constant and ignore the come-up time.

4. The time–temperature relationships of a conduction-heating food in a no. 1 picnic can (211×400) insulated by other cans at both ends when processed at a retort temperature of $240°F$ and cooled by $60°F$ water gave the following data:

$$f_h = 35.0 \text{ min}, \quad j = 1.8, \quad T_0 = 140°F$$

Assuming ten *C. botulinum* spores ($D_{250} = 0.2 \text{ min}$) per mL as the initial concentration in the container, determine the process time necessary to achieve a 10^{-9} probability of survival in the whole container.

5. A food company is thermally processing 307×306 cans. The process time t_p at $240°F$ is 70 min, with a 10-min come-up time for the retort. The result of the heat penetration test is given below. Calculate total sterilizing value of the process by Ball's method.

Time (min)	Temperature (°F)
0	166.0
5	170.0
10	195.0
15	221.5
20	232.5
25	233.4
30	234.4
35	235.2
40	236.0
45	236.6
50	237.1
55	237.5
60	237.9
65	238.4
70	238.5

6. A 500-mL tray is filled with 400 mL of liquid food and 100 mL of air or headspace. The easy-peel lid is sealed to the tray at atmospheric pressure and room temperature (72°F). The tray is retorted at 250°F.

   a. What is the partial pressure of air and partial saturated water vapor pressure at the equilibrium temperature of 250°F?

   b. For the tray to remain undamaged, what should the recommended overriding air pressure during (i) heating and (ii) cooling be?

   c. How much will the tray distort if the retort overriding air pressure is 24.3 psig at an equilibrium retort temperature of 250°F?

6.5 SUGGESTED READINGS AND REFERENCES

1. D. B. Lund, 1975. "Heat processing". In (M. Karel, O. R. Fennema, and D. B. Lund, eds.), *Principle of Food Science, Vol. 2: Physical Principles of Food Preservation*, New York: Marcel Dekker.

2. C. R. Stumbo, 1973. *Thermobacteriology in Food Processing*, 2nd ed., New York: Academic Press.

3. C. O. Ball and C. W. Olson, 1957. *Sterilization in Food Technology*, New York: McGraw-Hill.

4. P. Fellows, 1988. *Food Process Technology, Principles and Practice*, Chichester, UK: Ellis Horwood Ltd.

7

BLANCHING AND FREEZING OF FOODS

7.1 BACKGROUND

7.1.1 Blanching

Blanching is a heat treatment process generally applied to fruits and vegetables prior to freezing, drying, or canning. Blanching is primarily carried out to inactivate enzymes before freezing or dehydration. Unblanched frozen or dried foods undergo relatively rapid changes in food quality such as color, flavor, texture, and nutritional value due to continuous enzyme activity.

In plant tissues, enzymes such as lipoxygenase, polyphenol oxidase, polygalacturonase, and chlorophenolase cause loss of nutrition, flavor, and texture. In addition, peroxidase and catalase are two of the most heat-resistant and widely distributed enzymes. Although they are not implicated as a cause of deterioration during storage, their activity is used to evaluate the effectiveness of blanching. If both these enzymes are inactivated, then it can be safely assumed that other significant enzymes are also inactivated. The heating time necessary to destroy catalase or peroxidase depends on the type of fruit or vegetable, the method of heat treatment, the size of the fruit or vegetable, and the temperature of the heating medium. In commercial blanching, the blanching time in boiling water at 100°C generally varies from 1.5 to 4 min as shown in Table 7.1.

Other heating mediums such as steam, hot air, or a microwave at a temperature other than 100°C can also be used. Although the blanching of vegetables is most often done in hot water or steam, and blanching of fruits is often done in calcium chloride solution to firm the fruit through the formation of calcium pectates, some other colloidal thickeners, such as pectin, carboxymethylcellulose, and alginates, also can be used to promote fruit firmness following blanching.

Table 7.1 Blanching times at 100°C in hot water for vegetables prior to freezing

Vegetable	Time (min)
Asparagus	
< 8 mm per butt	2
8–15 mm per butt	3
> 16 mm	4
Beans, green and wax	
Small	1–1.5
Medium	2–3
Large	3–4
Beets	
Small, whole	3–5
Diced	3
Broccoli	2–3
Corn	2–3
Peas	1–1.5
Spinach	1.5

For frozen fruits that are not to be further heated after thawing, blanching is not used to avoid undesirable changes in texture and flavor. However, in such cases, enzymatic activities are controlled by other preservation techniques such as packing or dipping fruits in sugar syrup, the addition of antioxidants, and the removal of oxygen from the container.

Prior to canning, blanching helps to remove tissue gases, increase the tissue temperature, cleanse the tissue, wilt the tissue to facilitate packing, and activate or inactivate enzymes. As the product generally receives a much more severe heat treatment during subsequent thermal processing in canning, enzyme inactivation is not the primary objective of blanching. Thus, the removal of tissue gases and preheating of tissue have a great influence on the final level of oxygen in the container and, therefore, directly influence the product storage life. In the case of green beans, blanching is used to activate pectin methyl esterase enzyme to prevent a quality defect known as "sloughing." A moderate heat treatment at 80°C for a few seconds prior to thermal processing activates the pectin methyl esterase enzyme, which deesterifies the pectin molecule to allow cross-linking with calcium ions. This cross-linking bonds the outer layers of tissue to the underlying structure, thus preventing sloughing.

Blanching reduces the number of contaminating microorganisms on the food surfaces and, hence, assists in subsequent preservation operations. Incomplete blanching of fruits and vegetables may cause more damage to foods than the absence of blanching itself. Heat, which is sufficient to disrupt tissues but not to inactivate enzymes, causes the mixing of enzymes and substrate.

7.1.1.1 Blanching Methods

The two most common and widely used commercial methods of blanching involve passing food through an atmosphere of saturated steam or a bath of hot water. Sometimes, the cooling stage may result in greater losses of product or nutrients than the blanching stage. Steam blanching results in higher nutrient retention, provided that cooling is by cold air or cold water sprays. Cooling with running water substantially increases leaching losses (washing of soluble compounds), but the food may gain weight and the overall yield may, therefore, increase, whereas air cooling may cause weight loss of the product due to evaporation. And this may outweigh any advantages gained by nutrient retention.

7.1.1.2 Steam Blanchers

A simple steam blancher consists of a mesh conveyer belt that carries food through a steam atmosphere. The residence time of the food is controlled by the speed of the conveyor. Typical equipment is 15 m long, 1 to 1.5 m wide, and up to 2 m high. In conventional steam blanching, there is often poor uniformity of heating in the multiple layers of food. The time–temperature combination required to ensure enzyme inactivation at the center of the bed results in the overheating of food at the edges and this results in losses in texture and other sensory characteristics of the food. Individual quick blanching (IQB) that involves blanching in two stages overcomes this problem. In the case of IQB, the food is first heated in a single layer to a sufficiently high temperature to inactivate enzymes. In the second stage, a deep bed of food is held for sufficient time to allow the temperature at the center of each piece to increase to the level sufficient for enzyme inactivation. This reduces the heating time from a conventional 3 min to about 75 s (25 s for heating and 50 s for holding). This increases the energy consumption from 25 to 30% as in conventional steam to 85 to 90% in IQB. The mass of product blanched per kilogram of steam increases from 0.5 kg per kilogram of steam in a conventional steam blancher to 6 to 7 kg per kilogram of steam with small particulate food such as peas, sliced carrots, etc.

Nutrient losses during steam blanching can be reduced by exposing food to warm air (65°C) in a short preliminary drying operation known as preconditioning. Surface moisture evaporates and the surface then absorbs condensing steam during IQB. Preconditioning and IQB have been reported to reduce nutrient losses by 81% in green beans, 75% for brussel sprouts, 61% for peas, and 53% for lima beans (Bomben *et al.*, 1973).

7.1.1.3 Hot-water Blancher

In a hot-water blancher, food is retained in hot water at 70 to 100°C for a specified time and then it is removed to a dewatering-cooling section. A reel blancher is a widely used method of blanching, in which food enters a slowly rotating cylindrical mesh drum partly submerged in hot water. The food is moved through the drum by internal flights. The speed of rotation controls the heating time. The development of hot-water blanchers based on the IQB principle reduces energy consumption and minimizes effluent production. In such a blancher, there are three sections: a preheating stage, blanching stage, and cooling stage as shown in Figure 7.1. The food travels on a single conveyer belt throughout each stage, and therefore, does not suffer any physical damage associated with the turbulence of a conventional hot-water blancher. The food is preheated with water that is recirculated through a heat exchanger. After blanching, a second recirculation system cools the food. The two systems pass water through the heat exchanger and, thus, simultaneously heat the preheated water and cool the water for the cooling section.

A conventional hot-water blancher is cheaper and more energy-efficient than steam blanchers. However, there are large losses of water-soluble components, including vitamins, minerals, and sugars. The loss of ascorbic acid is used as an indicator of food quality and, therefore, the severity of blanching. When blanching peas, green beans, and broccoli, generally, 15 to 25% loss of ascorbic acid loss has been observed in a water blanch-water cool system.

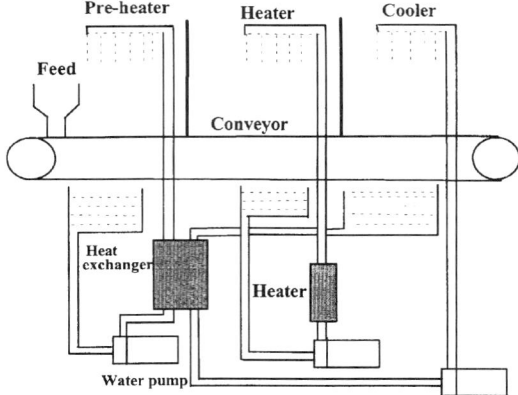

Figure 7.1 Schematic diagram of a hot-water blancher.

7.1.2 Freezing

During freezing, the temperature of the food falls below its freezing point, and a proportion of the water changes from a liquid to solid state to form ice crystals. The concentration of the dissolved solute increases in the food due to the immobilization of water to ice, thus lowering its water activity. Preservation of frozen food is, therefore, achieved by a combination of low temperature, low water activity, and in some cases, pretreatment such as blanching that significantly reduces the rate of chemical, biochemical, and microbiological activities. The major frozen foods commercially available are fruits such as strawberries, raspberries, blackcurrant, either whole or puréed or as juice concentrate; vegetables such as peas, green beans, sweet corn, sprouts, potatoes; fish fillets and seafoods such as cod, shrimp, crab; meats such as beef, lamb, poultry; baked foods such as bread, cakes, fruit pies; prepared food such as pizzas, desserts, ice creams, cook-freeze dishes.

In a freezing operation, the product is exposed to a temperature much lower than the desired final product temperature to remove first the sensible heat and later the latent heat from the food to form ice crystals. In fresh food, respiration heat, termed heat load, is also removed. The latent heat of other food components such as fat also must be removed before they solidify. However, most foods contain a large amount of water and other components require a relatively small amount of heat for crystallization. The temperature at which ice formation begins is called the initial freezing point of the product. The initial freezing and crystallization process taking place within a food product is different from those occurring during the freezing of pure water.

If the temperature is monitored at the slowest cooling point at the thermal center of a food, a characteristic curve is obtained, as shown in Figure 7.2. The *AB* portion of the curve indicates that food is cooled below its freezing point T_f (with the exception of pure water) that is always below 0°C. At point *B*, the water remains liquid, although the temperature may be below the freezing point. This phenomenon is known as supercooling and may be as much as 10°C below the freezing point. During *BC*, the temperature rises rapidly to the freezing point as the ice crystals begin to form and latent heat of crystallization is released. During *CD*, heat continues to be removed from the food as latent heat and ice crystals are formed. The freezing point is depressed due to an increase

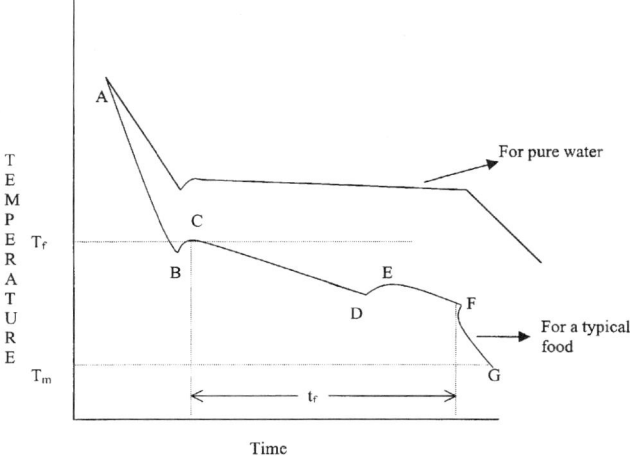

Figure 7.2 Comparison of freezing curves for pure water and a typical food with one major solute: t_f = freezing time, T_f = freezing temperature, T_m = temperature of freezing medium.

in solute concentration in the unfrozen liquid and the freezing point falls. A major amount of ice is formed during this part of the freezing operation. During *DE*, one of the solutes becomes supersaturated and crystallizes out. The latent heat of crystallization is released and the temperature rises slightly to the eutectic temperature of that solute. During *EF*, the crystallization of water and solute continues. Finally, during the *FG* portion of the curve, the temperature of the ice-water mixture falls to that of the freezer. A proportion of the water remains unfrozen and it depends on the food type, its composition and storage temperature.

7.1.2.1 Calculation of Freezing Time   Estimation of the freezing rate and, therefore, freezing time is most valuable for selecting and designing a freezing process. During freezing, heat is conducted from the interior of a food to the surface and then removed by the freezing medium. The freezing time depends on factors such as the following:

1. Size and shape of the product.
2. Thermal conductivity of the food material.
3. Area of the food available for heat transfer.
4. Surface heat transfer coefficient of the medium.
5. Temperature difference between the food and freezing medium.
6. Type of packaging film in the case of packaged food.

It is difficult to define the freezing time precisely. Effective freezing time is the time required to lower the temperature of a food from an initial value to a predetermined final temperature at the thermal center. The effective freezing time measures the time that food spends in a freezer and is used to measure the throughput of a manufacturing process. The nominal freezing time is the time between the surface of the food reaching $0°C$ and the thermal center reaching $10°C$ below the temperature of the first ice formation. Nominal freezing time is used to estimate the product damage. The calculation of the freezing time is complicated due to a difference in the freezing point and ice crystal formation within different regions of a piece of food, and changes in the density, thermal conductivity, specific heat, and thermal diffusivity with a reduction in the temperature of the food. For simplicity, Plank's equation and Cleland and Earle's (1979) equation have been found most suitable to predict freezing time. The main assumptions involved in the estimation of freezing time are:

1. Freezing begins with all water in the food in an unfrozen state.
2. Loss of sensible heat is ignored.
3. Heat transfer takes place slowly under steady-state conditions.
4. The freezing front maintains a similar shape to that of food; for example, in a rectangular block, the freezing front remains rectangular.
5. There is a single freezing point.
6. The thermal conductivity and specific heat of the food are constant when unfrozen and then change to a different constant value when the food is frozen.

A typical case of one-dimensional freezing of an infinite product slab of a thickness L being cooled by convection in an environment of constant temperature T_m is shown in Figure 7.3. The frozen layer grows with time and at time t, a thickness x of the frozen layer is formed on both sides. The initial freezing temperature is constant at T_f. The unfrozen center is also at T_f. If in time dt, a layer of product thickness dx freezes, then the rate of heat removal is given by Eq. (7.1):

$$q = A\lambda\rho\left(\frac{dx}{dt}\right) \tag{7.1}$$

where A is the surface area, λ the latent heat of freezing, and ρ the density of frozen product. At steady state, the heat given off at the freezing front of the product must be removed by conduction through the frozen layer of thickness x followed by its removal by convection from the outside surface. The heat transfer by conduction inside the product

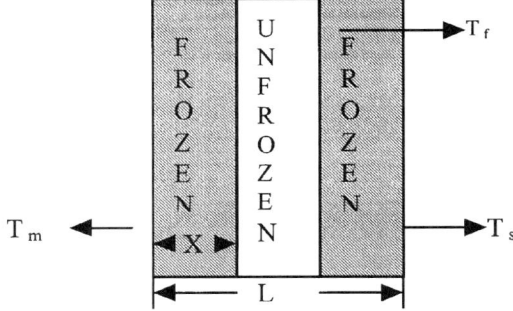

Figure 7.3 Schematic illustration of one-dimensional freezing of a product section used to derive Plank's equation: L = thickness of the slab, T_s = surface temperature, T_m = temperature of freezing medium, T_f = freezing temperature, x = thickness of frozen section.

can be described by Eq. (7.2):

$$q = \frac{k_f A(T_f - T_s)}{x} \qquad (7.2)$$

where k_f is the thermal conductivity of the frozen material and T_s the surface temperature. The convective heat transfer at the surface can be shown by Eq. (7.3):

$$q = hA(T_m - T_s) \qquad (7.3)$$

where h is the convective heat transfer coefficient at the product surface and T_m the medium temperature. By eliminating surface temperature T_s and combining Eqs. (7.2) and (7.3), heat transfer in series is given by Eq. (7.4):

$$q = \frac{A(T_f - T_m)}{\dfrac{x}{k_f} + \dfrac{1}{h}} \qquad (7.4)$$

Equating Eqs. (7.1) and (7.4), and rearranging the terms and integrating between $t = 0$ and $x = 0$ to $t = t_f$ and $x = L/2$, yield the relationship in Eq. (7.5):

$$\int_0^{t_f} (T_f - T_m)dt = \lambda \rho \int_0^{L/2} \left(\frac{x}{k_f} + \frac{1}{h} \right) dx \qquad (7.5)$$

Integrating and solving Eq. (7.5) give Plank's equation to estimate freezing time as shown by Eq. (7.6):

$$t_f = \frac{\lambda \rho}{(T_f - T_m)} \left(\frac{L}{2h} + \frac{L^2}{8k} \right) \qquad (7.6)$$

A general form of Plank's equation is given in Eq. (7.7):

$$t_f = \frac{\lambda \rho}{(T_f - T_m)} \left(\frac{P \cdot L}{h} + \frac{R \cdot L^2}{k} \right) \qquad (7.7)$$

where L is the characteristic length, which is the total thickness in the case of an infinite slab, the diameter in the case of a long cylinder or sphere, or the smallest dimension of a rectangular brick. The P and R constants of various geometry are shown in Table 7.2.

For brick-shaped objects, P and R values are obtained from the graph shown in Figure 7.4 prepared by Ede (1949), where β_1 and β_2 are the ratios of the two longest sides divided by the shortest ones. This would provide a value of either P or R. The other values for P and R are obtained by interchanging β_1 and β_2.

Plank's equation gives a satisfactory estimation of freezing time, as long as the product is initially at its freezing temperature. There are other approaches to account for the initial product temperature above its freezing temperature and the final product temperature

Table 7.2 *P* and *R* values for various geometries used in the calculation of freezing time

Geometry	P	R
Infinite slab	1/2	1/8
Infinite cylinder	1/4	1/16
Sphere	1/6	1/24

below the freezing point. One of the modified equations proposed by Nagoaka *et al.* (1955) is shown in Eq. (7.8) for SI units:

$$t_f = [C_{pu}(T_i - T_f) + \lambda X_w + C_{pf}(T_f - T)][1 + 0.008(T_i - T_f)]$$

$$\times \left[\frac{\rho}{(T_f - T_m)} \left(\frac{P \cdot L}{h} + \frac{R \cdot L^2}{k_f} \right) \right] \tag{7.8}$$

where T_f = freezing point of a product; X_w = water weight fraction, wet basis; T_i = initial product temperature; T_m = temperature of freezing medium; T = final product temperature; C_{pu} = specific heat of unfrozen food; C_{pf} = specific heat of frozen food; λ = latent heat of fusion; k_f = thermal conductivity of frozen material; ρ = density of frozen material.

When metric units are used, a factor of 0.0045 will replace 0.008. When the product to be frozen is contained in a package, the convective heat transfer h in Eq. (7.7) is replaced

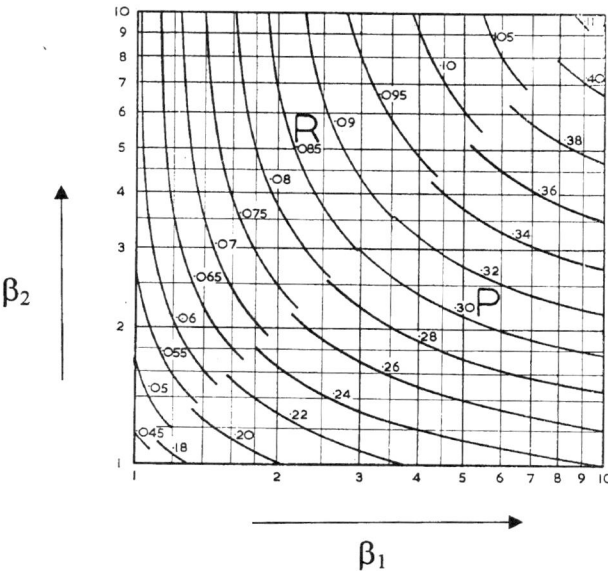

Figure 7.4 Values of *P* and *R* for a brick-shaped product (based on Ede, 1949).

by an overall heat transfer coefficient U to account for the resistance to heat flow offered by the packaging material. For a packaged system, the value of U is given by Eq. (7.9):

$$U = \frac{1}{x_p/k_p + 1/h} \tag{7.9}$$

Cleland and Earle (1979) have modified Plank's equation to a dimensionless form as shown by Eq. (7.10):

$$N_{FO} = \frac{P}{N_{Bi}N_{st}} + \frac{R}{N_{St}} \tag{7.10}$$

where N_{FO} = Fourier number, $\alpha t/L^2$; N_{Bi} = Biot number, hL/k; N_{St} = Stefan number, $C_{pf}(T_f - T_m)/\lambda$; N_{Pk} = Plank's number, $C_{pu}(T_i - T_f)/\lambda$; α = thermal diffusivity, $k/\rho C_p$. P and R values can be calculated by using the following relationships for a slab, cylinder, and sphere.

For a slab:

$$P = 0.5072 + 0.2018N_{Pk} + N_{St}(0.3224N_{Pk} + 0.0105/N_{Bi} + 0.0681)$$
$$R = 0.1684 + N_{St}(0.274N_{Pk} + 0.0135)$$

For a cylinder:

$$P = 0.3751 + 0.0999N_{Pk} + N_{St}(0.4008N_{Pk} + 0.0710/N_{Bi} - 0.5865)$$
$$R = 0.0133 + N_{St}(0.0415N_{Pk} + 0.3957)$$

For a sphere:

$$P = 0.1084 + 0.0924N_{Pk} + N_{St}(0.2310N_{Pk} - 0.3114/N_{Bi} + 0.6739)$$
$$R = 0.0784 + N_{St}(0.0386N_{Pk} - 0.1694)$$

7.1.2.2 *Example Exercise*

A 0.30 m × 0.60 m × 0.90 m rectangular block of lean beef initially at 15°C is to be frozen by immersion freezing in liquid refrigerant $R - 12$ (−29.8°C) to a final temperature of −15°C. Compute the freezing time for the following cases: (1) the meat block is unpackaged, (2) the meat block is packaged in 1.0-mm-thick cardboard, and (3) the meat block is immersed in liquid nitrogen at −196°C.

Solution: Since the initial and final temperatures of the product are different, the use of Nagaoka et al.'s (1955) modification is therefore more appropriate. The physical and thermal properties of the product being frozen are as follows: $X_w = 0.68$; $C_{pu} = 3.5$ kJ/kg·K; $T_i = 15°C$; $T_f = -1.7°C$; $\lambda = 332.7$ kJ/kg {(411 − 78.3), Appendix B, Table B.4}; $C_{pf} = 2.05$ kJ/kg·K; $T = -15°C$; $\rho = 1050$ kg/m³; $T_m = -196°C$ or −29.8°C; $h = 170$ W/m² K; $k_f = 1.1$ W/m·K.

For the first case, in order to obtain the shape factors P and R, we calculate β_1 and β_2:

$$\beta_1 = 0.60/0/30 = 2.0 \quad \text{and} \quad \beta_2 = 0.90/0/30 = 3.0$$

We find the values of P and R as 0.275 and 0.078, respectively, as shown in Figure 7.4. Substituting the values into Eq. (7.8), we get

$$t_f = [3.5(15 + 1.7) + 0.68(332.7) + 2.05(-1.7 + 15)] \cdot [1 + 0.008(15 + 1.7)][1050/(-1.7$$
$$+ 29.8)][(0.275)(0.30)/0.170 + (0.078)(0.30)^2/1.1 \times 10^{-3}] = 14{,}844 \text{ s} \quad \text{or} \quad 4.1 \text{ h}$$

For the second case, when the meat is packaged, h is replaced with U in Eq. (7.8). The value of U is obtained from the following data:

$$X_p = 1 \times 10^{-3} \text{ m}$$
$$k_p = 0.04 \text{ W/m} \cdot \text{K}$$
$$U = 1/[(0.001/0.04) + (1/170)] = 32.38 \text{ W/m}^2 \cdot \text{K}$$

Substituting for h as $0.0323 \text{ kW/m}^2 \cdot \text{K}$ in Eq. (7.8), we obtain

$$t_f = [3.5(15 + 1.7) + 0.68(332.7) + 2.05(-1.7 + 15)]$$
$$\cdot [1 + 0.008(15 + 1.7)][1050/(-1.7$$
$$+ 29.8)][(0.275)(0.30)/0.0323 + (0.078)(0.30)^2/1.1 \times 10^{-3}]$$
$$= 42{,}180.6 \text{ s} \quad \text{or} \quad 11.7 \text{ h}$$

For the third case, when the meat slab is immersed in liquid nitrogen at $-196°\text{C}$,

$$t_f = [3.5(15 + 1.7) + 0.68(332.7) + 2.05(-1.7 + 15)]$$
$$\cdot [1 + 0.008(15 + 1.7)][1050/(-1.7$$
$$+ 196)][(0.275)(0.30)/0.170 + (0.078)(0.30)^2/1.1 \times 10^{-3}] = 2146 \text{ s} \quad \text{or} \quad 0.6 \text{ h}$$

7.2 LABORATORY

7.2.1 Objectives

The objectives of this lab exercise are to:

1. Evaluate the effectiveness of blanching operations in fruits and vegetables.
2. Compare experimentally determined freezing times with those obtained from predictive equations.
3. Determine the effect of packaging on freezing time.
4. Examine the effects of blanching and freezing on the quality of fruits and vegetables.

7.2.2 Materials

1. Fruits and vegetable such as apples, potatoes, and brussels sprouts.
2. Food product such as beef cubes or frankfurter, and fruit juices such as apple, orange, etc.
3. Weighing balance, knife, trays.
4. Hot-water or steam blancher.
5. 1% Catechol solution.
6. Hunter Lab calorimeter or Macbeth color eye or other color-measuring instrument.
7. Food blender.
8. Petri dish, test tubes, etc.
9. A bath containing either liquid nitrogen or a mixture of 50/50 ethylene glycol/water to a level that will allow complete immersion of the product.
10. Thermocouples.
11. Data logger.
12. Polyethylene shrink film, vacuum seal heater, hot air blower, aluminum cans of small size (211 × 300), can sealer.

7.2.3 Procedure for Estimating Freezing Time

Apples or potatoes can be used to measure freezing time in fresh fruits and vegetables. Orange juice or apple juice can be used to estimate freezing time in packaged can.

1. Select two apples or potatoes as identical as possible. Measure each fruit diameter and determine the fruit tissue density.
2. Insert and secure a thermocouple near the geometric center of each apple or pear. Shrinkwrap one apple or potato in a plastic film.
3. Connect thermocouples to the recorder and program the recorder to record product temperature at a desired interval of time, say, every 30 or 60-s interval.
4. Record the initial product and freezing medium or blast freezer temperatures in Data Sheet 7.1.
5. Place the product in a blast freezer ($-40°C$) to initiate the freezing process.
6. Record the time-temperature data to freeze the apple to $-10°C$.
7. Put a thermocouple in the center of an aluminum can.
8. Pour orange or apple juice into the can and attach the thermocouple to the data logger.
9. Record the initial juice and freezing medium temperature.
10. Start the freezing process by immersing the can in the liquid freezing medium or in a blast freezer.
11. Immediately, start measuring the temperature-time data on the data logger.
12. Freeze the juice up to $-10°C$, as in the case of apples or pears.
13. Estimate the density of the orange or apple juice.

7.2.3.1 Preparation of Fruits for Blanching and Freezing

1. Weigh out 3 to 4 kg each of apples or potatoes.
2. Peel and core the apples, reweigh them and hold in cold water until ready for processing.
3. Slice the apples or potatoes into pieces before various treatments. Save some slices as a control for a comparison test.
4. Blanch 500 g of fruit slices in a steam or hot-water blancher for 2 min.
5. Cool in slush/ice for 1 min, drain and weigh.
6. Package one-half the slices in polyethylene, remove the air, and seal.
7. Save the other half of slices for testing and color evaluation.
8. Add 500 g of apple slices to a 35% sucrose solution containing 0.5% ascorbic acid. Make sure all exposed slice surfaces are coated with syrup.
9. Package and repeat step 7.
10. Submerge another 500 g of apple or potato slices in 0.25% $NaHSO_3$ solution for 2 min, contained in a cheesecloth. During soaking, agitate to ensure the exposure of all surfaces to the solution. Drain and hold for half an hour at room temperature.
11. Package one-half of the treated apple or potato slices in the polyethylene bag, remove the air, and seal. Save the other half of slices for sensory and color evaluation.
12. Freeze the apple slices by immersing them in liquid nitrogen and keep frozen for 4 to 5 days.
13. Prepare a homogenate of 100 g of fruit sample in 100 mL of 0.25% sodium bisulphite solution. Place a portion of homogenate into an appropriate cuvette and measure color parameter L, a and b for the blanched and unblanched (control) sample.
14. Perform a catechol test to test the enzymatic activity of the enzyme polyphenol oxidase.
15. Perform a sensory test to evaluate the color, texture, and flavor of the fruit immediately after treatment and on frozen samples after 5 days.
16. Compare the quality of the frozen sample treated with various treatments; report all the data in Data Sheet 7.2.

7.2.4 Results and Discussion

1. Plot temperature versus time data for each product.
2. Record the thermophysical property data and experimental freezing times on Data Sheet in 7.1.
3. For the computation of freezing time, use Eq. (7.8) (Nagoaka et al., 1955) for apples or potatoes and Eq. (7.7) (Plank's equation) for the juice. The freezing times of apple is defined as the time required to decrease the temperature from its initial temperature to $-10°C$. For juice, on the other hand, it may be defined as the time spent to bring down the temperature essentially at the freezing point $(-2°C)$.

4. Use a Microsoft$^{(R)}$ Excel spreadsheet or any other program to compute the freezing time. Show at least one sample calculation.

5. Discuss the results in terms of the effect of packaging on the freezing times. How does the freezing time differ in apple with or without polyethylene wrap?

6. How would freezing time be affected if the juice was frozen in a cardboard container rather than an aluminum container?

7. What are the likely sources of error in estimating the freezing times?

8. Discuss why Plank's equation is used to compute freezing times for juice and Nagaoka's equation for apples.

9. Discuss what effect on the freezing time you would expect, if air were trapped inside the package.

10. Discuss the effect of blanching and freezing on the quality of frozen and thawed apples or potatoes in terms of color, flavor, and texture.

11. Discuss the problems associated with overblanching the samples.

7.3 SUGGESTED READINGS AND REFERENCES

1. A. J. Ede, "The calculation of freezing and thawing of foodstuffs'. *Modern Refrig.* 52:52 (1949).

2. A. C. Cleland and R. L. Earle, "Prediction of freezing times for foods in rectangular packages." *J. Food Sci.* 44:964 (1979).

3. J. S. Nagoaka, D. H. Taylor, and N. J. Downes, 1955. "Experiments on the freezing of fish in an air blast freezer." In *Proceedings 9th International Congress on Refrigeration* Vol. 2, Paris.

4. J. C. Bomben, W. C. Dietrich, D. F. Farkas, J. S. Hudson, and E. S. de Marchena, "Pilot plant evaluation of individual quick blanching for vegetables." *J. Food Sci.* 38:590 (1973).

5. J. D. Selman, 1987. "The blanching process." In *Developments in Food Preservation*, *Vol. 4* (S. Thorne, ed.), Barking, Essex, UK: Elsevier Applied Science.

6. P. Fellows, 1988. "Freezing." In *Food Processing Technology Principles and Practice*, Chichester, UK: Ellis Horwood Ltd.

7. D. R. Heldman and R. P. Singh, 1981. "Thermodynamics of food freezing." In *Food Process Engineering*, Westport, CT: AVI Publishing Co.

DATA SHEET 7.1

Estimating the freezing times of products

Date: _____ Group: _____

Freezing medium: _____

Property	Product 1	Product 2
Dimensions		
Initial freezing point		
Water content X_w		
Initial product temperature T_i		
Freezing medium temperature T_m		
Final product temperature T		
Specific heat of unfrozen product C_{pu}		
Specific heat of unfrozen product C_{pf}		
Latent heat of fusion λ		
Density of frozen product ρ		
Thermal conductivity of frozen product k_f		
Thermal conductivity of packaging material k_p		
Convective heat transfer coefficient h		

Shape factors
  P
  R		

Computed freezing times by Plank's or Nagoaska's equation
  Packaged
  Unpackaged		

Experimental freezing times
  Packaged
  Unpackaged		

DATA SHEET 7.2

Effect of freezing on the quality of frozen and thawed apple or potato

Particulars	Day 0 (before freezing)	Day 0 (after freezing)	Day 5 (after thawing)
Color parameters Control sample L a b			
Blanched sample L a b			
Sample treated with 35% sucrose L a b			
Sample treated with 0.25% $NaHSO_3$ L a b			
Sensory evaluation (flavor, texture, and appearance) Control			
Blanched			
Sample treated with 35% sucrose			
Sample treated with 0.25% $NaHSO_3$			

8

ULTRA-HIGH-TEMPERATURE THERMAL PROCESSING

8.1 BACKGROUND

Ultra-high-temperature (UHT) treatment when coupled with aseptic packaging provides better-quality foods than its traditional retort/canning counterpart. In ordinary canning processes, the food product is packaged and then subjected to heat treatment as shown in Figure 8.1A. In aseptic processing, the food and packages are sterilized separately and the packages are then filled under aseptic conditions as shown in Figure 8.1B.

Aseptic processing has several advantages over conventional canning:

- *Better quality.* In addition to destroying bacteria and spores, thermal processing destroys vitamins and reduces the quality of foods. As temperatures increase, the rate

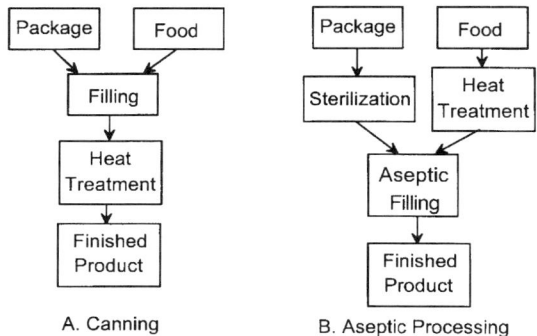

A. Canning   B. Aseptic Processing

Figure 8.1   Comparison of canning versus aseptic processing.

of bacterial destruction increases faster than that of vitamin loss and quality degradation. Therefore, processing at higher temperatures for shorter times can bring about the required bacterial reduction with a minimum loss of quality.

- *Lighter, cheaper packaging.* Packaging materials used for aseptic processing do not have to withstand high temperatures and the pressure of retorting. As a result, they are usually lighter and cheaper than the cans and bottles used in canning. The weight reduction results in lower manufacturing and transportation costs.
- Heat used to process food can be reclaimed efficiently, thereby reducing energy costs.

8.1.1 Food Sterilization

Several systems are available for the sterilization of food prior to packaging.

8.1.1.1 *Plate Heat Exchangers*
As shown in Figure 8.2, in a plate heat exchanger, stainless steel plates are stacked with a small gap between them. The product enters through a hole in one corner and exits from the opposite corner. The heating (or cooling) medium follows an opposite path. Gaskets are arranged so that the product and medium pass through alternate spaces between the plates. The spaces are kept very narrow so that a thin layer of film is formed that heats faster. In addition, plates are usually corrugated to induce turbulence and increase the surface area, which further speeds the heating process. Because of the narrow spacing, this method of heating is best suited to low-viscosity fluids with little or no particulate matter such as juice, milk, thin sauces, etc. Plate heat exchangers are more prone to fouling because of a narrow product flow path than other types of heaters, but they can be easily disassembled for cleaning.

Plate heat exchangers usually contain several sections as shown in Figure 8.3:

- *Heating section.* In the heating section, steam or hot water is used to raise the product to the proper temperature for sterilization. The product then passes through a holding

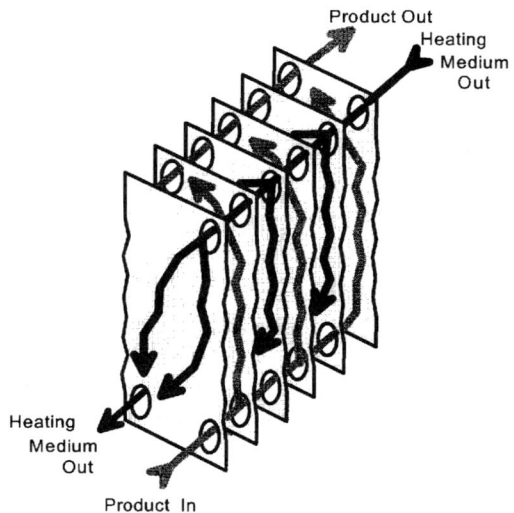

Figure 8.2 Schematic cutaway of a plate heat exchanger.

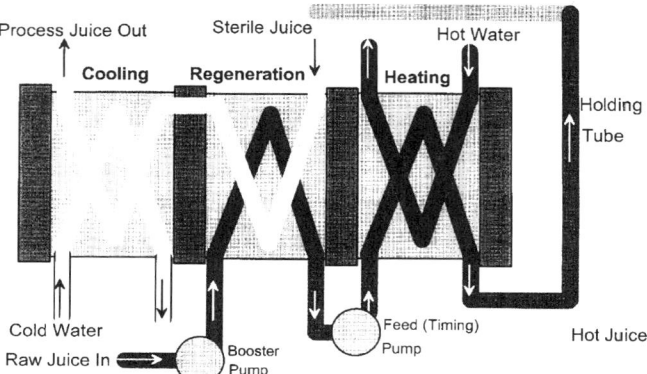

Figure 8.3   View of product and hot-water flow into various sections of the plate heat exchanger.

tube where the sterilizing temperature is maintained for a sufficient time to achieve the desired PNSU.

- *Regeneration section.* In order to conserve energy, a regeneration section is used to recover as much heat as possible from the hot product and transfer it to the incoming raw product.
- *Cooling section.* In the cooling section, cold water is used to reduce the temperature to a suitable temperature for packaging or storage.
- The flow rate through this system is critical and is controlled by a positive displacement rotary feed pump, usually placed between the regenerative section and the heating section.
- If a leak were to develop in the regenerative section, the product could pass between the sterile side to the raw side. In order to protect the sterile product from contamination, the sterile side must be held at a higher pressure than the raw side so liquid will flow through a leak only from the sterile to the raw side. This pressure differential is created by the feed pump and regulated by controlling the rotational speed of the booster pump, in order that the sterile side of the regeneration section be maintained at a higher pressure than the raw product side.

8.1.1.2   *Single Spiral-Tube Heat Exchanger*   In this type of heat exchanger, the product is forced under high pressure through a helical tube located inside a shell. The heating (or cooling) medium passing through the surrounding shell heats the product. This type of unit is suitable for low- and medium-viscosity liquids with little or no particulate matter such as ice cream mix, various kinds of cream, thick sauces, etc.

8.1.1.3   *Single Tube-in-Tube Heat Exchanger*   The food passes through an inner tube and the medium passes through a surrounding tube as shown in Figure 8.4. This system is suitable for liquids containing particulates up to $\frac{3}{4}$ in. diameter, including diced fruit, soups, purées, etc.

8.1.1.4   *Multiple Tube-in-Tube or Shell-in-Tube Heat Exchanger*   In this type of heat exchanger (Fig. 8.5), the product passes through several tubes within the same

Figure 8.4 Tube-in-tube heat exchanger.

outer tube. The smaller-diameter tubes increase the rate of heating, but limit use to a product of low viscosity with very small particulates.

8.1.1.5 *Scraped Surface Heat Exchanger* In this type of unit, the product passes through a cylinder that is inside another cylinder containing the heating/cooling medium as shown in Figure 8.6. Scraper blades rotating inside the inner cylinder scrape the product off the wall. This removes fouling, brings fresh product in contact with the heated surface, and transports heat into the product. This type of heat exchanger is more expensive, but is more suitable for medium- to high-viscosity liquids and handles smaller particulates.

8.1.1.6 *Steam Injection* In this system, steam is sprayed into the product. After processing, evaporative cooling is used to remove excess water.

8.1.1.7 *Steam Infusion* This process is similar to steam injection, except that the food is sprayed into the steam. A typical steam infusion system is shown in Figure 8.7. Again, evaporative cooling is used to remove excess water.

8.1.1.8 *Double-cone Processor* For some products that consist of particulates in a liquid, the particulates require a longer processing time than the liquid. Shortening the particulate processing is unsafe and lengthening the liquid processing reduces quality. In

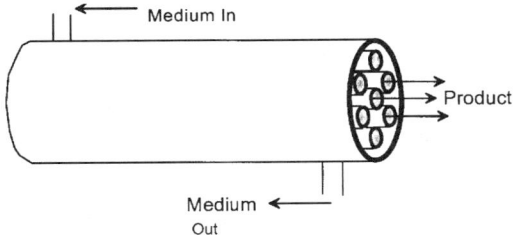

Figure 8.5 View of multiple tube-in-tube or shell-in-tube heat exchanger.

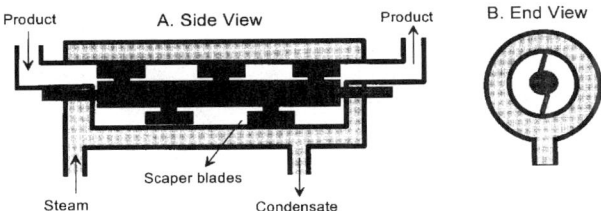

Figure 8.6 Schematic view of scraped surface heat exchanger.

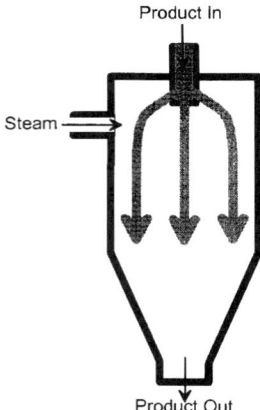

Figure 8.7   Operation of steam infusion system.

these circumstances, the liquid and particulates must be processed separately. The liquid can be processed in a conventional heat exchanger. For the particulates, AVI has developed the Jupiter double-cone aseptic processing vessel, as shown in Figure 8.8. The product is loaded into the chamber and tumbled by rotating the unit. Heat is provided by steam or hot water in an outside jacket or by injecting steam directly into the food chamber. A preheated liquid or stock is sometimes added to the food chamber. After heating, the particulates are removed and combined with the liquid portion of the food before packaging.

8.1.2   Typical UHT Processing System

A typical simplified diagram of an aseptic processing unit such as you may use to process fruit juice in your pilot plant is shown in Figure 8.9. In this example, the food is heated and cooled with a plate heat exchanger.

1. Raw juice is loaded into a balance tank.
2. It enters a deaerator where air is removed by a vacuum pump.
3. A centrifugal booster pump transports the juice into the regeneration section of a heat exchanger where heat is recaptured from processed juice in order to partially heat the new juice.

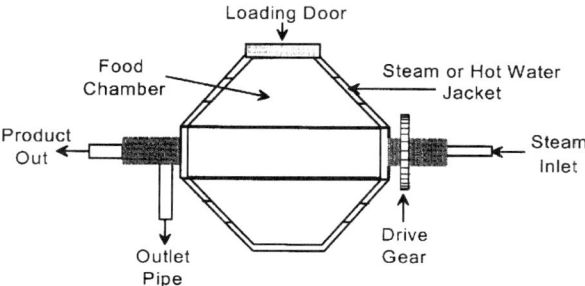

Figure 8.8   Schematic view of a Jupiter double-cone aseptic processing vessel.

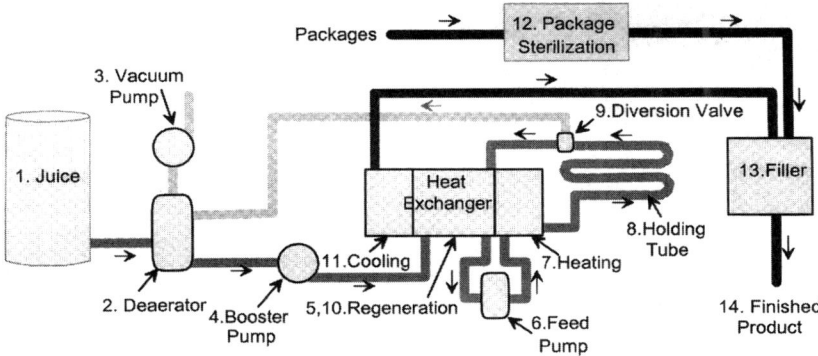

Figure 8.9 Detailed view of an aseptic processing system.

4. A positive displacement feed pump pushes the juice at a controlled volumetric flow rate into the heating section, where hot water is used to bring the juice to the proper processing temperature.

5. The heated juice passes through a holding tube where it is kept for the time necessary for proper processing. The time in the tube is determined by the volumetric flow rate and flow characteristics of the juice and by the length of the tube as described in a later section. Furthermore, to prevent the formation of air pockets in the holding tube, it is required to slope upward at a rate of $\frac{1}{4}$ in. per foot.

6. At the end of the holding tube, the juice is checked to see if it has maintained the required processing temperature. If it has dropped below the desired level, a diversion valve routes the juice back to the beginning of the system so that it can be reprocessed.

7. Returning to the regeneration section, the hot juice gives up heat to heat the incoming juice.

8. The juice is then cooled close to the storage temperature in the cooling section.

9. In the mean time, packages are being sterilized on the other side of the processing machine.

10. Finally, the sterile juice is packaged in the sterile packages by an aseptic filler.

In addition to what is shown in Figure 8.9, such systems also include arrangement for a cleaning in place (CIP) system that circulates detergent solution and rinsing hot water.

8.2 ENGINEERING BACKGROUND

8.2.1 Holding Tube Design

After heating the product to the proper temperature, it must be held at that temperature for a period of time necessary to achieve the desired probability of nonsterile unit (PNSU). This is generally obtained by passing the product through a holding tube whose length is calculated so that the fastest-moving particle in the product stream will remain for the necessary time. This section reviews the information needed to calculate the holding time and design a tube that will achieve that time.

8.2.1.1 Heat Treatment

1. *PNSU.* When the average microbial count drops below one per unit (can, package, liter, etc.), we refer to it as the PNSU.

   > **Example 1**   *For C. botulinum spores, the goal is usually to have a PNSU of 10^{-9} or less. In other words, one unit in 10^9 would be contaminated at this PNSU.*

2. *Log cycle reduction.* A microbial count is reduced to 1/10th of the starting value, during one log cycle reduction. In N log cycles of reduction, the count is reduced to $1/10^N$ times the starting value. If the microbial count per package and the desired PNSU are both known, the number of log cycles of reduction is given by

$$N = \log(N_0) - \log(\text{PNSU}) \tag{8.1}$$

   where N = the number of log cycles of reduction to be achieved and N_0 = the initial count per unit.

   > **Example 2**   *If a can has a count of 3500 and a PNSU of 10^{-9} is sought, the number of log cycle reductions should be*
   >
   > $$N = \log(3500) - \log(10^{-9}) = 3.5 - (-9) = 12.5$$
   >
   > *The final count will be reduced to $1/10^{12.5}$ times the initial count.*

3. *Required process time.* If the D_T valule is known for an organism, the required time at temperature T is given by

$$F_T = ND_T \tag{8.2}$$

   where D_T = the time required at temperature T for the count to be reduced to 1/10th of its starting value, N = the number of log cycles of reduction to be achieved, and F_T = the required time.

4. *Determining D_T.* If z and D_{250} are known for a microorganism, D_T can be computed by

$$D_T = D_{250} 10^{(250-T)/z} \tag{8.3}$$

   > **Example 3**   *C. botulinum as $D_{250} = 0.2$ min and $z = 18$. At $240°F$, we have*
   >
   > $$D_{240} = 0.2^{(250-240/18)} = 0.72 \text{ min}$$

To achieve 12.5 log cycles of reduction at 240°F would require a heating time of

$$F_{240} = 12.5(0.72) = 8.9 \text{ min}$$

8.2.1.2 Fluid Dynamics When a fluid passes through a pipe, the velocity within the pipe varies from 0 at the wall to a maximum in the center as shown in Figure 8.10. The velocity profile across the pipe depends on the rheological properties of the fluid and the flow characteristics of the stream.

1. *Dimensions.* We will use the following coordinate systems when referring to fluid flow:
   - *X dimension.* In the direction of flow.
   - *Y dimension.* At right angles to the direction of flow.
2. *Flow characteristics.* Flow can be either:
   - *Laminar.* All fluid flows in the X direction. There is no lateral movement in the Y direction. Since the flow can occur at different velocities, we define the flow like the movement of parallel plates sliding over each other. Very little mixing occurs in this type of flow.
   - *Turbulent.* Fluid moves in different directions in different locations with considerable lateral movement. This produces considerable mixing.
3. *Local velocity.* In either type of flow, the velocity will vary from place to place. We call the velocity at any given location the local velocity. The x component of local velocity is defined by

$$v_s = \frac{dx}{dt} \tag{8.4}$$

4. *Shear rate.* As you cross the stream (in the Y direction), the velocity changes. The rate of change of the velocity in the x direction with a change of the y position is called the shear rate and defined by

$$\gamma = \frac{dv_x}{dy} \tag{8.5}$$

5. *Shear stress.* In order for shear to take place, the fluid in two layers must be subjected to opposing forces as shown in Figure 3.1. This force will be spread out

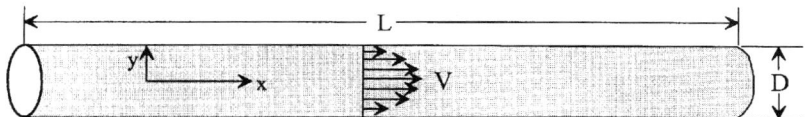

Figure 8.10 Two-dimensional pattern view of the velocity profile in a tube.

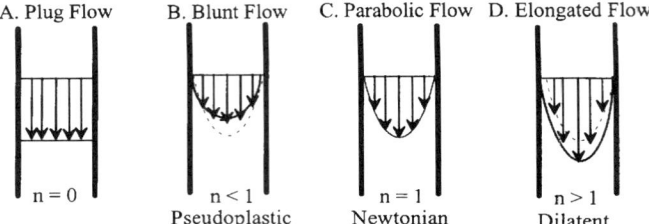

Figure 8.11 View of velocity profiles for Newtonian, pseudoplastic, and dilatant types of fluid while flowing in a tube.

over the area of contact between the two layers. Shear stress measures the force per unit area, that is,

$$\tau = \frac{F}{A} \qquad (8.6)$$

6. *Newtonian viscosity.* For water and similar fluids, the shear stress required is proportional to the shear rate, that is,

$$\tau = \mu\gamma \qquad (8.7)$$

7. *Non-Newtonian viscosity.* For many food products, the apparent viscosity changes as the shear rate increases. This behavior can frequently be approximated with the power law equation:

$$\tau = m\gamma^n \qquad (8.8)$$

where $m =$ the consistency coefficient and $n =$ the flow behavior index.
- When $n = 0$, the left side of Eq. (8.8) reduces to a constant and the shear stress is independent of the shear rate.
- When $n = 1$, Eq. (8.8) reduces to Eq. (8.7) and $m = \mu$. We call this a Newtonian fluid.
- When $n < 1$, the fluid appears to thin as the shear rate increases and we refer to it as a pseudoplastic fluid.
- When $n > 1$, the fluid appears to thicken as the shear rate increases and we refer to it as a dilatant fluid.

8. *Velocity profiles*
- *Plug flow.* When $n = 0$, the local velocity is constant across a pipe and we call it a plug flow as shown in Figure 8.11A. This happens with very viscose materials.
- *Blunt flow.* For pseudoplastic fluids, the velocity profile varies less across the pipe than in parabolic flow as shown in Figure 8.11B.
- *Parabolic flow.* The velocity profile of a Newtonian fluid is parabolic as shown in Figure 8.11C.
- *Elongated flow.* For dilatant fluids, the velocity profile varies more across the pipe then in parabolic flow as shown in Figure 8.11D.

9. *Newtonian Reynolds number.* For flow of a Newtonian fluid through a pipe, the Reynolds number is computed by

$$Re = \frac{\rho D v_{\text{avg}}}{\mu}$$

(8.9)

where ρ = the density of the fluid, D = the diameter of the pipe, v_{avg} = the average velocity of the fluid, and μ = the viscosity of the fluid.
- If $Re < 2100$, flow will usually be laminar.
- If $Re > 4000$, flow will usually be turbulent.
- Between 2100 and 4000, flow will be in transition between the two patterns.

10. *Generalized Reynolds number.* For fluids that obey the power law (Eq. 8.8), the generalized Reynolds number is computed by

$$Re = 2^{3-n} \left(\frac{n}{3n+1}\right)^n \left(\frac{\rho D^n v_{\text{avg}}^{2-n}}{m}\right)$$

(8.10)

where m = the consistency coefficient and n = the flow behavior index.

8.2.1.3 Flow Rates and Velocities

1. *Mass flow rate.* This can be determined by timing the filling of a bucket from the pipe. The rate is then

$$\dot{m} = \frac{m}{t}$$

(8.11)

where m = the mass of fluid in the bucket and t = the time required to collect that mass.

2. *Volumetric flow rate.* This can be determined by timing the filling of a bucket from the pipe. The rate is then

$$\dot{G} = \frac{V}{t}$$

(8.12)

where V = the volume collected and t = the time required. It can also be computed from the mass flow rate by

$$\dot{G} = \frac{\dot{m}}{\rho}$$

(8.13)

where ρ = the density of the fluid.

3. *Average velocity.* This can be computed from the volumetric flow rate as

$$V_{\text{avg}} = \frac{\dot{G}}{A}$$

(8.14)

where A = the cross-section area of the pipe (for a circular pipe, $A = \pi r^2$, where r = the radius of the pipe).

Example 4   *A holding tube is found to discharge 12.6 kg of water in 32 sec. The inside diameter of the pipe is 1.78 in. The average velocity of the fluid is computed as follows:*

$$\dot{m} = \frac{12.6 \text{ kg}}{32 \text{ s}} = 0.39 \text{ kg/s}$$

$$\dot{G} = \frac{0.39 \text{ kg/s}}{1000 \text{ kg/m}^3} = 3.9 \times 10^{-4} \text{ m}^3/\text{s}$$

$$V_{\text{avg}} = \frac{3.9 \times 10^{-4} \text{ m}^3/\text{s}}{\pi \left(\frac{(1.78 \text{ in.})(0.0254 \text{ m/in.})}{2} \right)^2}$$

$$= \frac{3.9 \times 10^{-4} \text{ m}^3/\text{s}}{1.61 \times 10^{-3} \text{ m}^2} = 0.24 \text{ m/s}$$

4. *Maximum velocity.* For a circular pipe, the maximum velocity occurs at the center of the pipe. This velocity can be computed from the average velocity as follows:
   · For a Newtonian fluid in laminar flow,

$$v_{\text{max}} = 2v_{\text{avg}} \tag{8.15}$$

   · For a fluid in turbulent flow,

$$v_{\text{max}} \approx 1.25v_{\text{avg}} \tag{8.16}$$

Example 5   *With the flow rate computed in the last example, what is the maximum velocity of water?*

$$Re = \frac{\rho D V_{\text{avg}}}{\mu} = \frac{(1000 \text{ kg/m}^3)[(1.78 \text{ in.})(0.0254 \text{ m/in.})](0.24 \text{ m/s})}{2 \times 10^{-4} \text{ kg/m} \cdot \text{s}} = 54{,}254$$

so flow is turbulent and the maximum velocity is

$$v_{\text{max}} = 1.25(0.24 \text{ m/s}) = 0.3 \text{ m/s}$$

8.2.1.4 *Holding Tube Length*   The required pipe length is computed by

$$L = v_{\text{max}} F_T \tag{8.17}$$

where V_{max} = the maximum fluid velocity in the pipe and F_T = the required holding time at temperature T.

Example 6 If the required holding time is 4.6 sec and the maximum velocity 0.3 sec, what length tube is needed?

$$L = (0.3 \text{ m/s})(4.6 \text{ s}) = 1.4 \text{ m}$$

8.3 LAB EXERCISE

8.3.1 Objectives

The objectives of this lab exercise are to:

1. Process foods using the UHT treatment system.
2. Determine the convective heat transfer coefficient and overall heat transfer coefficient of the UHT heat exchanger in the heating section.
3. Study the effect of viscosity and flow rates on the heat transfer coefficients.

8.3.2 Equipment

A schematic diagram of typical laboratory scale UHT/HTST Lab-25 DH is shown in Figure 8.12. It is designed to process fluid products at temperatures ranging from 170 to 295°F and pressure up to 350 psig. Specific design features include two tubular heaters. The first heater also known as a preheater or heater 1 uses hot water to heat products before it enters the final heater or heater 2. The product temperature in heater 1 is controlled by adjusting the steam pressure used to generate hot water. A bypass line is also provided for

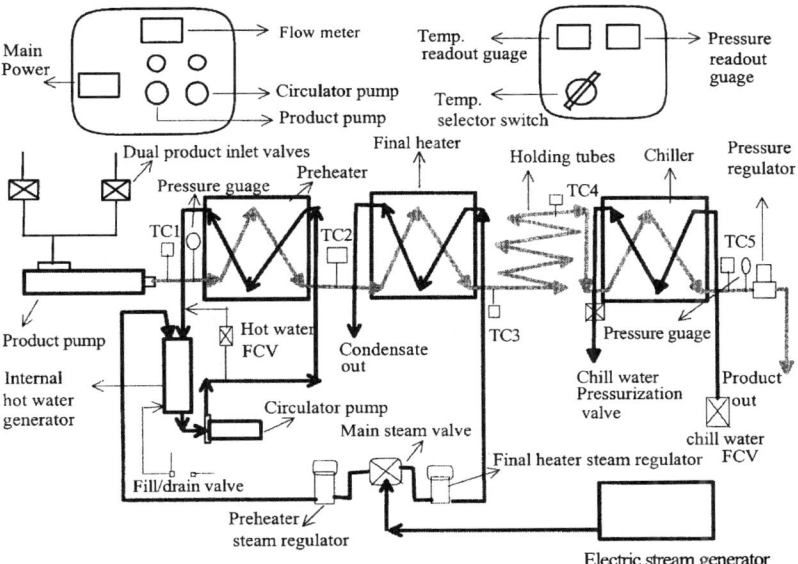

Figure 8.12 Typical schematic view of UHT/HTST lab-25 DH processing system.

adjusting the heater temperature. The heater 2 uses direct steam as the heating medium. A number of holding tubes are used to obtain a desired holding time. This holding time can be also varied by controlling the pump speed. The product is cooled in the chiller to less than 45°F.

The system is equipped with a spring-loaded pinch-type back pressure valve located at the end of the heaters and chiller. It is used to control product back pressure, which prevents back flashing (boiling) of the product during processing. Generally, it is useful to maintain an approximately 75 to 100 psi back pressure in the system to avoid boiling. The product pressure immediately before the back pressure valve is displayed on the digital readout.

Various thermocouples are located in the system to measure the temperature of the product at various stages of the processing, it is displayed on a digital readout by selecting the proper channel. Two dual product inlet valves feed the product to the pump. The product pump is a variable-speed positive displacement pump to control the product flow rate.

8.3.3 Procedure

1. Turn on the main power supply.
2. Close the cold-water flow control valve.
3. Close the cooler pressurization valve.
4. Open the hot-water flow control valve.
5. Open the fill/drain valves (set to the filling position). The water will start flowing to the internal hot-water generator tank.
6. Turn on the hot-water circulator pump and let the hot-water loop flush free of air for approximately 1 min.
7. Close the fill/drain valve (set to the running position).
8. Feed the product pump with water.
9. Open the product inlet valves and product back pressure valves.
10. Turn on the product pump and let product lines flush free of air.
11. Set the flow rate and adjust back pressure by adjusting the needle valve to approximately 75 psi.
12. Set the sterilization temperature.
13. Close the hot-water flow control valve and then open $\frac{1}{8}$ of a turn to start.
14. Open the preheater steam regulator valve to adjust steam pressure in between 5 to 10 psi to regulate the product heating temperature in the preheater.
15. Open the final steam regulator 2 to adjust steam pressure in heater 2 in order to set the final product temperature.
16. Hot water should be circulated in between 250 to 265°F for 30 min to ensure the sterility of the entire system.
17. Open the cold-flow control valve and cooler pressurization valve.
18. Allow the product of sterile water to cool down to a desired level.
19. Adjust back pressure and system conditions such as the flow rate, holding times, reheating, final heating and cooling temperatures, etc.

20. The final processed product may either collect at the filling station or the back pressure valve outlet near the sink.
21. Once the raw product processing is finished, rinse the system with water and then a standard CIP procedure may be adopted to clean the system.

8.3.4 Materials

Low-viscosity product such as apple juice or milk, or a high-viscosity product such as tomato juice, cream, etc.

8.3.5 Procedure

1. Record the product flow rate (L/min or kg/min) in Data Sheet 8.1 for low viscosity product and in Data Sheet 8.2 for high viscosity product. Maintain the same flow rates for low- and high-viscosity products for a comparison of overall heat transfer coefficients and holding tube design in both cases.
2. Once the steady-state conditions have been achieved, record product inlet temperature (TC1), product outlet temperature after preheater or heater 1 (TC2), product outlet temperature after heater 2 (TC3), product temperature at the end of holding tube (TC4), product outlet temperature after cooling section (TC5) for a low- and high-viscosity product.
3. Similarly, record the hot-water inlet temperature in the preheater (TC6), hot-water outlet temperature (TC7), steam inlet temperature before the heater 2 (TC8), chilled-water inlet temperature (TC9), and chilled-water outlet temperature (TC10).
4. Record the outer and inner diameters of the UHT preheating, heating, and cooling section tube, length of preheater tube (heater 1), heater 2, cooling tube, and type of flow in heat exchanger during heating/cooling, whether countercurrent or of the parallel type (Data Sheets 8.1 and 8.2).

8.3.5.1 Assumption

1. The thermal properties of the product such as specific heat and density may be assumed as constant during various processing sections.
2. We maintain a holding time of 5 s for each low- and high-viscosity product.
3. Assume that the absolute viscosity of the product remains constant during UHT processing.

8.3.6 Calculations

1. Find out the properties of the product at the average inlet and outlet temperature of the heater such as density ρ, specific heat C_p, and absolute viscosity μ.
2. Determine the product average velocity $v =$ product mass flow rate/tube cross-sectional area.

3. Calculate the Reynolds number by using Eq. (8.18):

$$\text{Reynolds number } (N_{\text{RE}}) = [\rho D v]/\mu \tag{8.18}$$

4. Estimate the maximum product velocity at the center of the tube:
   a. If Reynolds number $(N_{\text{RE}}) < 2100$, then it is a laminar flow and maximum velocity = average velocity/0.5.
   b. If Reynolds number $(N_{\text{RE}}) > 10,000$, then it is a turbulent flow and maximum velocity = average velocity/0.8.
5. Holding tube length = maximum velocity × holding time.
6. Log mean temperature difference:
   a. Depending on whether product flow, and heating or cooling-medium flow are of the countercurrent or parallel flow types (Figs. 8.13 through 8.15), estimate the ΔT_1 and ΔT_2. Calculate the log mean temperature difference by using

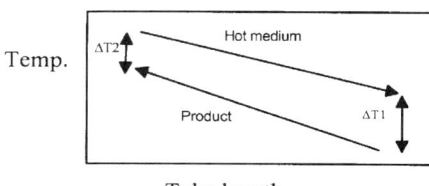

Figure 8.13   Countercurrent-type flow in a heat exchanger.

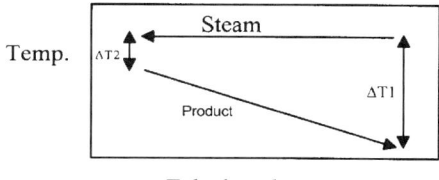

Figure 8.14   View of countercurrent-type flow when steam is used as a heating medium.

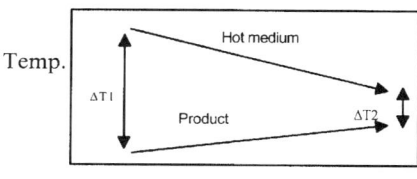

Figure 8.15   Schematic view of co-current-type flow in a heat exchanger.

Eq. (8.19):

$$LMTD = \left(\frac{\Delta T_1 - \Delta T_2}{\ln \dfrac{\Delta T_1}{\Delta T_2}} \right) \qquad (8.19)$$

7. Calculate the overall heat transfer rate by using Eq. (8.20). The overall heat transfer coefficient can be estimated based on either the outer area (A_o) or inner tube area (A_i):

$$q = U_i \cdot A_i \cdot (LMTD) \qquad (8.20)$$

where $A_i = \pi D_i L$, and $D_i =$ inner diameter of the tube (m) and $L =$ length of the tube.

8. The heat transfer rate q can be estimated by the following relationship.

$$q = m \cdot C_p (Tp_e - Tp_i)$$

where $q =$ KJ/min or J/s, $m =$ mass flow rate (kg/s), $C_p =$ specific heat of the product, $Tp_e =$ product temperature at the heater outlet, $Tp_i =$ product temperature at the heater inlet.

9. Mass flow rates of hot water in the preheater or cooling water are calculated by energy balance equations. Heat gained or lost by the product = heat lost or gained by the heating/cooling medium.

10. Amount of steam used per hour in the heater can be calculated by assuming that only the latent heat of stream is used in heating the product:

$$\text{Steam used} = \frac{q}{\lambda} \qquad (8.21)$$

where $q =$ heat transfer rate (KJ/h) and $\lambda =$ latent heat of steam (KJ/kg).

8.3.7 Lab Report

1. Determine the overall heat transfer coefficients in the preheater, final heater, and cooling section of the UHT processing plant.
2. How the overall heat transfer coefficients and log mean temperature difference differ for low- and high-viscosity products?
3. Estimate the hot-water flow in the preheater, cooling-water flow rate in the cooling section and amount of stream required per hour in the final heater of the UHT processing plant.
4. Determine the length of the holding tube required to hold each product for a minimum period of 5 s. How can holding time be doubled without increasing the holding tube length?
5. Label the hot-water and cooling-medium temperature at various sections of the UHT/HTST heat exchanger system.

6. Discuss the effects of flow rates and viscosity on the heat transfer coefficients. How the overall heat transfer coefficients would change if flow types were changed from the countercurrent to parallel flow type.

7. Summarize the results in Data Sheet 8.3.

8. Discuss the advantages and disadvantages of UHT heat exchanger over retorting.

8.4 SUGGESTED READINGS AND REFERENCES

1. P. Fellows, 1988. *Food Process Technology, Principles and Practice*, Chichester, UK: Ellis Horwood, Ltd.

2. H. G. Kesseler, 1981. *Food Engineering and Dairy Technology*, Freising, Germany: Verlag A. Kesseler.

3. R. J. Swientek, "Free falling film UHT system 'sterilizes' milk products with minimal flavour change." *Food Proc.* 44:114 (1983).

4. S. Palaniappan and C. E. Sizer, "Aseptic process validated for foods containing particulates." *Food Tech.* 51:60 (1997).

DATA SHEET 8.1

UHT system type: _____

Outer diameter of heating and cooling section tube: _____ mm

Inner diameter: _____ mm

Length of preheater tube (heater 1): _____ m

Length of heater 2 tube: _____ m

Length of cooling section tube: _____ m

Type of flow during heating/cooling: _____

Type of flow during cooling: _____

Low-viscosity product: _____

Manufacturing and label information: _____

Flow rate: _____

1. Product temperature data at steady state

Time	TC1	TC2	TC3	TC4	TC5
5 min					
10 min					
15 min					
Average					

TC1 = product inlet temperature.

TC2 = product outlet temperature after preheater.

TC3 = product outlet temperature after final heater.

TC4 = product temperature at the end of holding tube.

TC5 = product outlet temperature after cooling section.

2. Heating/cooling-medium temperature data at steady state

Time	TC6	TC7	TC8	TC9	TC10
5 min					
10 min					
15 min					
Average					

TC6 = hot-water inlet temperature in preheater.
TC7 = hot-water outlet temperature.
TC8 = steam inlet temperature in final heater.
TC9 = chilled-water inlet temperature.
TC10 = chilled-water outlet temperature.

3. Thermo-physical properties of low-viscosity product

Particulars	Preheater	Final Heater	Cooling Section
Inlet temperature ($^\circ$C)			
Outlet temperature ($^\circ$C)			
Average temperature ($^\circ$C)			
Density (kg/m^3)			
Specific heat (KJ/kg·K)			
Apparent viscosity (Pa·s)			

DATA SHEET 8.2

High-viscosity product: _____

Manufacturing and label information: _____

Product flow rate: _____

1. Product temperature data at steady state

Time	TC1	TC2	TC3	TC4	TC5
5 min					
10 min					
15 min					
Average					

TC1 = product inlet temperature.

TC2 = product outlet temperature after preheater.

TC3 = product outlet temperature after final heater.

TC4 = product temperature at the end of holding tube.

TC5 = product outlet temperature after cooling section.

2. Heating/cooling-medium temperature data at steady state

Time	TC6	TC7	TC8	TC9	TC10
5 min					
10 min					
15 min					
Average					

TC6 = hot-water inlet temperature in preheater.

TC7 = hot-water outlet temperature.

TC8 = steam inlet temperature in final heater.

TC9 = chilled-water inlet temperature.

TC10 = chilled-water outlet temperature.

3. Thermo-physical properties of high-viscosity product

Particulars	Preheater	Final Heater	Cooling Section
Inlet temperature (°C)			
Outlet temperature (°C)			
Average temperature (°C)			
Density (kg/m^3)			
Specific heat (KJ/kg·K)			
Apparent viscosity (Pa·s)			

DATA SHEET 8.3

Comparison of UHT processing parameters

Particulars	Low-viscosity Product	High-viscosity Product
Preheater		
Overall transfer coefficient U_i (W/m^2·K)		
Hot-water flow rate (Kg/min)		
Final heater		
Overall transfer coefficient U_i (W/m^2·K)		
Steam flow rate (Kg/h)		
Cooling section		
Overall transfer coefficient U_i (W/m^2·K)		
Hot-water flow rate (Kg/min)		
Length of holding tube (m)		

9

MEMBRANE PROCESSING
OF LIQUID FOODS

9.1 BACKGROUND

Membrane processing is a growing field in food processing. Separations are based on molecular weight size and can be very discriminating. Essentially, fluids are passed over semipermeable surfaces that selectively pass components of the fluid without chemical or physical alterations. An obvious advantage is the recovery of substances without alterations. Another advantage is the fine separation of components difficult to achieve by other means. Membrane processing has made some former food waste products, such as whey proteins and waste water, recoverable and recyclable. Membranes are also being used to preconcentrate milk solids as a step in the cheese-making process.

Membrane separations can be performed at ambient or low temperatures, thus preventing temperature-sensitive ingredients from damage. It improves product quality by retaining flavors and vitamins and minimizing protein denaturation. High selectivity makes membrane separation a more efficient method. Membrane processing is a capital- and energy-efficient method too. Usually, membranes with 25 kJ of energy are required for the removal of 1 kg of water, compared to about 500 kJ using evaporation. Compact and modular design makes the membrane separation system easy to install and space-saving. Membrane separation is also referred to as the cold sterilization process. Generally, the pore size of membrane is smaller than that of bacteria, yeast, and mold and, therefore, the permeate is bacteria-free. However, their retentate is often full of bacteria.

9.2 PRINCIPLES OF MEMBRANE PROCESSING

Generally, there are three main membrane techologies: namely, hyperfiltration or reverse osmosis, ultrafiltration, microfiltration. They may be distinguished by the size of the particles or molecules they are capable of retaining.

9.2.1 Hyperfiltration

In hyperfiltration, pore size is the smallest (0.0001–0.001 μm) of all three systems, and even small dissolved molecules such as salts are retained by the membrane. At this molecular level, high pressures are required on the order of 10 to 50 bar because osmotic forces also come into play. The hyperfiltration system is additionally used for dewatering purposes. For example, water purification by the desalination of sea water.

9.2.2 Ultrafiltration

The pore size of ultrafiltration membrane ranges from 0.001 to 0.1 μm. It is a selective separation process in which suspended solids, colloids such as proteins, emulsified solids such as fat protein complex, and dissolved macromolecules with a molecular weight of approximately 10,000 to 100,000 are retained by the membrane. The molecule that does not pass through the membrane is called retentate or concentrate. Lower-molecular-weight dissolved materials such as solvents, salts, sugars, and water pass through the membrane under a driving force of relatively low hydrostatic pressure (1–10 bar) and this stream is known as the permeate, as shown in Figure 9.1. Ultrafiltration is generally used in the concentration and fractionation of large molecules from cheese whey and milk.

9.2.3 Microfiltration

Microfiltration membranes have a pore size 0.1 to 10 μm and only very large macromolecules such as large fat globules, large proteins, and suspended particles such as microbial cells are held back. Sometimes, microfiltration is used for the selective

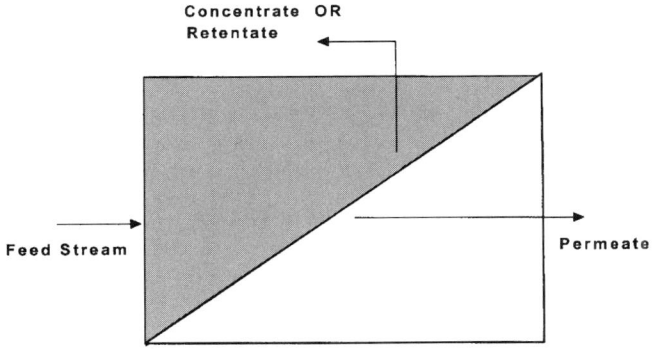

Figure 9.1 Schematic diagram depicting the principle of ultrafiltration membrane.

separation of molecules by choosing a specific membrane size. For example, milk protein caseins can be retained in the milk while filtering out the whey proteins from skim milk. Microfiltration is generally used in the clarification and separation of pharmaceutical products, beer, wine, and soft drinks.

9.3 MEMBRANE CONFIGURATION

9.3.1 Dead-end Membrane

Dead-end membrane filtration is one of the simplest and most general methods of filtration. The liquid mixture flows at right angles to the membrane barrier and cake layer that is formed by deposited particles as shown in Figure 9.2. In this type of membrane configuration, the cake layer increases and the flux rate decreases with time. Filter aids and flocculants are generally used to improve filtration efficiency.

9.3.2 Cross-flow Membrane Filtration

In this process, the feed flows tangentially over the membrane surface as shown in Figure 9.3. The particles deposited on the filter medium are swept away by the feed. The highly turbulent velocity of the feed maintains a clean surface with a high efficiency. This type of configuration is most desirable, since it usually maintains a constant cake-layer thickness and, thus, a constant rate of flux through the membrane.

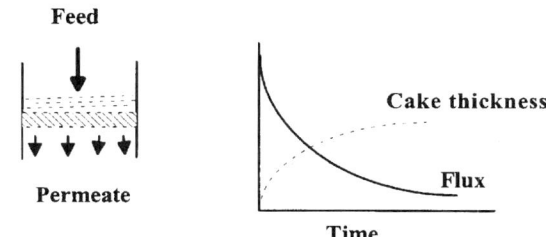

Figure 9.2 Dead-end membrane filtration configuration.

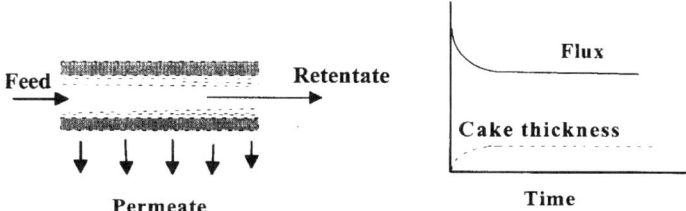

Figure 9.3 Cross-flow membrane filtration configuration.

9.4 TYPES OF MEMBRANES

9.4.1 Organic Membranes

Organic membranes are generally prepared from cellulose by acetylation, that is, reaction with acetic anhydride, acetic acid, and sulphuric acid. These are known as cellulose triacetate, proprionate, or ester. There are several advantages to the use of cellulose acetate: high flux and high salt rejection properties, for instance. They are also easy to manufacture. However, they have many disadvantages. For example, they have a narrow temperature range, as many manufacturers recommend a maximum operating temperature of 50°C. The membrane material, cellulose acetate, is highly biodegradable under acidic conditions. They have a restricted pH range of 3 to 9. They have a limited life time. Cellulose acetate has poor resistance to chlorine, being oxidized readily, leading to a poor operating life time.

9.4.2 Synthetic Membranes

These membranes are made of synthetic polymers such as nylon, polyvinylidene fluoride, polyurethane, polysulphone, polyester, etc. Polysulphone is very popular for ultrafiltration applications because it has a wide temperature application; a typical temperature of up to 75°C can be used. Sulphones can be exposed to a wide pH range, from 1 to 13. It has fairly good resistance to chlorine for sanitation purposes. However, other sanitizers are now currently being recommended. A wide range of pore sizes are available in synthetic membranes for ultrafiltration purposes, ranging from 0.001 to 0.02 μm.

The main limitations of polysulphone membranes are their apparent low-pressure limits. A flat sheet membrane has a pressure limit of 100 psi and hollow fiber membrane of up to 25 psi.

9.4.3 Inorganic Mineral Membranes

These membranes are made from inorganic mineral materials such as α-alumina (ceramic), silica, stainless steel, carbon, zirconium, etc. They are formed by the deposition of inorganic solutes onto a microporous support that is reusable. These mineral or ceramic membranes are extremely versatile and have none of the disadvantages associated with polymeric membranes. Both membrane and support materials possess a high degree of resistance to chemical and abrasion degradation. They tolerate a wide range of pH and temperature ranges. A schematic of the ceramic membrane is shown in Figure 9.4. The main advantages of these membranes include excellent mechanical properties to withstand high pressure from 600 to 1200 psi. They also withstand a wide temperature range, from 0 to 300°C. They can be sterilized with steam. They are highly resistant to chemical organic solvents, a corrosive medium, and abrasion degradation. They are additionally available in a wide range of pore sizes, from 0.01 to 1.4 μm. They have a long life span, are not subjected to compaction, and are easy to replace or alter.

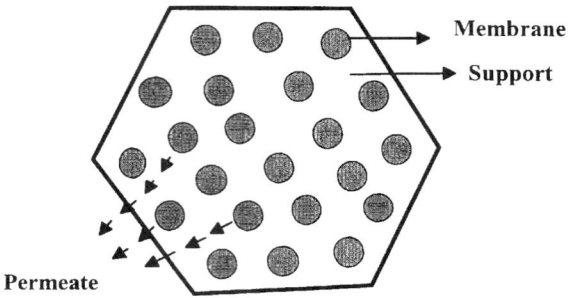

Figure 9.4 Typical ceramic-type membrane filtration configuration.

9.5 MEMBRANE MODULE CONFIGURATIONS

9.5.1 Tubular Filter

In the tubular filter configuration, the unit is placed in a shell-and-tube configuration similar to hollow-fiber devices, but the tube diameters are larger, generally 2.5 to 5.0 mm. The flow through all the tubes can be either in parallel, or in series or of the mixed type. Tubular units are capable of handling feedstreams and slurries containing large particles. They operate under turbulent flow conditions with a Reynolds number greater than 10,000. The open tube design and turbulent flow makes the tubular filter easy to clean by standard clean in place (CIP). However, high-pressure drop and high flow rates require high-energy consumption. Tubular units also have the lowest surface-area-to-volume ratios of all module configuration, resulting in high floor space requirements to install equipment.

9.5.2 Hollow Fiber

In this case, the membrane exists in the form of a self-supporting tube, with a dense "skin" layer inside the tube. The unit employs a multitude of fibers with an internal diameter of 0.25 to 1 mm. Fibers have a cross-sectional thickness of about 0.2 mm and are arranged in a shell-and-tube geometry. Feed enters the center of the tubes at one end of the unit, is concentrated, and exists at the other end. Permeate moves radially outward as shown in Figure 9.5.

A hollow fiber unit generally operates in a laminar flow region. Pressure drops are typically 5 to 20 psi. This combination of pressure drops and flow rates makes hollow fibers somewhat more economical in terms of energy consumption. Hollow fibers also have the highest surface-area-to-volume ratio. The hollow fibers have a "back-flushing" capability, as fibers are self-supporting. However, one of the disadvantages of a hollow fiber is its limitation to withstand high operating pressure. Shear rates are very high near the wall in a hollow fiber configuration. Moreover, small diameters make the fibers susceptible to plugging at the cartridge inlet. Replacement costs are also high: Even if only a single fiber out of 50 to 3000 in a bundle bursts, the entire cartridge must be replaced.

9.5.3 Spiral Wound

A spiral wound configuration is one of the most compact and inexpensive designs available. Imagine a sheet of plastic mesh 1 mm thick. On each side of the mesh is a

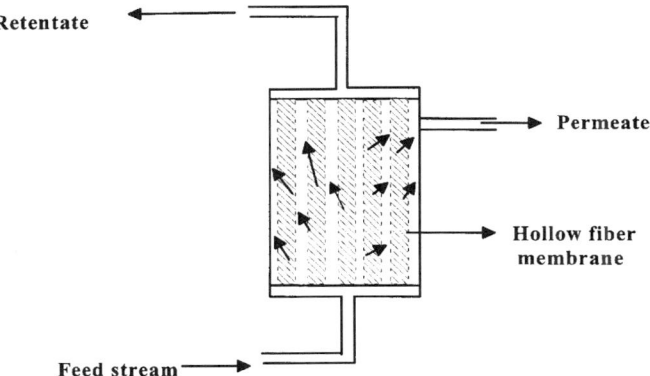

Figure 9.5 Hollow fiber membrane filtration configuration.

membrane and thin layer of absorbent material to serve as a permeate collector. If this assembly is rolled into a spiral, a spiral wound module is formed as shown in Figure 9.6. Spiral wound modules are flat-sheet-arranged to form a narrow slit for fluid flow. A mesh spacer controls the feed channel height. A narrow channel height allows more membrane surface area that can be packed into a given pressure vessel. Larger channel height is considered more desirable, as it minimizes pressure drops and reduces the feed channel plugging, however, slightly reducing the surface-area-to-volume ratio.

In a spiral wound configuration, the laminar flow profile is substantially increased by mesh spacers. However, one of the problems with these mesh spacers is the creation of dead space. This may cause particles to become lodged in the mesh network, resulting in a cleaning problem. Overall, the combination of low flow rates, moderate pressure drops, and relatively high turbulence makes this configuration one of the more economical modules in terms of energy consumption.

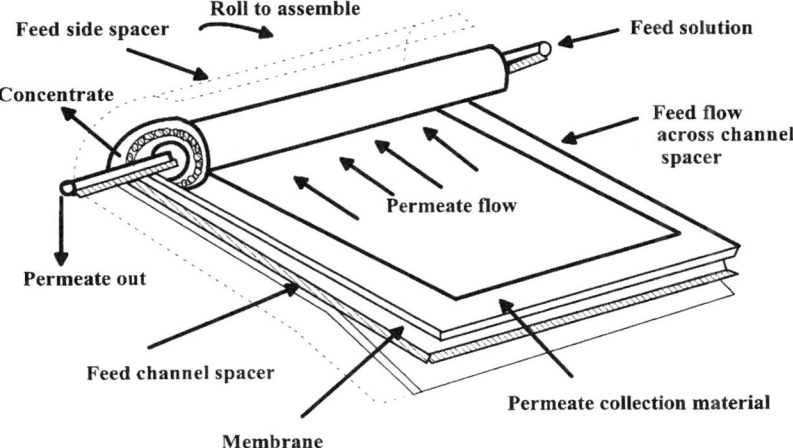

Figure 9.6 Schematic diagram indicating the working principle of a spiral wound membrane filtration configuration.

9.5.4 Plate and Frame Membrane

The plate and frame module is one of the earliest configuration based on a simple flat sheet configuration. The membrane is placed on both sides of a porous injection-molded plastic disk. A membrane spacer is placed on one of the membranes to serve as a permeate collection place. The space between the sheets ranges from 0.25 to 2.5 mm. This membrane-spacer-plate assembly is stacked one part on top of the other around a central tie-rod. These plates are hydraulically clamped together between two endplates to form a tight fit like a plate heat exchanger. Gaskets around the edges are used to prevent leaks.

Feed is pumped parallel to the membrane, usually under turbulent flow conditions into the bottom endplate, flowing up through holes on the periphery of the spacer and then radially onward across the membrane. The retentate leaves the stack through the top endplate. Permeate goes through the membrane and a porous support, and is removed through an outlet connection that is connected by plastic tubing.

9.6 LAB EXERCISE

Understanding the membrane system is essential to its proper functioning and longevity. The user must match the membrane and system to the job. The user must also observe the constraints of pressure, temperature, and transmembrane flux and use proper cleaning and storage procedures.

9.6.1 Objectives

The objectives of this exercise are to:

1. Microfilter food products such as whole milk, skim milk, and orange juice. As an alternative to food products, the microfiltration of water mixed with blue dextran can be carried out using a membrane.
2. Observe and determine some key components of membrane processing and fouling such as transmembrane pressure, flux rate, membrane resistance, etc.

9.6.2 Materials

1. Milk (preferably skim milk), or water mixed with blue dextran.
2. A membrane filtration system of any configuration module with known characteristics and dimensions. A typical diagram of a UF membrane setup is shown in Figure 9.7.
3. Plastic pails.
4. Stopwatch.
5. Weighing scale.

9.6.3 Procedure

1. Determine the specifications of the membrane, such as length, diameter, and effective membrane area, etc. Typical specifications for a Romicon membrane are: fiber cartridge type PM 10, internal diameter $(2r) = 43$ mils (1.0922 mm), length of module $(L) = 21$ in. (0.5334), and effective area $(Ae) = 5$ ft^2/per module.

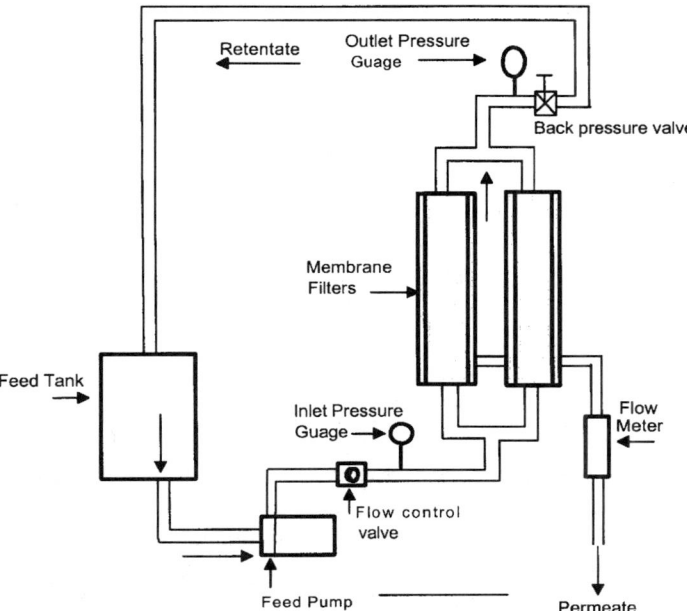

Figure 9.7 Typical experimental membrane filtration system.

2. Fill the tank with milk or water mixed with blue dextran. Adjust the flow control valve on the system.

3. Measure the mass flow rate of fluid through the filtration system by determining the time taken to collect some amount of the water sample delivered through the pump. Use a stopwatch to measure the time accurately. Weigh this sample and calculate the mass flow rate (remember to repeat this step each time you adjust the transmembrane pressure).

4. Open the valve that leads to the module and adjust the valve on the downstream section to obtain a suitable pressure drop. For the permissible ranges of pressure drop in relation to the temperature of operation, please observe the chart provided on the membrane module.

5. Record the inlet and outlet pressure gauges and calculate the average internal pressure as

$$P_{avg} = \frac{P_{in} + P_{out}}{2} \tag{9.1}$$

6. Calculate the transmembrane pressure as the difference between the average pressure and atmospheric pressure:

$$P_{TM} = P_{avg} - P_{atm} = P_{avg} - 0 = P_{avg} \tag{9.2}$$

(since all are gauge pressures).

7. Record at known and regular time intervals, the pressure, flow rates, and permeate flux in Data Sheet 9.1.

8. The observed flux J_v can be related to the applied transmembrane pressure by an expression of the form

$$J_v = \frac{P_{TM}}{R_c + R_m} \tag{9.3}$$

where J_v = flux or filtration rate per unit area, P_{TM} = transmembrane or the differential pressure drop across the filter media, R_m = resistance due to the membrane alone, R_c = resistance due to the cake buildup on and in the membrane, and is related to the terms shown in Eq. (9.4):

$$R_c = \frac{\alpha \omega V_t (P_{TM})^\beta \eta}{A} \tag{9.4}$$

where β = compressibility of the cake (deformation of the deposited particles), η = viscosity of the liquid, ω = concentration of particles in the fluid stream per unit volume of fluid, V_t = volumetric throughput up to any given time, α = constant dependent on the packing density of the particles deposited in the cake, A = membrane area over which the particles are deposited.

In a typical situation, where the concentration of blue dextran is low, the resistance offered by the cake layer formed is negligible (or it can be easily inferred). The expression for the flux is reduced to Eq. (9.5):

$$J_v = \frac{\Delta P_{TM}}{R_m} \tag{9.5}$$

9. For each applied transmembrane pressure, calculate the average flux over each time interval and calculate the membrane resistance.
10. Record the data in the appropriate part of Data Sheet 9.1. Record the final weight of the retentate and permeate. Calculate the concentration factor defined by the expression below and include them in your report:

$$CF = \frac{\text{Feed}}{\text{Retentate}} \tag{9.6}$$

11. Repeat the above procedure for two other applied transmembrane pressures.

9.6.3.1 *Calculations*   The total effective area of a typical hollow fiber membrane is calculated as follows:

$$Ae = n*A$$
$$A = 2\pi rL$$

where Ae = total effective area, n = number of tubes in the bundle, A = area of one tube, r = radius of the tube, L = length of the tube.

The Reynolds number NR_e of the flow inside the tube can be calculated as

$$NR_e = \frac{\rho D \langle V \rangle}{\mu} \tag{9.7}$$

where NR_e = Reynolds' number, $\langle V \rangle$ = average velocity (m/s) [$\langle V \rangle$ = volumetric flow rate/$(n\pi r^2)$], ρ = density of the material (kg/m^3), μ = dynamic viscosity of the fluid (mPa·s), D = diameter of the tube (m).

9.6.4 Report

Write a report with an introduction, procedure results, a discussion, and a conclusion. Include the following:

1. A complete, neat, accurate, well-labeled diagram (not a picture) of the system.
2. Tables of data collected.
3. Compute inherent resistance.
4. Plot permeate flux versus time curves for the three different applied transmembrane pressures.
5. A graph of the Reynolds' number (for water blue dextran) versus the transmembrane pressure. Determine the flow pattern or regime under which the membrane system was run (i.e., laminar, transition, or turbulent flow).
6. Plot a curve between intrinsic membrane resistance versus the P_{TM} for the blue dextran plus water system. (Use the average flux over the run for each applied transmembrane pressure in the calculation procedure.) On what factors do you think this resistance will depend? Briefly mention their significance in your report.
7. Plot the average flux over the total time of the filtration run versus P_{TM} for the blue dextran solution. Draw a tangent to the flux versus P_{TM} curve [at the intermediate pressure point you have selected and calculate the slope of the tangent line (slope)$^{-1}$ = resistance]. Do you think this exercise is of practical significance when filtering any fluid? Explain.
8. Discuss the above and their implications on food processing. Also discuss microfiltration versus ultrafiltration for fluids such as milk, juice, and for processing waste, etc. Which system would you use for each and why? If we used a ceramic membrane, how does this compare to membranes of other materials? Use your knowledge of chemistry and microbiology.

9.7 SUGGESTED READINGS AND REFERENCES

1. M. Cheryan, 1997. *Ultrafiltration Handbook*, 2nd ed., Lancaster, PA: Technomic Publishing Co.
2. P. Fellows, 1988. *Food Process Technology, Principles and Practice*, Chichester, UK: Ellis Horwood Ltd.
3. H. G. Kesseler, 1981. *Food Engineering and Dairy Technology*, Freising, Germany: Verlag A. Kesseler.

DATA SHEET 9.1

Temperature: _____

Initial feed weight: _____

Initial feed flow rate: _____

Retentate weight: _____

Permeate weight: _____

Membrane type: _____

Internal diameter: _____

Length of module: _____

Effective area/module: _____

Number of module: _____

Estimation of Reynolds number (type of flow) in membrane filtration corresponding to various transmembrane pressure

Transmembrane pressure P_{TM} (kPa)	Density ρ (kg/m^3)	Diameter D (m)	Velocity v (m/s)	Viscosity μ (Pa·s)	$NRe = (\rho DV)/\mu$

Estimation of transmembrane pressure, permeate flux, and membrane resistance

Flow and Pressure Data							
Number of Readings	Time (min)	P_{in} (psi or kPa)	P_{out} (psi or kPa)	P_{avg} (psi or kPa)	Permeate Weight (kg)	Permeate flux $(m^3/m^2 \cdot min)$	Membrane Resistance $(kPa \cdot min/m)$
1							
2							
3							
4							
5							
1							
2							
3							
4							
5							
1							
2							
3							
4							
5							

10

EVAPORATION CONCENTRATION OF LIQUID FOODS

10.1 BACKGROUND

Evaporation is a unit operation that is used for the partial removal of water from liquid foods by boiling. The separation of water or concentration of solids are achieved by a difference in volatility between water (solvent) and solute. The preconcentration of foods like fruit juice, milk, and coffee is desirable prior to drying, freezing, or sterilization to reduce weight and volume. An increase in solids by evaporation reduces water activity, such as in jams or molasses, and hence helps in preservation. Evaporation can also be used to develop flavor and color in a product, such as in caramelized syrups for baked goods.

During evaporation, latent heat is transferred from the heating medium to the food to raise its temperature to the boiling point. The rate of evaporation is determined by the rate of heat transfer to the foods and the rate of mass transfer of vapors from the foods. Evaporation is often carried out under vacuum conditions to increase the rate of evaporation and to reduce the boiling point of the solution so that heat degradation of the product is minimized.

10.1.1 Description of a Typical Batch Evaporator

A schematic diagram of a simple evaporator-type kettle is shown in Figure 10.1. The various features of the kettles are as follows:

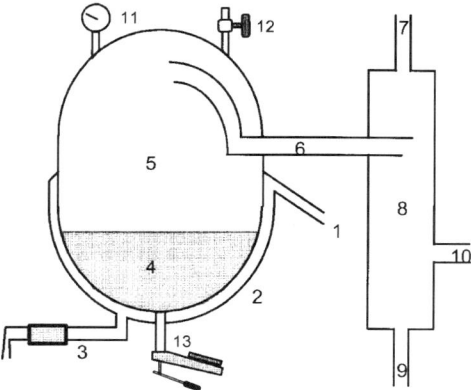

Figure 10.1 Schematic diagram of an evaporator kettle. 1, steam supply inlet; 2, outer jacket; 3, condensate outlet; 4, product; 5, vapors; 6, vapors' outlet; 7, cold water inlet; 8, condenser chamber; 9, condenser pump inlet; 10, vacuum pump inlet; 11, pressure gauge; 12, vacuum release valve; 13, sampling valve.

1. Steam from the main supply enters the jacketed body of the kettle at point 1.
2. The outer jacket where it condenses gives up heat to the juice in the kettle.
3. The condensed steam drains through a trap that maintains the steam pressure. The discharge is directed to the floor.
4. The juice is heated by the condensing steam and evaporates at a temperature that depends on the applied vacuum and the concentration of sugar molecules dissolved in it. As the sugar concentration increases, the boiling point increases.
5. The vapors from the juice occupy the outer open space of the kettle.
6. Vapor outlet that allows the vapors to pass out of the kettle and be piped to the condenser.
7. Cold water enters the top of the condenser.
8. Condenser chamber where cold water mixes with the vapors, condensing them to liquid. Since condensation reduces volume, a partial vacuum is created, reducing the pressure in the kettle and thereby the boiling temperature.
9. Inlet to a pump that removes the mixture of condensed juice vapors and cooling water, and discharges them onto the floor.
10. Inlet to the vacuum pump that is used to increase the vacuum in the condenser and further reduce boiling temperature.
11. A pressure gauge at the top of the tank reads the pressure in the evaporation chamber.
12. A valve at the top of the tank that is opened to let air enter and reduce the vacuum in the evaporation chamber.
13. A sampling valve at the bottom of the evaporation chamber to allow collection of samples for analysis while the evaporator is running.

10.1.2 Load Calculation

Enthalpy is a measure of the heat content of a system at constant pressure. It has units of energy (joules or BTUs). The enthalpy of a system changes when either heat or work is exchanged between the system and its surroundings. The following enthalpy relationships are of importance to evaporation processes:

1. The total enthalpy needed for an evaporation process is

$$H_{Total} = H_{Sensible} + H_{Latent} \tag{10.1}$$

   where $H_{Sensible}$ = the energy required to raise the temperature of the juice to the boiling point and H_{Latent} = the energy required to vaporize a required amount of water at its boiling point.

2. The sensible heat required to raise the temperature of a batch of juice from its starting temperature to the boiling point can be estimated with the equation

$$H_{Sensible} = m_{Juice} C_p (T_{Avg} - T_{Initial}) \tag{10.2}$$

   where C_p = the specific heat of the juice (this can be determined from its composition), m_{Juice} = mass of juice that is heated, $T_{Initial}$ = initial temperature of the juice, T_{Avg} = the average boiling point of the juice if there is a significant boiling point elevation during the process; otherwise, use the initial boiling point.

3. The latent heat consumed by evaporation can be estimated from the equation

$$H_{Latent} = m_v h_{fv} \tag{10.3}$$

   where m_v = the mass of water evaporated and $h_{fv} = h_v - h_f$ = the latent heat per unit mass of water, obtained from saturated steam tables. (Assume that the entire process takes place at the average vapor pressure inside the evaporator).

4. The average rate of heat transfer between condensing steam and juice during the experiment is

$$Q = \frac{H_{Total}}{t} \tag{10.4}$$

   where t = the process time required for the transfer of H_{Total}.

5. To provide the needed heat of evaporation, thermal energy is transferred from the condensing steam through a stainless steel wall to the evaporating juice. The rate of this heat transfer is computed by

$$Q = UA(T_s - T_b) \tag{10.5}$$

   where Q = the rate of heat transfer in watts, Btu/h, or other power units, U = the overall heat transfer coefficient in appropriate units, A = the area through which the heat is being transferred in m^2, ft^2, etc. (see below), T_s = the temperature of the steam, T_b = the temperature of the juice.

6. If a hemispherical tank is partially filled with a liquid as shown in Figure 10.2, the area of contact between the liquid and tank is given by the equation

$$A = 2\pi RD \tag{10.6}$$

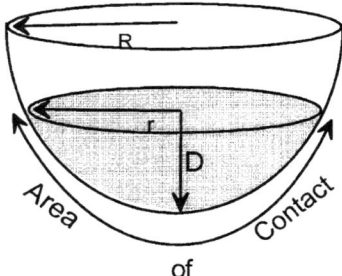

Figure 10.2 Partially filled hemisphere kettle for determining surface area. R = radius of the kettle; r = radius of the top surface of the liquid, D = depth of the liquid food filled for evaporation.

where R = the radius of the hemisphere and D = the depth of the liquid.

If the radius (or diameter) of the surface of the liquid is measured rather than the depth, the depth can be computed as

$$D = R - \sqrt{R^2 - r^2} \qquad (10.7)$$

where r = the radius of the top surface of the liquid (when $r < R$). The average depth D of the liquid over the process time can be used to approximate the average area of heat transfer A for the computation of the overall heat transfer coefficient U in Eq. (10.5).

10.1.3 Other Calculations

1. In principle, every kg of steam condensed at 100°C should provide enough energy to evaporate approximately 1 kg of water. However, because of pressure differences, sensible heat, and heat losses, this relationship can be less than 1 at 100°C. The steam economy of a process is defined as the ratio of the mass of water evaporated to the mass of steam used, that is,

$$SE = \frac{m_E}{m_S} = \frac{m_0 - m_F}{m_C} \qquad (10.8)$$

where m_E = mass of water evaporated, m_S = mass of steam used, m_0 = initial mass of product, m_F = the final mass of the product, m_C = the mass of the condensed steam that can be collected.

2. The concentration factor is defined as the following ratio:

$$CF = \frac{m_0}{m_F} \qquad (10.9)$$

where m_0 = initial mass of the juice and m_F = final mass of the concentrate.

3. In some cases, the boiling point elevation during evaporation will be significant. The boiling point elevation of a solution is defined as

$$\Delta T_b = 0.51M \tag{10.10}$$

where ΔT_b = increase in boiling point and M = molality of solution.

4. The molality of a solution is defined as

$$M = \frac{\text{g-moles of solute}}{\text{kg of solvent}} = \frac{\text{g of solute/MW of solute}}{\text{kg of solvent}} \tag{10.11}$$

Example.   *Determine the boiling point elevation of a 30% (w/w) sucrose solution.*

Solution.   Use a basis of 100-g solution. This solution will contain 30 g of sucrose and 70 g of water. The MW of sucrose = 342.3.

$$\text{Modality} = (M) = \frac{\text{g-moles of solute}}{\text{kg solvent}} = \left[\frac{\left(\dfrac{30 \text{ g sucrose}}{342.3 \text{ g/g-mole}}\right)}{70 \text{ g solvent}}\right]\left(\frac{1000 \text{ g}}{\text{kg solvent}}\right) = 1.252$$

$\Delta T_b = (0.51°\text{C/Molality}) (1.252 \text{ Molality}) = 0.638°\text{C}$, which is relatively small because of sucrose's high molecular weight.

10.2   LABORATORY

10.2.1   Objectives

The objectives of this lab exercise are to:

1. Observe the effects of the evaporation process on product quality.
2. Make an energy balance of the system.

10.2.2   Equipment and Materials

- Hamilton kettle evaporator or equivalent.
- 25 to 30 L of single-strength apple juice, maple sap, if available.
- Temperature probe, an extra thermometer.
- Refractometer.
- Seven sample containers (vials), masking tape, and black pen.
- Water activity meter.
- Stopwatch.
- Two small and two large buckets, funnel, and boots from pilot plant.
- Small weighing balance.

10.2.3 Experimental Procedure

10.2.3.1 Before the Run

1. Weigh the container with the juice. Empty the juice into the vacuum kettle. Weigh the empty container and calculate the weight of the feed.
2. Take an initial sample of the product for comparison with later samples.
3. Assign the following tasks during the experimental run:

   a. Monitor the steam temperature (230–240°F). This is generally provided with two valves and a pressure-temperature gauge as shown in Figure 10.3. Temperature for this experiment is regulated by rotating the steam valve and thereby controlling the steam pressure.

   b. Monitor the vacuum by watching the gauge at the top of the tank and adjusting the valve at the top of the tank as needed. Try to maintain a constant vacuum at a value somewhere between 10 and 20 in Hg. Record the vacuum at 10-min intervals.

   c. Allow a few minutes for the steam to flush out the water, then collect all steam condensate throughout the experiment. Empty the condensate into a larger bucket as necessary and, at the end of the experiment, weigh the total amount of condensate collected. *Take precaution and use gloves when working with the steam condensate.*

   d. Begin timing when the juice begins to boil (time 0). Take a product sample every 10 min for 1 h. To draw a sample from the evaporator, raise the handle of the sampling valve to allow the apple juice into the collection chamber (Fig. 10.4).

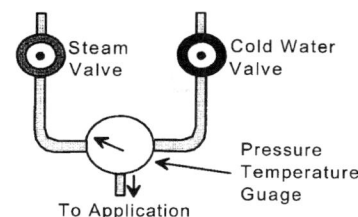

Figure 10.3 Typical steam-water mixing system.

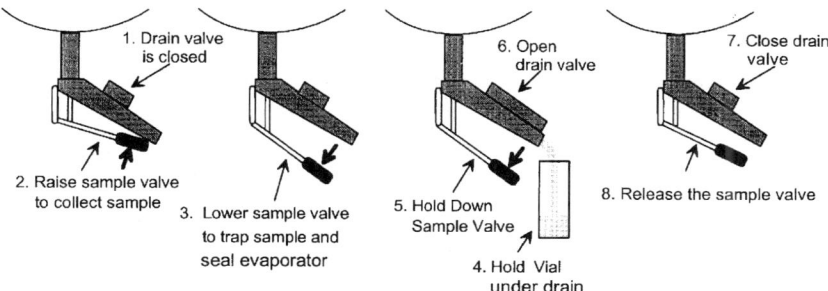

Figure 10.4 Sequence of operations of the sample port to collect samples.

Push down the handle to trap the sample and seal off the chamber. Hold the handle down firmly, place a vial under the sample drain, and open the drain valve to collect a sample. Close the drain valve before releasing the sample valve. Collect any extraneous juice in another bucket and include them in calculating product loss. Be sure to record the vacuum pressure, temperature of the vapors, condensate, and product in Data Sheet 10.1.

10.2.3.2 Start Up

1. Clamp the cover shut.
2. Attach the inlet water and steam hoses.
3. Check to make sure that the valve to the condenser is half-open. Turn on the inlet water.
4. Turn on both the vacuum pump and cooling water removal pump.
5. Close the vacuum release valve on the top and drain the valve to seal off the kettle from the atmosphere.
6. Turn off the water inlet and vacuum pump, when the vacuum gauge reads approximately 15 in the Hg vacuum.
7. Make sure that the steam valve to the evaporator jacket is full open. Turn on the steam and allow it to reach about 220°F. Monitor the steam every 10 min and compensate for deviations from the specified temperature.
8. Keep the vacuum between 10 and 20 in Hg. Adjust if necessary by opening the valve on the top of the evaporator.

10.2.3.3 During the Run

1. Take the first sample when boiling begins (time 0). After that, take samples at 10-min intervals for 1 h.
2. Determine the Brix, refractive index, pH, and water activity of the sample.
3. Evaluate the color and aroma of the sample.

10.2.4 Shut Down

1. Shut off the steam.
2. Allow the water to run through the system for approximately 5 min, then shut off the water to the condenser.
3. Open the valve on the top of the evaporator.
4. Shut off the water pump.
5. Collect the remaining juice from the sampling chamber at the bottom of the evaporator.
6. Measure the final product weight.
7. Disconnect the water and steam hoses.

10.2.4.1 After the Run

1. Continue to measure sample properties.
2. Weigh each product sample that was collected during the experiment and use these weights to calculate product loss.
3. Use a refractometer to measure °Brix and the refractive index of the product samples. Keep all samples at a constant temperature in a water bath maintained at 25°C. Note that the refractometer is calibrated for use at 20°C (68°F). Use the information provided to correct the sucrose concentration readings for the actual temperature.
4. Record qualitative observations such as product aroma, color, or stickiness.

10.2.4.2 Clean Up

1. Turn on the water and then the steam (lukewarm, about 130–140°F).
2. Rinse the evaporator with soapy water.
3. Turn on the pump for a few seconds.
4. Release the soapy water from the bottom of the evaporator through the collection chamber.
5. Rinse.

10.2.5 Lab Report

1. *Introduction.* Include a short inroduction to evaporation as a process.
2. *Objectives.* State the objectives of the lab exercise.
3. *Materials and Methods.* Draw the flow process diagram for a mass and energy balance, labeling all mass and energy inputs, outputs, and process conditions. Include a table of the operating conditions used (e.g., initial product weight, vacuum level, steam temperature, etc.). This should be sufficiently detailed to allow one to reproduce the experiment.
4. *Results.* Include the following specific items:
   a. A mass and energy balance of the experiment (insert the calculation details in an appendix). Assume negligible losses to the environment.
   b. The total energy required of this process.
   c. The average overall heat transfer coefficient considering sensible and latent heat transfer during the process.
   d. The overall steam economy of the process.
   e. The concentration factor.
   f. The initial and final calculated boiling point elevations. Assume that all solids are sucrose (MW = 342.3).
   g. Plots of refractive index (RI), °Brix, and H_2O evaporation rate versus time.
   h. A table of product quality versus time (Data Sheet 10.2).
   i. Is the vacuum measured on the gauge consistent with the boiling temperature for the saturated steam at that pressure?

5. *Discussion*.

   a. Discuss the results that you obtained. Include a description of how the product characteristics and product quality changed with time. Are they linear or nonlinear? Discuss the significance of boiling point elevation, steam economy, and the appropriateness of the vacuum kettle for apple juice concentration. How could this process be improved?

   b. Was the product heated to a supersaturation condition? Discuss the product quality at different concentrations that would have been obtained during the process.

10.3 SUGGESTED READING AND REFERENCES

1. P. Fellows, 1988. "Evaporation." In *Food Processing Technology, Principle and Practice*, Chichester, UK: Ellis Horwood Ltd.

2. APV, 1997. *Evaporator Handbook*, 3rd ed. Rosemont, IL: APV Crepaco Inc.

3. R. P. Singh, and D. R. Heldman, 1984. "Evaporation." In *Introduction to Food Engineering*, New York: Academic Press.

DATA SHEET 10.1

Product

Initial weight: _____

Final weight: _____

Losses (samples): _____

Losses (other): _____

Initial product temperature: _____

Time to reach boiling point: _____

Average boiling temperature: _____

Condensate

Final weight: _____

Hamilton vacuum kettle

Diameter at initial filled level: _____   Depth: _____

Diameter at final filled level: _____   Depth: _____

Observations during the experimental run

Time (min)	Steam Condensate		Evaporator		Juice Sample	
	Temperature (°F)	Weight (kg)	Vacuum (in Hg)	Temperature (°F)	Weight (kg)	Temperature (°F)
0[a]						
10						
20						
30						
40						
50						
60						

[a] This would include condensate for preheating to the initial boiling point.

DATA SHEET 10.2

Qualitative evaluation of the product

Time (min)	°Brix	RI	pH	a_w	Color/Aroma
Original Quality					
0					
10					
20					
30					
40					
50					
60					

RI = refractive index.
a_w = water activity.

11

PRINCIPLES OF EXPERIMENTAL DESIGN

11.1 BACKGROUND

In the food industry, product developers and process engineers often use experiments to develop new products and processes and to improve existing ones. The experiments are performed to learn how one set of variables affects another one. For example, an experiment might be performed to learn how different types of sugar affect texture or how processing temperature affects the moisture content of a product? The information obtained makes it possible to manipulate one or more variables in order to control others. In a highly competitive industry, a premium is placed on getting useful information as quickly and economically as possible, and therefore, experiments should be designed with this in mind.

Experiments can be done on a "Let us try this and see what happens" basis or they can be carefully planned. Most good researchers do some of each: ad hoc experiments to try out hunches and carefully designed experiments to gather information and confirm ideas. In the design of experiments (DOE), statistical principles are used to design experiments that will yield the right information in the shortest time and at the least cost. Its use leads to the rapid and efficient development of new and improved products and processes. In this chapter, we attempt to introduce students to some of the basic principles of DOE.

This chapter is divided into two parts:

1. A brief introduction to some key concepts of DOE.
2. A set of problems so you can practice using these concepts.

See Appendix C for how to apply these concepts in greater detail in industrial research.

11.2 BASIC PRINCIPLES

In a comparative experiment, two or more treatments are performed. These treatments will differ in some way that is selected by the experimenter. After applying the treatments, some variable is measured and these measurements are compared to see if the treatments affected the measured variables.

11.2.1 Types of Design

DOE can be divided into three phases:

11.2.1.1 Treatment design.   The goal of any experiment is to answer some specific question about the system of interest. In the treatment design phase, treatments and methods of measurements are selected to get information that will best answer the questions. In this chapter, two types of design are discussed: a one-variable experiment and the two-way factorial experiment.

11.2.1.2 Experimental design.   Experimental results are subject to uncertainty due to many variations. Ideally, any differences that are measured in an experiment are the result of the applied treatment. In practice, many other variables may also affect the results and this leads to the uncertainty. In a well-designed experiment, the magnitude of the uncertainty will be both small and predictable, and the experiment will require a minimum of time and expense. The experimental design phase deals with this problem.

11.2.1.3 Analysis design.   After the experiment has been performed, the experimenter must examine the data to see how it answers the questions. This usually involves the preparation of tables and graphs to help decipher the effects being studied. It also involves estimating the effects of interfering variables and determining whether they alone might have produced the observed effects. The conclusions drawn will be much more convincing if analysis design had been taken into consideration before the experiment was conducted.

11.2.2 Treatment Design

11.2.2.1 Objectives   The first step in designing any experiment is to state the goals as specifically as possible. Usually, the goals involve determining the effect of one or more variables on other variables. Here are some examples of various types of objectives.

- *Screening.* Does variable A affect variable B?

   Example 1   *Does the type of starter culture A used affect the yield B of cheddar cheese?*

- *Screening.* Which of several variables affects variable B?

   Example 2   *Which of the following variables has a practical effect on the cheese texture B: protein content of milk A_1, fat content of milk A_2, cheddaring time A_3, amount of salt added A_4, strain of starter culture A_5, and quantity of starter culture A_6?*

- *Quantifying*. What is the size of the effect of variable *A* on variable *B*?

Example 3 *By how much is bulk density B reduced if a new type of blender is used (A)?*

Example 4 *How many organisms B are killed per minute of exposure to heat at 250°F A?*

- *Trends*. What are the direction, shape, and rate of change of the trend exhibited by variable *B* with a change in variable?

Example 5 *If baking temperature increases A, does the loaf volume B increase, decrease, or remain unchanged? If there is an increase or decrease, is it linear? Does it pass through a maximum or minimum? Does it level off? What is the rate of change? etc.*

- *Optimization*. What level of variable *A* (or of several variables) gives the best level of variable *B*?

Example 6 *What combination of baking time A₁, baking temperature A₂, and sugar concentration A₃ yields the lightest angel food cake B?*

- *Sensitivity*. Which variables must be carefully controlled if variable *B* is to remain constant?

Example 7 *In order to maintain uniform overrun B of ice cream, which of the following variables must be controlled carefully and which can be allowed some freedom to vary: freezer temperature A₁, feed rate A₂, air pressure A₃, scraper rotation rate A₄, fat concentration A₅?*

- *Theory*. Is the relationship between variable *A* and *B* consistent with a particular theory?

Example 8 *Does the heating rate B of a food increase exponentially with time A?*

11.2.3 One-variable Experiment

One-variable experiments are run to learn how one experimental variable affects one or more response variables. In these experiments, treatments are simply selected levels of the experimental variable. Depending on the experimental variable selected, treatments can differ either qualitatively or quantitatively.

11.2.3.1 *Experimental Design Terms*

- *Experimental variable*. A variable that the experimenter manipulates.
- *Level*. A specific value, either qualitative or quantitative, of the experimental variable that is selected for the experiment.
- *Treatment*. In a one-variable experiment, each treatment is a level of the experimental variable.
- *Response variable*. A variable that the experimenter measures after performing the treatments to see how it is affected by the experimental variable.

Example 9 *Bread is baked at three different temperatures (the experimental variable), and the volume (the response variable) of the resulting loaves is measured to see how it is affected by baking temperature. The three temperatures are the levels*

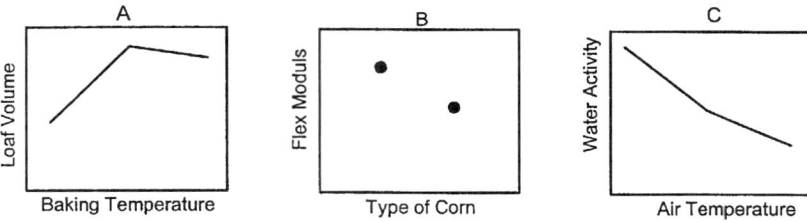

Figure 11.1   Experimental results of one-variable experiment on response variable.

of the experimental variable and, hence, the treatments of this experiment, which in this case differ quantitatively. If the results looked like those in Figure 11.1A, we could conclude that increasing temperature increases volume up to a point, after which volume levels off.

Example 10   *Tortilla chips are made from two different types of corn flour (the experimental variable) and the flexural modulus (the response variable) of the resulting chips is measured to see how it is affected by flour type. The two types of flour are the two levels of treatments in this experiment. In this case, the treatments differ qualitatively. If the results looked like those in Figure 11.1B, we could conclude that the first type of flour produces the higher modulus.*

Example 11   *Milk is dried using three different air temperatures. The water activity of the powder is measured to see how it is affected by the temperature. This is another example of quantitative treatments. If the results looked like those in Figure 11.1C, we could conclude that increasing air temperature decreases water activity.*

11.2.4 Two-way Factorial Experiments

In a two-way factorial experiment, two experimental variables are chosen and, for each, two or more levels are selected. There will be a treatment for each combination of levels. Although factorial experiments are usually larger than one-variable experiments, they usually produce a lot more information in the time they take.

11.2.4.1 Experimental Terms

- *Factor.* When more than one experimental variable is manipulated in a single experiment, these are called experimental factors.
- *Treatment.* In a two-way factorial experiment, a treatment is a combination of levels of the two factors. If there are n_1 levels of factor 1 and n_2 levels of factor 2, then there will be n_1 times n_2 treatments.
- *Interaction.* If the effect of one factor on the response variable is different for different levels of the other factor, we say the factors are exhibiting an interaction. Interactions can only be discovered with factorial experiments, which gives them an important advantage over one-variable experiments.

Example 12 *Bread is made using two different types of flour (factor 1) and baked for three different lengths of time (factor 2). The loaf volume of the bread is measured to see how it is affected by various combinations of these factors. There will be a total of $2 \times 3 = 6$ treatments, one for each combination of flour type and baking time. For example, if we select a high-gluten and a low-gluten flour and bake for 40, 50, and 60 min, the six treatments will be:*

1. *High-gluten for 40 min*
2. *High-gluten for 50 min*
3. *High-gluten for 60 min*
4. *Low-gluten for 40 min*
5. *Low-gluten for 50 min*
6. *Low-gluten for 60 min*

The three main possible outcomes of this experiment are shown in Figure 11.2. In both A and B, high-gluten flour produces larger volumes, and increasing time increases the volume. However, in A the effect of time is the same for each flour (parallel lines) so we say the factors act independently; there is no interaction. One can easily discuss the effect of either factor without taking the other one into account. In Figure 11.2B, increasing time has a greater effect on high-gluten flour than low-gluten flour, indicating that there is an interaction between the factors. Under these circumstances, it is meaningless to discuss the effect of baking time without specifying the type of flour being used. In Figure 11.2C, there is an even greater interaction between the factors. Here, with high-gluten flour, loaves increase in volume with increasing time. On the other hand, loaves made with low-gluten flour actually decrease in volume with increasing time. In fact, at the shorter baking time, the low-gluten flour actually has the greater volume.

11.2.5 Experimental Design

When you examine results such as those in Figure 11.2, you should always ask whether the observed differences were due to the applied treatments or some other, possibly unknown, variable. The goal of experimental design is two-fold:

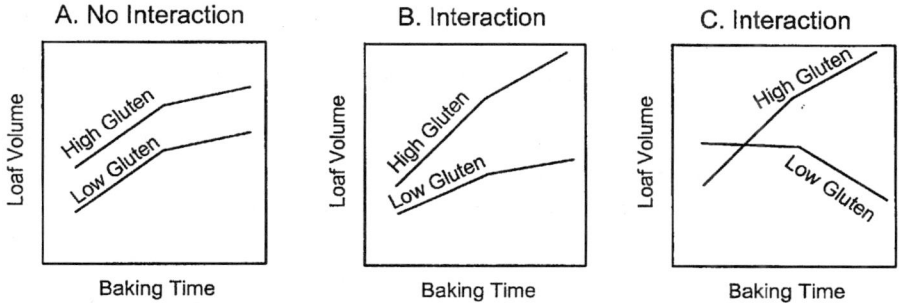

Figure 11.2 Experimental results of factorial experiment with or without interactions.

- To reduce the effects of other variables that might affect the response variable, so that, as much as possible, the observed differences can be attributed to just the treatments.
- To provide a way to estimate the amount of variation produced by other variables so you can decide whether the observed differences between treatments are really due to the treatments.

11.2.5.1 New Experimental Terms

- *Interfering variable.* Any variable other than the experimental variable that has an effect on the response variable. It is also called a nuisance variable. For example, in an experiment to study the effect of baking time on bread volume, oven temperature, mixing time, weight of dough per loaf, and ingredient variation are possible interfering variables.
- *Experimental error.* Variation in the response variable that can be attributed to interfering variables and not treatments (not to be confused with mistakes). For example, variation in the weight of dough could produce variation in loaf volume. The following sections describe a few techniques that can be used to reduce the effect of experimental variables.
- *Hold constant.* Many potentially interfering variables can be held constant throughout the experiment. In this way, they have the same effect on each treatment and do not affect comparisons between treatments. For example, you should carefully weigh each loaf and maintain a uniform oven temperature throughout an experiment. Unfortunately, not all interfering variables can be controlled so other techniques are needed.

11.2.5.2 Replicate   Whenever possible, each treatment should be repeated two or more times. Each repetition of a treatment is called a replicate. A complete repetition of a treatment is a replicate. Replication is not the same as making repeated measurements on one treatment. Replication has the following effects:

1. The average of several replicates of a treatment will usually be closer than any single trial to the "true" effect of that treatment.
2. By observing the variation among replicates of the same treatment, it is possible to estimate the magnitude of experimental error. This allows determination of whether differences in a response variable observed between treatments is really due to the treatments or just the interfering variables.
3. By randomly assigning experimental material to different replicates, we can reduce biases that might produce misleading results.

Three questions should be answered about replicates when designing experiments:

- What is a proper replicate?
- How many replicates are needed?
- How should material for replicates be assigned to treatments?

11.2.5.3 What Is a Proper Replicate   A common error in experimental design is to improperly replicate treatments. Each replicate of a treatment must completely repeat the

entire treatment, not simply the measurements. To illustrate what this means, consider the following procedure that could be used in a variety of bread-baking experiments.

> *Example 13* *A general procedure, as shown in Figure 11.3, is to mix batches of dough, then divide each batch into several loaves. After baking, the loaves are sliced. Circles are then cut from each slice and their mechanical properties measured. Thus, there are four subdivisions of the experimental material: batches, loaves, slices, and circles.*

When performing an experiment using this procedure, should each replicate be a separate batch, separate loaf, separate slice, or separate circle? The answer depends on the purpose of the experiment. To see how to decide, we introduce the notion of an "experimental material and experimental unit."

- *Experimental material.* The material to which the experimental treatments are to be applied. Examples include dough, Guinea Pigs, cheese, etc.
- *Experimental unit.* A subdivision of the experimental material to which one replicate of one treatment is to be applied. An experiment with t treatments, each replicated r times, requires t times r experimental units.

To properly replicate and experiment, you must specify the experimental unit in such a way that any two experimental units can be assigned to two different treatments. Each replicate of a treatment must use a separate experimental unit. We cannot measure the same experimental unit twice. Consider each measurement to be a separate replicate. To see what this means, let us look at some examples of bread experiments based on the procedure outlined in Example 13.

> *Example 14* *The goal is to determine the effect of an ingredient on bread texture. Each treatment will be a different ingredient. The appropriate experimental unit is the batch since any two batches can have different ingredients. The loaf, or any subdivision of it, cannot be an experimental unit because two loaves from the same batch cannot have different ingredients and cannot, therefore, be part of different*

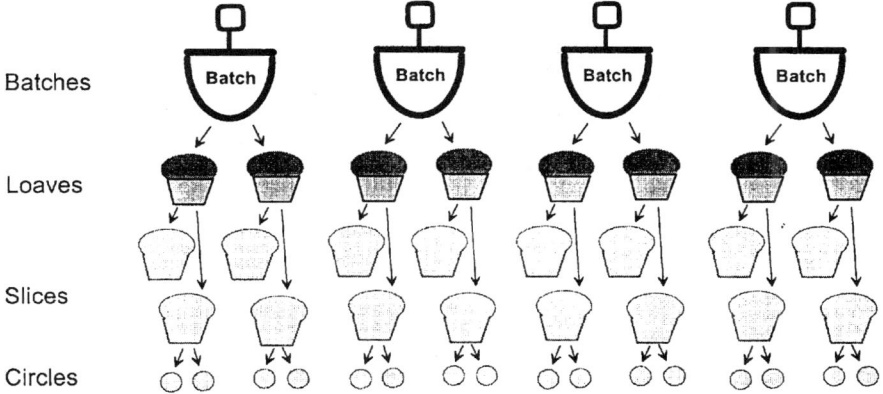

Figure 11.3 Experimental plan for the subdivision of experimental material.

treatments. This means that to properly replicate this experiment, you must make two or more batches for each treatment (ingredient).

Example 15   *The goal is to determine the effect of baking time on bread texture. Each treatment will require a different baking time. The appropriate experimental unit is the loaf since two loaves can always be assigned to different baking times even if they come from the same batch. The slice cannot be an experimental unit because two slices from the same loaf cannot be baked at different times. This means that to properly replicate this experiment, you must bake two or more loaves for each baking time.*

Example 16   *The goal is to determine the effect of toasting time on bread texture. Each treatment will require a different toasting time. The appropriate experimental unit is the slice since two slices can always be assigned to different toasting times even if they come from the same loaf. The circle cannot be an experimental unit because two circles from the same slice cannot be toasted at different times. This means that to properly replicate this experiment, you must toast two or more slices for each toasting time.*

Example 17   *The goal is to determine the effect of measurement method on measurement of bread texture. Each treatment will require a different measuring method. The appropriate experimental unit is the circle since two circles can always be assigned to different measurement methods even if they come from the same slice. This means that to properly replicate this experiment, you must measure two or more circles by each method.*

11.2.5.4 *Assigning Experimental Units*   Once we have determined the proper experimental unit for an experiment and the number of replicates to use, it is important that we assign the units to treatments in a way that will minimize the effects of interfering variables and, hence, experimental error. We will discuss just two of the many ways this can be done. Let us first define the new term, *bias*.

- *Bias*. Experimental units will differ from each other as a result of interfering variables. If experimental units are assigned to treatments so that the interfering variables alone produce treatment differences, we say the experiment is biased. For example, in a comparison of the effects of diets on rat growth, if we assign overweight rats to one diet and lightweight rats to another, we may conclude that the diet is producing different growth rates when, in fact, the results are due to the initial weight of the rats. Such an experiment is biased.

11.2.5.4.1 *Completely random design* (*CRD*).   This design reduces biases by assigning experimental units to treatments in a completely random manner:

1. For an experiment with *t* treatments and *r* replicates of each treatment, assemble *tr* experimental units.
2. Number the units from 1 to *tr*.

3. Randomize the order of the numbers from 1 to *tr*. This can be done by placing numbered slips of paper in a box, shaking them up, and then drawing them one at a time. (A random number table or computer program can also be used. See Appendix C for more details.)

4. Assign units with the first *r* random numbers to treatment 1, assign the next *r* to treatment 2, etc.

5. Note that in this design, all units are randomized together.

11.2.5.4.2 Randomized complete block design (RCB). This design reduces the bias produced by one interfering variable by grouping and the randomization of another interfering variable. It is done as follows:

1. For an experiment with *t* treatments and *r* replicates of each treatment, assemble *tr* experimental units.

2. Select a troublesome interfering variable.

3. Place experimental units into *r* groups of *t* units each. The groups are called blocks.

4. Group them in such a way that, within each block, all units are as similar as possible with respect to the selected interfering variable. For example, all units in a block could come from a single batch of dough, or be experimented on the same day, or be of similar initial weights or use ingredients from the same supplier, etc.

5. Within each block, number units from 1 to *t*.

6. Obtain *r* different random arrangements for the numbers 1 to *t*, one for each block.

7. For each block, assign the unit with the first random number to the first treatment, etc.

8. In the final design, each block should contribute one unit to each treatment and each treatment should receive one unit from each block.

In each of the following examples, the goal is to measure the effect of baking temperature on loaf volume so that the appropriate experimental unit is the loaf. One problem facing the experimenter is the fact that average loaf volume may vary from batch to batch, even within the same treatments so this must be considered in the experimental design. In the following examples, we will use three treatments with two replicates of each treatment. The loaf is the experimental unit. The following examples show how the experiment could be conducted as a completely random design or randomized complete block design.

Example 18 (CRD) *One batch of bread is made and divided into six loaves that are then numbered from 1 to 6. Numbers are drawn randomly from a box in the order 1, 4, 6, 2, 3, 5. This sequence leads to the assignment shown in Figure 11.4A. Since only randomization is used to assign units to treatments, this is a completely random design.*

Example 19 (RCB) *Two batches of bread are made and each is divided into three loaves. Loaves from the same batch are similar so they form a block. Within each block, loaves are numbered from 1 to 3. The following two sets of random numbers are drawn from a box: 1, 3, 2 and 3, 1, 2. These sets lead to the assignments shown*

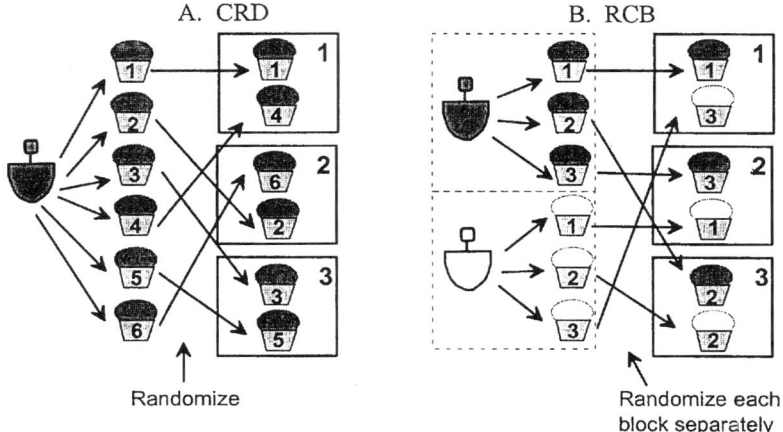

Figure 11.4 Experimental plan for a completely random design (CRD) and randomized complete block design (RCB).

in Figure 11.4B. This procedure guarantees that each treatment contains exactly one loaf from each batch. Since randomization is done separately in each block, this is a randomized complete block design.

11.2.6   Analysis Design

The analysis of the data should accomplish two things:

- Display the data, either in tables or graphs, in such a way so that it can be meaningfully interpreted.
- Test whether the differences observed in these displays are the result of the treatments or just the result of interfering variables.

11.2.6.1 Graphs of Data   Some examples of graphs made from one-variable experiments are shown in Figure 11.1. They are made as follows:

- Scale the horizontal axis to represent the experimental variable.
- Scale the vertical axis to represent one of the response variables.
- For each treatment, compute the mean of the responses variable and plot it on the graph.
- If the experimental variable is quantitative, connect the points with lines to suggest trends.

The examples of graphs made from two-way factorial experiments are shown in Figure 11.2. They are made as follows:

- Scale the horizontal axis to represent one of the factors. If one factor is qualitative and the other quantitative, this axis will usually represent the quantitative factor. Otherwise, pick the factor with the most levels if there is a difference.
- Scale the vertical axis to represent the response variable.
- For each treatment, compute the mean of the responses and plot it as a point on the graph.
- Connect the points with lines as shown in Figure 11.2 so that the means are grouped according to levels of the other factor. For example, if the second factor has two levels, there will be two connecting lines. Label the lines to indicate the levels.

11.2.6.2 *Analysis of Variance*

As we pointed out earlier, you should always ask whether the differences between responses to different treatments are really due to the treatments or whether they could be due to just interfering variables. The analysis of variance is a statistical technique used to answer this question.

- Perform an analysis of variance (ANOVA) to test whether observed differences are due, in part, to treatments or whether they could be entirely due to interfering variables. Instructions for performing ANOVAs on Minitab are given in Appendix C.

Example 20 *The following is an analysis of variance of a one-variable experiment with no blocks. There were three treatments, each with four replicates:*

```
ANALYSIS OF VARIANCE ON Volume
    SOURCE   DF      SS      MS      F       p
    Treat.    2   1326.2   663.1   7.03   0.014
    ERROR     9    848.8    94.3
    TOTAL    11   2174.9
```

- For one-variable experiments, check the P value for treatments. If it is close to or less than 0.05, it is very probable that at least two treatments differ significantly. By this, we mean that the differences observed were greater than interfering variables alone could produce, so the treatments have been shown to have an effect on the response variable. It makes sense to interpret the data. The smaller the P value, the more convincing the results. A P value much larger than 0.05 means that this experiment has failed to separate the effects of treatments from those of interfering variables and no conclusions about treatment effects can be confidently drawn.

Example 21 *In the last example, the* P *value for treatments is 0.014. Since this is quite small, we can be fairly confident that the treatments are affecting the response variable.*

- For factorial experiments, look at the interaction first. If it has a P value of around 0.05 or less, it means that there is an interaction. This tells us that the factors are having an effect but that the effects of one factor will depend on the levels of the other factor. This must be taken into account when interpreting the data.

Example 22 *The following shows an analysis of variance for an experiment with two factors, flour type and temperature, that has been replicated with blocks:*

```
Analysis of Variance for loaf Volume
  Source        DF       SS       MS        F       P
  Block          1    216.75   216.75     6.17    0.056
  Flour          1      4.08     4.08     0.12    0.747
  Temp           2    906.50   453.25    12.89    0.011
  Flour*Temp     2    481.17   240.58     6.84    0.037
  Error          5    175.75    35.15
  Total         11   1784.25
```

We examine the interaction first. Since the P value for interaction is fairly small (0.037), we conclude that the effect of temperature must be different for each flour. The effect of temperature alone is also significant ($P = 0.011$) but because of the interaction, we cannot interpret it independent of flour type.

- If there is no interaction, or if after you examine it, you decide the interaction is small, look at the P values for each factor to determine if either or both are affecting the response. You can then interpret the plots for any factor that is significant.

Example 23 *The following is an analysis of variance for an experiment on the effect of metal used in a baking pan and temperature on loaf volume. This is a two-way factorial experiment using a completely random design, so there is no line for blocks:*

```
  Source        DF       SS       MS        F       P
  Metal          1     24.08    24.08     0.29    0.611
  Temp           2   1028.17   514.08     6.14    0.035
  Flour*Temp     2    656.17   328.08     3.92    0.082
  Error          6    502.50    83.75
  Total         11   2210.92
```

In this case, the interaction is not quite significant ($P = 0.082$) so an interaction may not exist and we can examine the factors independently. The type of metal is not significant ($P = 0.611$), but temperature ($P = 0.035$) is having an effect.

11.2.6.3 Interpretation   Once you have finished the statistical analysis, you can go on and interpret the data. It is important to understand that the statistical analysis simply tells you which differences can be interpreted and which cannot be interpreted. It is not in itself an interpretation. The interpretation describes how the experimental variables affect the response variables. In making your interpretation, be guided by the objectives of the experiment and by the statistical tests. Clearly, your interpretation of the results can only be as good as your knowledge of the underlying physics and chemistry of the baking process. Statistics is not a substitute for science.

11.3 PROBLEMS

1. Farm-raised fish are cut into two fillets each and cooked. The cooked fillets are then cut into cubes so that they can be analyzed for protein content. What is the appropriate experimental unit for each of the following experiments? Justify each selection. In each case, how many experimental units will you require to replicate each treatment five times? What is the minimum number of fish required to supply these units?

   a. To compare two methods of measuring protein.

   b. To compare the effect of two fish diets on protein content.

   c. To compare two marinades used in cooking on the protein content.

2. You are designing an experiment to compare the effects of Fe^{++} versus Fe^{+++} in the diet on the growth rate of rats. You also want to study the effects of various levels of vitamin X over the range from 10 to 50 units and see if the effect of vitamin X is influenced by the type of iron. Although you are not interested in the effect of initial weight, you anticipate that the initial weight of the rats will also affect growth weight and would like to design the experiment so as to minimize its effect on the experimental results. The initial weight of available rats is shown in Figure 11.5.

   a. Describe an appropriate treatment design for this experiment, including the response variable and brief descriptions of the factors levels and treatments. Justify your choice of levels.

   b. List several variables that you could hold constant in all treatments.

   c. Select and justify a good experimental design for this experiment? From the available rats (Fig. 11.5), select the rats you need and assign them to treatments, clearly describing and/or demonstrating each step as you do this and listing any sets of random numbers that you use. Color-code the rats in the figure, indicating to which treatment each is assigned.

3. An experiment is performed to determine the effect of frying temperature and frying time on the flex modulus of tortilla chips. It was impossible to perform enough replicates of all treatments on a single day and since it was thought that day-to-day variation would have a noticeable effect, on each of three successive days, each of

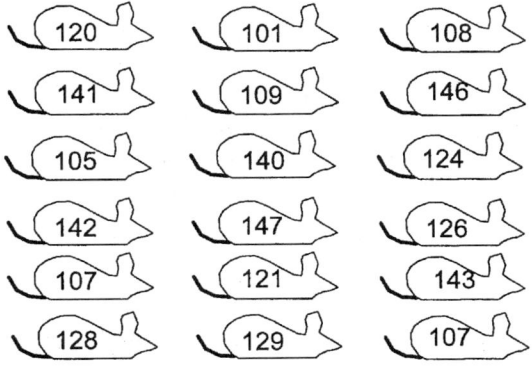

Figure 11.5 Weights of available rats to study the effects of Fe^{++} and Fe^{+++} on the growth rate of rats.

the six treatments were performed once. The following 18 moduli (psia) were obtained:

	350°F			400°F		
Day	1.5 min	2.0 min	2.5 min	1.5 min	2.0 min	2.5 min
1	767	879	1,065	1,143	1,600	2,271
2	1,344	1,261	1,612	1,019	1,564	2,167
3	570	635	1,278	937	1,242	1,658

a. Select an appropriate ANOVA and perform the analysis using Minitab or any other suitable program. (See Sect. C.5.1 in Appendix C.)
b. Use your ANOVA to determine which plot(s) should be examined (main effect and/or interactions).
c. Draw only the selected plot(s) and draw confidence limits (See Sect. C.5.2 in Appendix C). Show the computation.
d. Interpret the results in practical language useful to the food scientist (See Sect. 11.2.6.3.) As much as possible, avoid statistical jargon, but make your explanation statistically correct.
e. How would your conclusions change with the following data? Show all calculations and plots. (*Note*: To make analysis easier, these data are the same as in the last problem except for the five boldfaced numbers. Simply change these numbers in the Minitab spread sheet or other computer program sheet and rerun the analysis.)

	350°F			400°F		
Day	1.5 min	2.0 min	2.5 min	1.5 min	2.0 min	2.5 min
1	**986**	879	1,065	1,143	**1,053**	2,271
2	1,344	1,261	**1,437**	1,019	**1,262**	2,167
3	**1,433**	635	1,278	937	1,242	1,658

4. You plan to compare the effects of three baking temperatures on the crust color of loaves of bread, using four replicates of each treatment so you need 12 experimental units. You consider three alternatives.

a. Make one batch and divide it into 12 loaves.
b. Make 12 batches and take one loaf from each batch.
c. Make three batches and take four loaves from each batch

Explain why alternatives a and b would both lend themselves to a completely random design experiment but alternative c would not.

11.4 SUGGESTED READING AND REFERENCES

1. G. W. Snedecor, and W. G. Cochran, 1980. *Statistical Methods*, Ames, IA: Iowa State University Press.
2. A. R. Fisher, 1990. In *Statistical Methods, Experimental Design, and Scientific Inferences* (J. H. Bennett, ed.), Oxford, UK: Oxford University Press.

12

SPRAY AND DRUM DRYING

12.1 BACKGROUND

Drying is the application of heat under controlled conditions to remove water from food. One purpose of drying is to extend the shelf life of foods by a reduction in water activity, which inhibits microbial growth and enzyme activity. The reduction in weight and volume upon drying also reduces the transport and storage costs and, for some types of food, provides greater variety and convenience for the consumer. However, drying also affects nutritive quality and food taste.

Drying involves the simultaneous application of heat and removal of moisture from foods. In some operations, such as in spray drying, air is used as a heating and moisture-removal medium. The capacity of air to remove moisture from a food depends on its temperature and the amount of water vapor already carried by the air (absolute humidity).

12.1.1 Drum Drying

In drum drying, slowly rotating hollow steel drums are heated internally by pressurized steam to 120 to 170°C. Drum dryers have high drying rates and energy efficiencies. They are suitable for slurries in which the particles are too large for spray drying. Therefore, drum drying is generally applicable to viscous and semisolid foods, such as cooked potatoes. Caution must be taken when drum drying heat-sensitive food materials in order to prevent scorching or a deterioration in product quality.

The configuration of a typical drum drier suitable for the laboratory is shown in Figure 12.1.

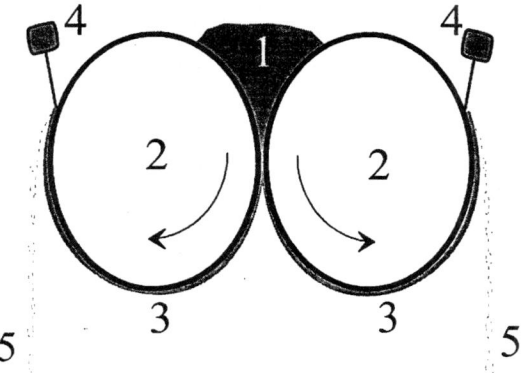

Figure 12.1 Drum drying operation. 1, product; 2, stainless drums; 3, thin product layer; 4, doctor blade; 5, dried product.

1. The product to be dried usually exists in the form of a thick paste or viscous fluid. A quantity of the product is placed above the drums.
2. Stainless steel drums that are heated by steam or other means have a thin gap between them. Rotation of the drums pulls the product through the gap.
3. This produces a thin layer of product that adheres to the drum and results in rapid drying.
4. Doctor blades are positioned to scrape the product from the drums.
5. The final dried product falls into collecting pans or conveyor belts below.

There are many types of drum dryers available. For example, the equipment may be classified according to:

- *Environmental drying pressure.* Chambers may enclose the drums so a vacuum can be pulled on the system while drying is being performed, or the equipment may be operated under atmospheric pressure.
- Number of drums in the process may be either one or two.

The food material can be applied to the drum by various devices, including the following:

- The rotating drum is dipped into a pool of material.
- Liquid material is fed at the drum gap (for double-drum dryers) at a rate that maintains a pool level between the drums.
- Material is sprayed onto the drum surface.
- The liquid is applied with a roller.

For double-drum dryers, the gap or clearance between parallel drums is set between 0.25 mm and 3 mm to regulate the film thickness that develops on the drum surface. The drums usually rotate in opposite directions. Efficient operation would not allow any dried material to remain on the drum surface. Before operation, the drum surface should be free of any deposits (e.g., grease) that would lower the rate of conductive heat transfer.

Optimum operating conditions can be achieved by control of several process variables, including:

- Solids content of feed that could be preconcentrated by evaporation before being applied to the drum surface.
- The feed may be preheated.
- The speed of drum rotation may be adjusted to control residence time.
- The temperature of the drum surface.
- The roughness of the drum surface.
- Flexible knifes (doctor blades) can be adjusted to fit securely against the drum surface.

Moreover, the steam supply should be of the highest quality (100%). Steam enters the interior of the drum and condenses on the interior wall. Proper operation of the drum dryer requires venting when the steam is introduced into the drum chamber. The condensate must be continually removed to maintain a high convective heat transfer coefficient between the steam and drum wall. If the drum chamber is pressurized (e.g., to have steam at, say, 300°F), the temperature of the product film will change during drying. The dynamics range is essentially between the boiling point of water and the maximum temperature that can be attained before removal by the knife.

The rate of heat transfer during this unit operation can be evaluated when the thermal resistance between the drum chamber and dried film is known. The outer surface area A of the drum(s) is the effective area of heat transfer, if we assume this surface is significantly larger than the drum wall thickness. A temperature difference exists between the interior drum surface and film temperature prior to removal. The overall drying rate can be expressed in the following equation:

$$\frac{dw}{dt} = \frac{UA\Delta T}{\Delta H_{fg}} \tag{12.1}$$

where dw/dt = the drying rate (kg water/s), A = the outer surface area of the drum (m^2) in contact with the product, U = the overall heat transfer coefficient (W/m^2°C), ΔT = the mean temperature difference between the inner drum surface and dried film (°C), H_{fg} = the latent heat of vaporization for the moisture removal in the food product (J/kg).

The overall heat transfer coefficient includes the combined effects of convective and conductive heat transfer between the steam and product according to the following equations:

$$R = \frac{1}{h_v A} + \frac{\Delta x}{k_w A} + \frac{1}{h_L A} \tag{12.2}$$

$$U = \frac{1}{RA} \tag{12.3}$$

where R = the total thermal resistance between the steam and product; h_v = the convective heat transfer coefficient of the steam boundary layer; Δx = the thickness of the cylinder wall; k_w = the coefficient of conductivity of the cylinder wall, approximately 15 W/m K for stainless steel; h_L = the convective heat transfer coefficient of the product; A = the surface area of the outside of the cylinder in contact with the product; U = the overall heat transfer coefficient. Values for the overall heat transfer coefficient are typically in the 1000 to 2000 W/m^2 K range.

12.1.2 Spray Drying

Spray drying is used in a wide range of applications from pharmaceutical products to foods, to detergents. Feed materials are usually in the form of a liquid solution, capable of being dispersed into a fluidlike spray. The fluid is atomized or dispersed as fine droplets that come into immediate contact with a flow of hot air or gas. These droplets provide an extensive surface area for heat and mass transfer. Therefore, evaporative cooling and short residence time maintain a low product temperature. This makes spray drying suitable for drying heat labile substances such as enzymes, blood plasma, and milk proteins. Essentially, sensible heat losses from the hot air provide the latent heat for evaporating liquid from the product.

Advantages

- Maintains quality and functional properties of the product (i.e., solubility of milk powder).
- Relative simplicity and ease of control of the system.
- Energy usage comparable to that of other drying methods.
- Preservation of most volatiles.

Disadvantages

- High initial investment.
- Hard to control final particle size.
- Requires a pumpable feed.
- Problems with product and dust recovery.

Spray drying involves four critical phases:

1. Atomization.
2. Droplet-air mixing.
3. Evaporation.
4. Recovery of dried product.

Atomization is the critical initial step and is achieved via a pressure nozzle, rotary wheel, or pneumatic nozzle. The atomizer controls the droplet size, size distribution, trajectory, and speed that in turn determine the final character of the dry particle. The contact between drying air and spray droplets can be conducted in either a co-current, countercurrent, or combined as mixed flow systems.

Cyclones are widely used in product recovery as they are highly efficient, require low maintenance, and are easily cleaned. A cyclone is a conical chamber into which a mixture of dry product and air enters tangentially at high speed, causing the mixture to spin. The heavier product is thrown to the sides of the cone and falls out at the bottom. The lighter air moves to the center and out the top.

One-stage drying is a process in which the final moisture content is achieved in the drying chamber alone. The resulting powder from this process is characterized by single particles of small size and high bulk density, which are difficult to disperse. To remove the last traces of moisture, high outlet temperatures are required and this may have a detrimental effect on powder quality.

Two-stage drying methods combine spray drying and fluidized bed drying. The advantage here is that all moisture is not driven off in the spray drying step. This allows for a lower outlet temperature.

Often, as is the case with milk powders, the initial product will be further processed with milk powders and other food ingredients. Agglomeration during storage and dispersibility during rehydration are of great concern to the processor. The hygroscopic properties of the initial milk powder cause the particles to rapidly reabsorb water back from the atmosphere and cake.

In commercial practice, the initial powder is rewet with a moist air stream and redried. This causes some agglomeration of particles, which improves its dispersion and solubility properties. The same concepts can be applied to other products. Agglomeration may also be used to a processor's advantage. Powders are often humidified and pressed into pelletlike forms as with soup stock or animal feed.

12.1.2.1 *Spray Dryer*

A schematic diagram of a spray dryer is shown in Figure 12.2. The features of the spray dryer include:

1. A tank reservoir holds the milk or other product to be dried.

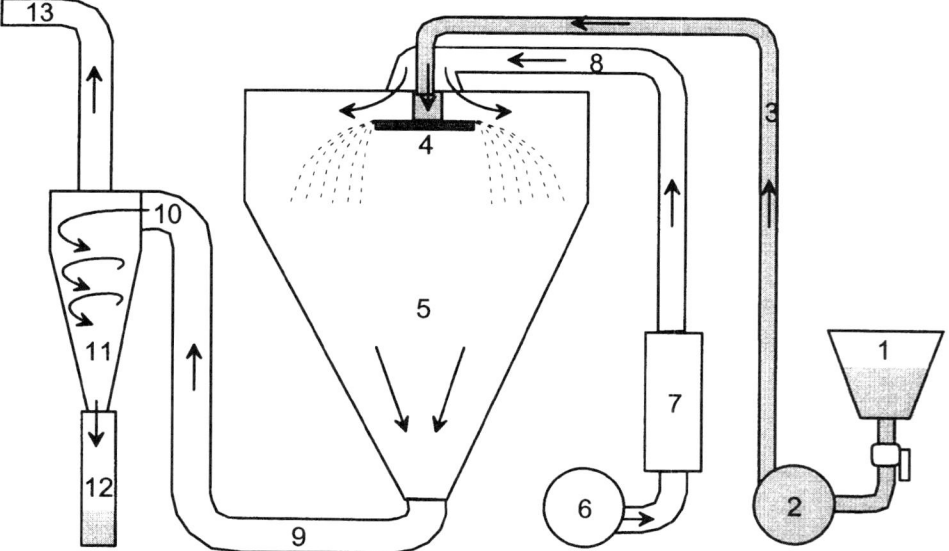

Figure 12.2 Spray dryer. 1, milk reservoir; 2, feed pump; 3, product feed pipeline; 4, atomizer; 5, drying chamber; 6, air fan; 7, air heater; 8, hot air duct; 9, a mixture of dried product and air-carrying duct; 10, cyclone separator; 11, heavy powder falling down; 12, product tank; 13, exhaust air.

2. A pump takes the liquid from the tank into the drying chamber.
3. A pipeline connects the pump outlet to the top of the dryer.
4. Atomizer, a spinning disk on the pressure nozzle that disperses it into very fine particles.
5. Drying chamber, where the liquid from the product particles evaporates into the surrounding air.
6. A fan forces air at high speed into the chamber.
7. It passes through a heater and its temperature is raised to approximately 150°C.
8. The hot air is carried to the top of the dryer where it enters above the spinning disk.
9. The mixture of dried product and hot air leaves the bottom of the chamber through a duct.
10. It enters a cyclone tangentially so that it starts spinning.
11. Spinning throws the heavier milk to the outside of the cyclone where it strikes the walls of the cyclone separator and falls down to the bottom of the chamber.
12. The powder is collected into a detachable cylinder.
13. The lighter air is displaced to the center and leaves through the top of the separator.

12.1.2.2 *Process Variables* The following variables can be controlled in this process.

- The solids content of the product before drying. Typically, milk is preconcentrated to 50% solids before spray drying.
- The flow rate of the product.
- The flow rate of the air.
- The inlet temperature or the air, typically between 150°C and 210°C.

12.1.2.3 *Product Characteristics* The following product properties can be measured:

- *Bulk density.* This is measured by filling a graduated cylinder with the dried powder and settling it in a standard manner, then measuring both its volume and weight. The density is computed using the normal formula:

$$\rho = \frac{m}{V} \tag{12.4}$$

where m = mass of powder and V = volume of powder after settling.

- *Moisture content.* Moisture content measures the total water in a product as a fraction (or percent) of the product's mass. It is measured by weighing a quantity of dried powder, then subjecting it to infrared radiation to drive off all residual moisture and watching its weight until it levels off. Moisture content (wet basis) is computed as

$$M_{wb} = \frac{m_b - m_a}{m_b} \tag{12.5}$$

where m_b = mass of powder before infrared treatment, and m_a = mass of powder after infrared treatment. For stability, this should be $< 4\%$ for milk powder. Moisture

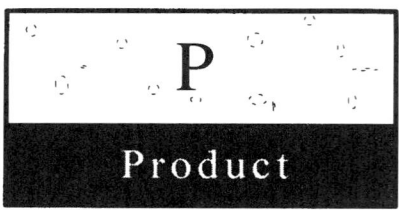 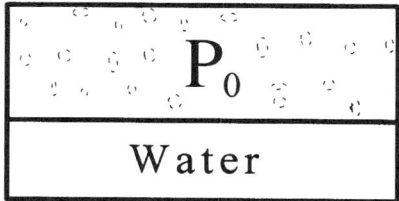

Constant temperature environment

Figure 12.3 Concept of water activity in a product when $P < P_0$.

can also be expressed on a dry basis, that is, the mass of moisture as a percent of the solids in the product. The equation is

$$M_{db} = \frac{m_b - m_a}{m_a} \tag{12.6}$$

- *Water activity a_w.* Water activity is a measure of interaction between water and other food components. This is measured with an instrument that measures the vapor pressure in the headspace above a sample. As illustrated in Figure 12.3, water activity is defined as

$$a_w = \left[\frac{P}{P_0}\right]_T \tag{12.7}$$

where $P =$ the partial pressure of water in the headspace over the product, and $P_0 =$ the partial pressure of water in the headspace over pure water for the same temperature T. Note that a_w will vary with temperature.

12.2 LAB EXERCISE

12.2.1 Objectives

The main objectives of this lab exercise are to:
- Demonstrate the application of drum drying as a dehydration process.
- Demonstrate spray drying as a dehydration process.
- Design a factorial experiment to evaluate the effect of both spray drying or drum drying process variables on product characteristics and product quality.

12.2.2 Materials and Methods

Products:

- Skim milk concentrated to 50% total solids for drying on a drum dryer and 40% total solids with a Niro atomizer spray drying system or similar type of equipment. The surface temperature of the drum dryer should be approximately 140 to 150°C and the inlet and outlet air temperatures of the spray dryer 180 and 90°C, respectively.

For bulk density measurements:

- Graduated cylinder.
- Top pan balance.
- Spatula.

For moisture content measurements:

- Infrared moisture balance, for example, Cenco or equivalent.
- Aluminum sample pans.

For water activity measurements:

- Water-activity-measuring instrument, for example, Decagon CX-2 or the equivalent.

12.2.3 Experimental Procedures

Operating instructions for the drum dryer, spray dryer, Cenco infrared moisture balance, and Decagon CX-2 water activity system are included in Appendix D. You will evaluate the powders obtained from the spray and drum drying processes:

Task 1: Measure bulk density.

Task 2: Measure moisture content.

Task 3: Measure water activity. First practice an experimental technique with the samples provided.

12.2.3.1 Drum Dryer Operation

1. Loosely attach the doctor blades and endplate.
2. Connect steam and electric supplies.
3. Start the drums and regulate them to the desired speed and direction. To change direction, the indicator must stop at 0 and the drums must stop completely before going in the opposite direction.
4. Slowly open the steam valve while the drums are turning.
5. Open the drain valve. When the drain has turned to steam (venting complete), close the drain valve and bypass valve.
6. Regulate the steam pressure with the pressure-regulating valve.
7. After the drums are thoroughly heated, turn off the steam and open the drain valve to release the pressure in the drums. When the pressure reaches zero, stop the drums and set the roller gap with the feeler gauge.
8. Never stop the drums when the steam is on or the drums are pressurized.
9. Start the drums, turn the steam on, and adjust the pressure, if necessary.

10. When the product is being dried and preheated, tighten the doctor blades and endplates; adjust the groove position between the drums and drum speed to get a uniform product.

11. Clean up when finished drying:

    a. Turn off the steam.

    b. Bleed off the pressure to 0.

    c. Stop the drums.

    d. Open the roller gap.

    e. Remove the doctor blades and endplates.

    f. Rinse the drums with water to cool.

    g. Clean the drums with detergent, sponges, and brushes.

    h. Rinse and sponge off the machine.

    i. Wash and rinse all disassembled parts in the sink and air dry.

    j. Rinse and mop-dry the floor.

12.2.3.2 *Spray Dryer Operation*

Prior to startup, doublecheck that all components are clean, dry, and properly installed and the powder buckets are in place. The damper at the outlet of the fan should be in the full open position. Ensure that all the individual components, such as the fan, atomizer, feed pump, etc., are working properly.

1. *Typical startup:*

    a. Make sure the inlet of the feed pump is connected to a reservoir of water and the outlet is connected to the atomizer inlet at the top of the dryer.

    b. Make sure a product collection cylinder is attached and properly seated at the bottom of the cyclone to the left of the dryer and the butterfly valve above the cylinder is open (vertical).

    c. Turn on the main power.

    d. Set the air temperature control to 180°C Settings on this control are only approximate. Check them against the dials on the front panel of the dryer.

    e. Turn on the air heater.

    f. Watch for the inlet temperature and when it reaches 100°C, turn on the atomizer.

    g. Wait until the atomizer reaches its speed (a minute or two), turn on the feed pump so that water is being sprayed into the dryer. Failure to achieve the proper speed will subject the atomizer to unnecessary loads. Set the pump to a relatively low speed to start.

    h. Adjust the temperature control until the inlet temperature reaches the desired level.

2. *Running operation:*

    a. Place the concentrated milk in the second reservoir.

    b. When ready, switch reservoirs so that the milk flows into the dryer, replacing the water. When this occurs, increase the speed of the feed pump slowly until it stays at the proper temperature. Changes in the product viscosity and the product itself will also affect the air temperatures.

  c. Once manufacturing of the dried powder begins, place the collection container underneath the cyclone separator duct.

  d. Knock out the dryer walls every 5 min with a rubber mallet. This should be done periodically, in order to knock down powder that may tend to rest on the tapered portion of the chamber. Some products will require more frequent knockdown than others, whereas some may require no knockdown at all.

  e. Switch the collection cylinder before changing temperature or other experimental settings.

  f. After changing settings, wait a few minutes for the old product to clear, then replace it with a clean cylinder.

  g. Observe both the inlet and outlet temperatures.

  h. Adjust the feed and temperature controller until the temperatures hold steady at the desired levels.

  i. Observe the drying closely in order to spot any potential problems.

3. *Shut down:*

  a. Turn off the air heater.

  b. When the air inlet temperature falls to 150°C, turn off the feed pump.

  c. When the inlet temperature falls to 80°C, turn off the atomizer.

  d. Open the side door and brush any residual powder off the inside walls of the dryer.

12.2.4 Design of Experiment

The experimental design includes the following.

12.2.4.1 Treatment Design

1. *Objective.* To determine the effect of some processing parameters such as initial solids content, air temperature, flow rate, or drum speed on the bulk density, moisture content, and water activity of dried milk powders.

2. *Factors.* Select one or two processing variables.

3. *Treatments.* Select appropriate levels of the processing variables.

12.2.4.2 Experimental Design

1. *Hold constant.* List variables that can be held constant.

2. *Replication.* Discuss how the experiment will be replicated. What are the appropriate number of replicates?

3. *Randomization.* Select completely random design (CRD) or randomized complete block design (RCBD).

12.2.5 Lab Report

1. Include a short introduction to drum and spray drying dehydration processes.

2. State the objectives of the lab exercise.

3. Include a table of the operating conditions used. This should be sufficiently detailed to allow one to reproduce the experiment. Report all the data in Data Sheets 12.1 and 12.2.
4. Include the specific results of the spray drying experiment:
5. Draw a flow diagram of the process, labeling all inputs, outputs, and process conditions.
6. Convert the experimental moisture contents from a wet to dry basis (Cenco moisture balance).
7. Plot the appropriate graphs to show the effect of experimental variables on bulk density, water activity, and moisture content. You can use the data given in Appendix D for spray drying processes conducted at different inlet temperatures.
8. Discuss and interpret the results in comparison with the quality of other products such as commercial agglomerated spray dried powder, instant coffee, freeze dried coffee, and cocoa mix.
9. Report the results in Data Sheet 12.3.

12.2.6 Problem

Using your experience in this laboratory, design a two-way factorial experiment to investigate some practical problem. Assume that, if necessary, you can make several runs on the equipment. Report the following:

1. Your experimental objectives.
2. Your treatment design, listing factors, levels, and treatments. Explain how the design meets your objectives.
3. Your experimental design, describing how each aspect of the design improves the experiment. Name specific variables wherever appropriate. Include descriptions of your experimental unit and your blocks (if any) and describe your plan for assigning replicates.
4. Your analysis design, including the response variable. Include descriptions of the graphs you would plot.
5. Describe the practical interpretation you would offer for each of various possible outcomes on quality factors.

See Chapter 11 on "Principles of Experimental Design" for guidance.

12.3 SUGGESTED READINGS AND REFERENCES

1. D. Heldman, and P. Singh, 1981. "Food dehydration." In *Food Process Engineering*, 2nd ed., Westport, CT: AVI Publishing Co.
2. M. Karel, O. R. Fennema, and D. Lund, 1975. "Dehydraqtion of foods." In *Principles of Food Science. Part II, Physical Principles of Food Preservation*, New York: Marcel Dekker.
3. P. Fellow, 1988. "Dehydration." In *Food Processing Technology*, Chichester, UK: Ellis Horwood Ltd.

DATA SHEET 12.1

Drum drying process

Process conditions: _____

Roller gap width: _____

Feed % total solids: _____

Steam pressure: _____

Rotation Speed (rpm)	Moisture (% wet basis)	Moisture (% dry basis)	Bulk Density (g/mL)	Water Activity	Powder Quality

Qualitative assessment of drum-dried product: look for color, aroma, presence of burned particles.

DATA SHEET 12.2

Spray drying process

Atomizer type: _____

Flow type: _____

Product recovery type: _____

Inlet Temperature (°C)	Outlet Temperature (°C)	Bulk Density (g/mL)	Moisture Wet Basis (%)	A_W	Product Quality

Qualitative assessment of spray-dried product: look for color, aroma, presence of burned particles.

Quality evaluation of commercial agglomerated spray-dried milk powder, instant coffee, freeze-dried coffee, and cocoa mix

Product	Moisture (% wet basis)	Moisture (% dry basis)	Bulk Density (g/mL)	Water Activity
Agglomerated spray-dried milk powder				
Instant coffee				
Freeze-dried coffee				
Cocoa mix				

13

CONVECTIVE DRYING
OF FOODS

13.1 BACKGROUND

The dehydration or drying of foods is a complex phenomenon involving momentum, heat, and mass transfer processes. All drying operations depend on the application of heat to vaporize the water or volatile constituents. The mechanism that controls the drying of a particulate product depends on its structure and drying parameters such as moisture content, product dimensions, heating medium temperature, surface transfer rates, and equilibrium moisture content.

All solid materials have a certain equilibrium moisture content when they come into contact with air at a particular temperature and humidity. Therefore, the materials tend to lose or gain moisture over a period of time to attain this equilibrium value. If the temperature or humidity of the air is changed, then moisture is either lost or gained until a new equilibrium value is attained. A few typical equilibrium moisture isotherms of some foods are shown in Figure 13.1. The equilibrium moisture curves depend on the environmental temperature for a particular food and its fibrous or colloidal structure. The moisture in a food solid is retained in two forms, namely, the so-called "bound" water or free water as shown in Figure 13.2. The bound water exerts an equilibrium vapor pressure less than that of free water at the same temperature. Moisture in the form of bound water may be retained in fine capillaries, or adsorbed onto the surface or within a cell or fibrous walls or in physical/chemical combination with the solid. Free water, on the other hand, exerts an equilibrium vapor pressure equal to that of pure water at the same temperature. The moisture in the form of free water may be retained in voids in food solids.

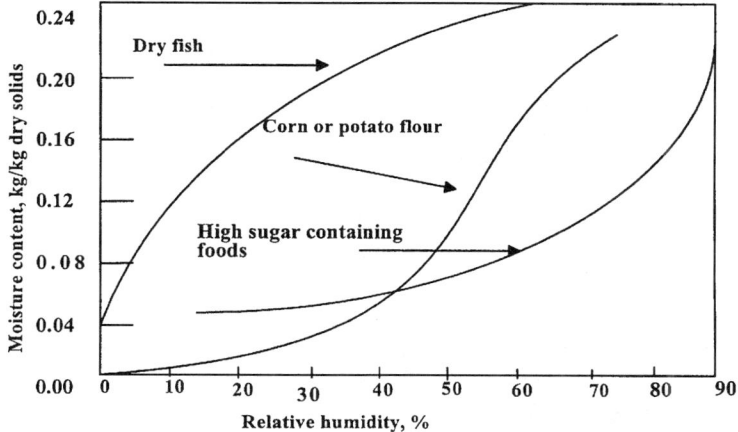

Figure 13.1   Some typical water sorption isotherms for dry food materials.

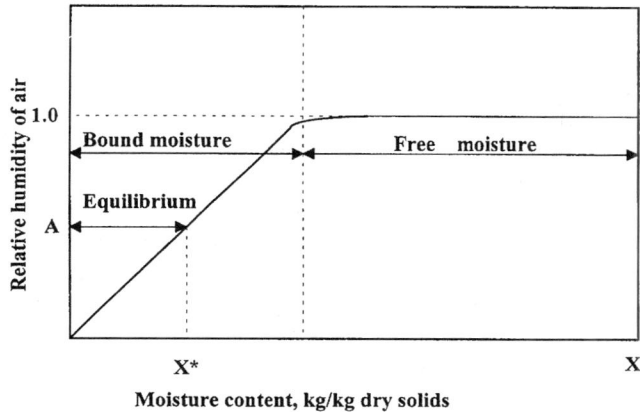

Figure 13.2   Typical curve showing the moisture equilibrium profile in a solid food material.

The distinction between "bound" and free water is a property of the specific food material under consideration. If a food is dried in air of relative humidity A (Fig. 13.2), mostly free water and partly "bound" water are removed. This consists of the free moisture above the equilibrium moisture content corresponding to the air condition. Therefore, the distinction between equilibrium and free moisture contents depends on the nature of the drying solids and conditions of the drying air. Based on the above considerations, the drying mechanisms can be classified into three categories:

1. Evaporation from a free surface, which follows the laws of heat and mass transfer from a moist object.
2. Liquid flow in capillaries.
3. Diffusion of liquid or vapor, which follows Fick's second law of diffusion.

Drying may involve various modes of heat transfer such as convection, conduction, or radiation. In convection drying, the heating medium, generally air, comes into direct contact with the solid food material and initiates diffusion of water vapors from and within the food material. Various oven, rotary, fluidized bed, spray, and flash dryers are typical examples of convection drying. In conduction drying, the heating medium, generally steam, is separated from the solid by a hot, conducting surface such as in a drum, cone, and trough dryers, any of which may be operated under vacuum. In radiation drying, heat is transmitted solely as radiant energy. Some dryers also use microwave energy to dry food materials at atmospheric pressure or vacuum.

13.1.1   Rate of Drying

The rate of drying of a material depends on properties of the material such as bulk density of the dried material, initial moisture content, and its relation to the equilibrium moisture content under drying conditions. It may be necessary to avoid the maximum rate of drying if it results in shrinkage, surface hardening, surface cracking, or other undesirable effects in the drying of food solids. A generalized rate of drying curve is shown in Figure 13.3, where the equilibrium moisture content is determined by the air condition. The initial $A*A$ portion of the curve represents the initial unsteady-state condition.

13.1.1.1   Constant Rate Period   The constant rate drying period (Fig. 13.3, section AB) is characterized by the evaporation of moisture from a saturated surface. It involves the diffusion of water vapor from a saturated surface of the material through an air film into the bulk of the air. Moisture movement within the solid is sufficient to maintain a saturated condition at the surface and the rate of drying is, therefore, controlled by the rate of heat transfer to the surface. The temperature at the surface remains constant and it approaches the wet bulb temperature. However, the rate of drying may be increased by additional heat transfer via conduction, or radiation, which raises the surface temperature above the wet bulb temperature. Constant rate drying is essentially equivalent to evaporation from a large body of water and is independent of the type of solids.

The rate of water evaporation is given by

$$N_c = K_g \cdot A(P_s - P_W) \tag{13.1}$$

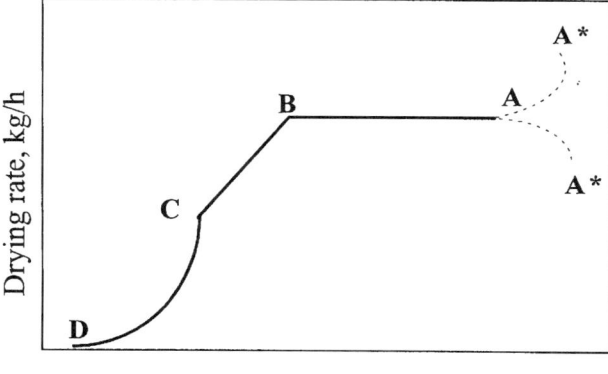

Figure 13.3   Typical drying rate curve for a hygroscopic food material (AB = constant rate period; B = critical moisture content; BC = first falling rate period; CD = second falling rate period).

where N_c = rate of water evaporation (kg/h), K_g = overall mass transfer coefficient for the gas film, P_s = vapor pressure of the water at the surface temperature, P_W = partial vapor pressure of the water in the air stream, A = surface area of the product exposed to drying. Vapor pressure terms in Eq. (13.1) can also be replaced with relative humidity terms as shown in Eq. (13.2):

$$N_c = K_g \cdot A(H_s - H_a) \tag{13.2}$$

where H_S = relative humidity of water vapor at the surface and H_a = relative humidity of drying air.

It is clear from Eq. (13.1) that $(P_S - P_W)$ determines the capacity of the air stream to take up moisture, as well as the driving force. Under the steady-state condition, the rate of evaporation is related to the heat transfer by forced convection, so that

$$N_c = \frac{dx}{dt} = K_g A(P_s - P_w) = \frac{hA(T_e - T_s)}{\lambda} \tag{13.3}$$

where x = moisture content of foods, dry basis at time t; h = convective heat transfer coefficient; T_S = wet bulb or surface temperature; T_e = air temperature; λ = latent heat of evaporation; $(T_e - T_s)$ = temperature driving force.

The heat transfer coefficient depends on the air velocity and direction of flow. If the mass flow rate of air stream is parallel to the drying surface, then the heat transfer coefficient can be estimated by the following relationship:

$$h = C \cdot G^{0.8} \tag{13.4}$$

where G = mass flow rate of air per unit area (kg/m$^2 \cdot$ s) and C = constant for the system and condition (C = 14.3 m^2/s$^2 \cdot$ K for parallel flow). In the case of perpendicular air flow, the heat transfer coefficient is given by

$$h = C \cdot G^{0.37} \tag{13.5}$$

where C = 24.2 m^2/s$^2 \cdot$ K.

13.1.1.2 *First Falling Rate Period*

At the end of the constant rate period, point B is termed the "critical moisture content." At this point, the surface of the solid is no longer saturated and dry spots appear. The outside wet area may, therefore, be reduced progressively, and the rate of drying falls off. Nonhygroscopic foods may have a single falling rate period, whereas hygroscopic foods have two falling rate periods. The drying rate depends on the factors affecting the diffusion of moisture away from the evaporating surface and rate of internal moisture movement. Point C represents a condition where the original surface film has evaporated completely, and beyond this point, the rate of drying is controlled by the rate of moisture movement through the solid.

The falling rate period is usually the longest period of a drying operation and in some foods such as grains being dried, where the initial moisture content is below the critical moisture content, the falling rate period is only part of the drying curve that is observed. In the falling rate period, the drying rate mostly depends on the air temperature and food bed thickness. It is unaffected by relative humidity (except in equilibrium moisture content) and velocity of air. The air temperature is, therefore, controlled during the falling rate period, whereas air velocity and temperature are more important during the constant rate period. In practice, foods may differ from these ideal drying curves due to shrinkage, change in temperature, and rate of moisture diffusion in different parts of the food.

13.1.1.3 Second Falling Rate Period   The second falling rate period C to D represents conditions where the drying rate is largely controlled by moisture movement within the solids and is independent of conditions outside the solid. Moisture transfer may occur by a combination of factors such as liquid diffusion, capillary movement, and vapor diffusion.

13.1.2 Time of Drying

Drying time in different stages of drying can be calculated as follows. Let W_0 = weight of dry solids and x = g moisture per g dry solid. The rate of moisture evaporation (N_a) is given by

$$N_a = -W_0 \frac{dx}{dt} \tag{13.6}$$

For constant drying rate period A to B, the drying time t_{AB} is given as

$$t_{AB} = \frac{W_0(X_A - X_B)}{N_C} \tag{13.7}$$

where X_A = moisture content/dry solids (kg water/kg dry solids) and X_B = moisture content/dry solids (kg water/kg dry solids).

For the first falling rate period (B to C), the drying time t_{BC} can be calculated if the moisture transfer rate N is represented by a straight line such as

$$N = m \cdot X + k$$

where m = slope of the line and k = intercept of the line. At $N_B = m \cdot X_B + k$, $N_C = m \cdot X_C + k$, so that $N_B - N_C = m(X_B - X_C)$:

$$t_{BC} = W_0 \frac{X_B - X_C}{N_B - N_C} \ln\left(\frac{N_B}{N_C}\right) \tag{13.8}$$

where X_C = moisture content/dry solids (kg water/kg dry solids). Equation (13.7) can be written in simplified form as

$$t_{BC} = W_0 \frac{X_B - X_C}{N_m} \tag{13.9}$$

where N_m is the log mean drying rate between B to C.

As an approximation, for most cases, the falling rate period between B to D can be taken as a straight line and, thus, the drying time from B to D can be calculated as shown in Eq. (13.10):

$$t_{BD} = W_0 \frac{X_B - X_D}{N_m} \tag{13.10}$$

where X_D = moisture content/dry solids (kg water/kg dry solids). The total drying time is therefore given by

$$t_{AD} = t_{AB} + t_{BD} = W_0 \left(\frac{X_A - X_B}{N_C} + \frac{X_B - X_D}{N_m} \right) \tag{13.11}$$

13.1.3 Moisture Content Wet or Dry Basis

The moisture content of a food can be represented on the basis of wet or dry mass of the product. Moisture content wet basis:

$$X_{\text{wet}} = \text{mass of water/initial mass of wet product} \quad \text{(kg water/kg food)}$$

Moisture content dry basis:

$$X_{\text{dry}} = \text{mass of moisture/mass of dry matter} \quad \text{(kg water/kg solids)}$$

$$\text{Mass of dry matter} = \text{mass of wet product} - \text{mass of moisture}$$

13.2 LAB EXERCISE

13.2.1 Objectives

The main objectives of this lab exercise are to:

1. Determine the moisture and drying rate histories of a food product in a wind tunnel dryer or forced convection air dryer.
2. Analyze the moisture loss rate as a function of moisture content to determine various drying rate periods, namely, the constant rate and falling rate periods.

13.2.2 Materials

1. Convective tunnel dryer or other forced air type of dryer.
2. Weighing balance.
3. Temperature-recording system.
4. Anemometer or other air-flow-metering system.
5. Food product in a different shape and size such as carrot cubes, thin circular carrot slice, or peas, etc.

6. Vacuum oven.
7. Aluminum dish.

13.2.3 Procedure

1. Measure the critical dimensions of the samples and calculate the surface area and volume.
2. Measure the sample initial weight and determine its initial moisture content and dry solids.
3. A tunnel dryer is comprised of a circular duct with an overall length of about 1.4 m. A window is provided at the front for the observation of the solid sample to be dried. The sample is inserted through the open end and is suspended in the center of the duct from a support wire coupled to a balance. Therefore, this instrument is designed to measure the sample weight loss more accurately.
4. Switch on the air blower and adjust it to the required flow rate by means of a damper. Compute the air flow measurements from an electronic anemometer.
5. Switch on the air heater and adjust it to the required temperature. Allow 10 min for the apparatus to attain steady-state conditions.
6. Measure and record the dry and wet bulb temperatures of the dry air. Measure the ambient air temperature.
7. Quickly insert the sample in the forced air oven or tunnel dryer and start the stopwatch and record the sample mass at time 0.
8. Measure the sample weight at short time intervals, initially every 1 min, gradually extending the period as drying progresses to about 5-min intervals and thereafter every 15 min, in case the tunnel dryer is provided with an electronic balance. Switch off the blower in order to obtain an accurate weight reading; this should be done rapidly so that air flow is reestablished without delay.
9. If a tunnel dryer is not available, place the sample in a forced air type of dryer and draw the sample after predetermined intervals to estimate the percent of moisture and dry solids.
10. Continue the drying process for at least 45 min.
11. Take the final mass of the sample after drying for 10 to 12 h to calculate the equilibrium moisture content X_e.
12. Record the data in Data Sheet 13.1.

13.2.4 Results and Discussion

1. Plot the sample mass versus time and estimate the equilibrium moisture content by extrapolation.
2. How much is the difference between the extrapolated equilibrium moisture content and experimental equilibrium moisture content.
3. Compute the moisture content X in the sample at various time intervals.
4. Calculate the drying rate $\Delta X / \Delta t$ and free moisture $X - X_e$.
5. Determine the rate of drying or moisture loss per min per kg of dry solids.

6. Plot the rate of moisture loss $\Delta X/\Delta t$ versus time and show various drying rate periods.
7. Plot $\Delta X/\Delta t$ versus the free moisture content $X - X_e$ and show various drying rate periods.
8. Plot $X - X_e$ versus time on semilog coordinates. Calculate the f and j parameters.
9. Parameter $j = (X - X_e)$ apparent$/(X - X_e)$ actual and $f = $ time to traverse one log cycle on $(X - X_e)$ versus the time graph.
10. Calculate the apparent moisture diffusivity D_m of the sample using the following equation:

$$f = \frac{2.303 \, R^2 \beta^2}{\pi^2 D_m} \tag{13.12}$$

where $R = $ radius of an equivalent sphere of equal volume $(V = 4/3\pi R^3)$ and $\beta = $ shape factor, using 1 for the sphere.

13.2 SUGGESTED READINGS AND REFERENCES

1. S. S. H. Rizvi, and G. S. Mittal, 1992. *Experimental Methods in Food Engineering*, New York: Van Nostrand Reinhard.
2. G. V. Jeffreys, and C. J. Mumford, 1982. *A Laboratory Course in Chemical Engineering: Mass Transfer Operations*, Birmingham, UK: University of Aston.
3. P. Fellows, 1988. *Food Processing Technology: Principles and Practice*, Chichester, UK: Ellis Horwood Ltd.

DATA SHEET 13.1

Drying characteristics of food

1. Moisture content

Initial sample mass: _____

Dried sample mass: _____

Moisture content, wet basis: _____

Moisture content, dry basis: _____

2. Equilibrium moisture content, dry basis: _____

3. Sample dimension

Sample dimensions: _____ m, _____ m, _____ m

Diameter of cylindrical slice: _____

Thickness of the slice: _____

Surface area: _____m^2

Volume: _____m^3

4. Air conditions

Hot air dry bulb temperature: _____

Hot air wet bulb temperature: _____

Hot air flow rate: _____

Time (min)	Sample Mass (g)	Moisture mass (g)	Moisture loss (g)	Drying rate $\Delta X/\Delta t$	Moisture content [g/g dry solids (X)]	Free moisture [g($X - X_e$)]

14

OSMOTIC DEHYDRATION
OF FOODS

14.1 INTRODUCTION

Removal of water is a major unit operation in food processing to stabilize foods by lowering water activity a_W. Air drying is commonly used to preserve fruits. However, air drying considerably reduces product quality, for example, tough texture, slow or incomplete rehydration, loss of juiciness, and unfavorable color and flavor loss. Freeze drying can be used to produce dried products of high quality, although at a higher cost. A newer technique, "osmotic dehydration," can be used as a means of food preservation. Osmotic dehydration is a water removal process based on the water and solubility activity gradient across a cell's semipermeable membrane. The application of osmotic treatments has been suggested as an intermediate step prior to drying or freezing to reduce product water load with a simultaneous improvement in quality. It has been proposed that osmosis is also a potential technique to produce intermediate-moisture foods.

Osmotic dehydration with osmotic syrup recycling requires two to three times less energy as compared to convection drying. At relatively low process temperatures (up to 50°C), it improves product color and flavor retention. Rate of water loss in osmotic dehydration depends on the concentration of the osmotic solution, contact time, process temperature, ratio of osmotic solution to food materials, and the exposed surface area.

However, osmotic dehydration's application in the food industry is restricted due to some associated problems. Simultaneous solute transfer into the foods can affect product quality. This so-called "candying" or "salting" may improve the taste and acceptability of some final products such as dried and sweetened fruit slices of banana, apple, pear, apricot, pineapple, etc. In other cases, however, extensive solute uptake spoils the taste and nutritional profile of the product. Leaching of natural sugar and acids in osmo-dehydrated foods also affects the taste by altering the natural sugar-to-acid ratio.

14.2 BACKGROUND

14.2.1 Mechanism of Osmotic Dehydration

Osmotic dehydration involves immersing high-moisture food materials in an osmotic solution, generally a solution of sugar or sodium chloride. Consider a system containing biological material and an osmotic solution containing sucrose, illustrated by two compartments as shown in Figure 14.1. A pseudo-membrane, generally the food material cell membrane, separates the internal solution in compartment 1 from the external solution in compartment 2. Two fluxes, the flux of water J_W and flux of sucrose J_{su}, are simultaneously taking place. The flux in a mass transfer process is a function of the difference in chemical potential ($\Delta\mu_w$ and $\Delta\mu_{su}$ for water and sugar, respectively).

The chemical potential of water is higher in the biological material and the chemical potential of sugar is higher in the osmotic solution. As a result, water flows out of the biological material and sugar may flow into the material, depending on the time of contact and membrane size. Therefore, two simultaneous countercurrent flows take place. Hence, osmotic dehydration has also been described as a water removal and solute impregnation soaking process. The removal of water from a fruit through the cell membrane is also considered a function of water activity. A lower water activity is maintained in the osmotic solution to remove water from the material of higher water activity across the cell membrane. A semipermeable cell membrane allows water to pass through more easily than solute. However, the process also includes the cases where there is no membrane such as gels and cheeses or where the cell membrane has been destroyed at a high temperature.

A simplified plant cell encountered in osmotic dehydration is a cylindrical equivalent consisting of two hollow co-axial cylinders with a membrane located between them as shown in Figure 14.2. It consists of three parts, namely, an intracellular volume, an extracellular volume, and a cell membrane. The intracellular membrane includes cytoplasm and a vacuole. The extracellular volume contains the cell wall and free space among individual cells. During the osmotic dehydration process, the solute diffuses into the extracellular volume and, depending on the solute characteristics, it may or may not penetrate through the cell membrane. The solute uptake during osmotic dehydration results

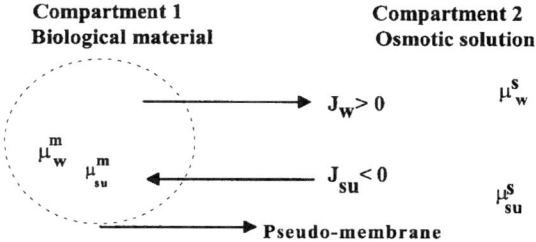

Figure 14.1 Schematic representation of a model for mass transfer during osmotic dehydration (μ_w^m = chemical potential of water in biological material; μ_{su}^m = chemical potential of sugar in biological material; μ_w^s = chemical potential of water in osmotic solution; μ_{su}^s = chemical potential of sugar in osmotic solution).

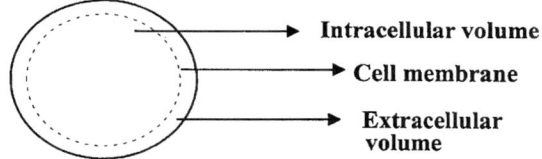

Figure 14.2 A schematic diagram of a simplified biological cell.

in the formation of a concentrated solids layer across the fruit surface, which decreases the driving force for water flow. The solute penetration is directly related to the solute concentration and is inversely related to the size of the sugar molecule. This process is carried out at a constant low temperature and, therefore, is considered isothermal and does not involve any phase change.

14.2.2 Selection of Solutes

The choice of solute and concentration of osmotic solution depend on several factors such as its effect on the organoleptic quality, final products' taste, its capacity to lower the water activity, solute solubility, permeability to the cell's membrane, preservative effect and its cost. Sucrose has been found to be one of the best osmotic agents because of its effectiveness, convenience, and desirable flavor. It is an effective inhibitor of polyphenol oxidase, prevents loss of volatile flavors, and is impermeable to most cell membranes. Its diffusivity is much lower than that of water, which results in little solid uptake in the tissue. However, its sweetness limits its application to vegetables.

Sodium chloride is an excellent osmotic agent because of its higher capacity to reduce water activity, resulting in a higher driving force during the water removal process. The salt driving force is much higher than for sucrose at the same concentration level. In some cases such as freeze-dried carrots, salt incorporation has shown a markedly improved effect during rehydration. Salt has been found satisfactory in the concentration range of 10 to 15%. However, its use is limited in the case of fruit dehydration. In some cases, a combination of sugar and salt has shown maximum results in terms of higher water loss, low solute gain, and good product flavor.

14.2.3 Processing Variables

The efficiency of a osmotic dehydration process depends on the rate and extent of water removal with a minimum cost. Food membrane structure, protopectin-to-soluble-pectin ratio, intercellular space, tissue compactness, entrapped air, etc. affect the osmotic dehydration process. The size and shape of foods also play an important role in mass transfer due to different surface areas or surface-to-thickness ratios.

Product pretreatment and process conditions change the material structural integrity, which affects the water loss and solids gain. Blanching, freezing/thawing, acidification, and high process temperatures all favor solids uptake and result in lower water loss. The acidification of a concentrated sugar solution increases water loss, probably by hydrolysis, and depolymerisation of pectin in fruits and vegetables tissues.

The molecular size of the osmotic solute has a significant effect. The smaller the solute size, the larger the depth and extent of solute penetration. Increased solute concentration results in increased water loss and solute gain up to a certain level. For example, a maximum concentration of about 60% in sugar solution is found most suitable. Process temperature has a significant effect. Higher temperature seems to increase water loss through the swelling and plasticizing of cell membranes.

Agitation ensures continuous contact of the product particles with the osmotic solution, resulting in higher water loss and lower solute gain during the first phase of osmotic dehydration. However, it seems to be of little importance during later stages of equilibrium. It has been reported that more than 50% of water loss takes place during the first hour of the process. Therefore, it is good practice to terminate the osmotic process at an early stage in order to limit solute uptake.

14.2.4 Kinetics of Osmotic Dehydration

The kinetics of osmotic dehydration are determined by estimating the rate of water removal and that of solid gain. Generally, higher rates of water removal take place within the first 60 min of osmotic dehydration due to a large driving force between the dilute fruit sap and osmotic solution. It is followed by lower rates due to the formation of a superficial solid surface layer, which reduces the driving force. The typical kinetics of osmotic dehydration are shown in Figure 14.3. The rate of solids gain also seems to behave in a similar fashion in most cases. However, it depends on the solute size and food membrane permeability.

14.2.4.1 Calculation of Osmotic Parameters In order to describe the kinetics of an osmotic dehydration process, total weight reduction *WR*, solid gain *SG*, and water loss *WL* based on initial dry matter content can be calculated as shown in Eqs. (14.1) through (14.3):

$$WR = \frac{(W_0 - W)}{S_0} \tag{14.1}$$

$$SG = \frac{(S - S_0)}{S_0} \tag{14.2}$$

$$WL = \frac{(W_0 X_0 - WX)}{S_0} \tag{14.3}$$

Equation (14.3) can be written as

$$WL = \frac{1}{S_0}[(W_0 - W) + (S - S_0)] \tag{14.4}$$

Equation (14.4) can be simplified as

$$WL = WR + SG \tag{14.5}$$

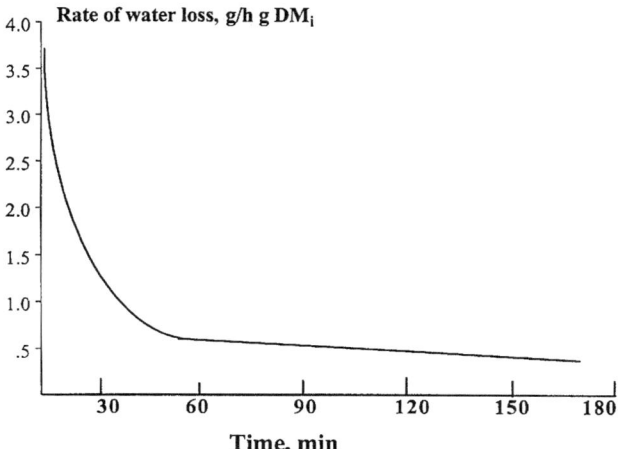

Figure 14.3 A typical graph illustrating the rate of water loss during osmotic dehydration of a biological material (DM$_i$ = Initial dry matter).

The water loss can also be calculated as

$$WL = \frac{(W_0 - W)}{S_0} + \frac{(S - S_0)}{S_0} \tag{14.6}$$

$$S_0 = W_0(1 - X_0) \tag{14.7}$$

$$S = W(1 - X) \tag{14.8}$$

where W_0 = initial weight of the material at time $t = 0$ (g), W = weight of the material at time t (g), S_0 = initial weight of dry matter DM_i in the material at time $t = 0$ (g), S = weight of dry matter in the material at time t (g), X_0 = initial weight fraction of water in the material at time $t = 0$, X = weight fraction of water in the material at time t.

The above equations are based on the assumptions that the amount of solid leaching out of the sample is negligible and, thus, the sample loses only water and picks up solute. Normalized moisture content (NMC) and normalized solid content (NSC) can be calculated as

NMC = total moisture at any time/initial total moisture

NSC = total solids at any time/initial total solids content

The kinetics of moisture and solids diffusion in osmotic dehydration generally follow an unsteady-state Fick's law of diffusion. The overall mass (solids) transfer coefficient $K(h^{-1/2})$ can be calculated by using a linear relationship between NSC and $t^{1/2}$ as shown in Eq. (14.9):

$$NSC = S/S_0 = Kt^{1/2} \tag{14.9}$$

The apparent effective moisture or soluble solids diffusivities can be calculated by a relationship shown in Eq. (14.10):

$$\ln\frac{M - M_e}{M_0 - M_e} = \ln\frac{8}{\pi^2} - \frac{D_a t\pi^2}{x^2} \tag{14.10}$$

where M_0 = initial moisture content (g/g DM_i), M = moisture at time t (g/g DM_i), M_e = moisture at equilibrium (g/g DM_i), D_a = effective diffusivity (m^2/s), x = characteristic length or thickness of an apple or fruit slice. The temperature effect on D_a can be evaluated by an Arrhenius type of relationship as shown in Eq. (14.11):

$$D_a = Ae^{-Ea/RT} \tag{14.11}$$

where E_a = activation energy (J/mol), R = gas constant (8.314 J·mol^{-1} K^{-1}), T = temperature ($^\circ K$), A = constant.

14.3 LAB EXERCISE

14.3.1 Objectives

The main objectives of this lab exercise are to:

1. Become familiar with the process of osmotic dehydration.
2. Determine the kinetics of the rate of water removal and rate of solute gain in fruits and vegetables during the osmotic dehydration process.

14.3.2 Materials and Methods

- Water bath.
- Glass beaker, 2-L capacity.
- Sugar.
- Apples, bananas, pineapple, etc.
- Weighing balance.
- Blotting paper.
- Vacuum oven.
- Stopwatch.
- Refractometer.
- Thermometer.

14.3.3 Procedures

14.3.3.1 Sample Preparation Wash, peel, and core the fruits. Cut the fruit into 5 to 8-mm-thick, wheel-shaped slices. Cut the midsection wheels into four quarters and trim them so they weigh exactly 5 g each. Note that the initial weight of each trimmed apple slice should be very close to obtain good results.

14.3.3.2 Osmotic Solution

Prepare an osmotic solution of 60% sugar concentration in deionized water. Pour the osmotic solution into three 2-L-capacity glass beakers.

14.3.3.3 Sampling Procedure

1. Place the beakers containing the osmotic solution in a water bath maintained at desired temperatures such as 20, 30, or 40°C.
2. Put about 16 to 20 pieces of apple slices (5 g each) or any other fruit slice in a petri dish. Keep them covered to avoid any loss of moisture during the experimental run.
3. Take out a control sample to estimate the water and initial dry matter content before osmotic dehydration.
4. Pour the fruit slices in the osmotic solutions maintained at 20, 30, and 40°C. The ratio of osmotic solution to fruit slice should be around 20 : 1 to maintain a constant osmotic solution concentration.
5. Record the osmotic process time.
6. Stir the fruit samples to maintain close contact between them and the osmotic solution.
7. Remove two treated fruit slices at 20-min intervals for up to 2 h.
8. Rinse out any superficially adsorbed osmotic solution in the deionized water and remove excess water from the rinsed sample with blotting paper.
9. Record the sample weight carefully to estimate weight reduction during the process. Record the data in the appropriate place on Data Sheet 14.1, 14.2, or 14.3.

14.3.3.4 Moisture For each sample, weigh about 3 to 4 g of the blended sample into predried aluminum dishes. Place the dishes in a vacuum oven (600 mm Hg) maintained at 75°C or in regular oven maintained at 105°C for a period of 24 h. Place the dried samples in a desiccator and reweigh to estimate their moisture content and dry solids gravimetrically.

14.4 LAB REPORT

1. Include a short introduction to the osmotic dehydration process.
2. State the objective of the experiment.
3. Include a list of materials and methods used.
4. Report the results on the attached Data Sheets 14.1 through 14.3.
5. Estimate the percentage of moisture and dry solids contents of the processed fruit slices and determine the total amount of water and dry solids at different time intervals.
6. Report the total weight reduction WR per unit weight of initial dry matter DM_i, solid gain per unit weight of initial dry matter, and water loss per unit weight of initial dry matter for each temperature (Data Sheets 14.1 through 14.3).

7. Plot the rates of water loss and of solid gain at different temperatures and discuss the effect of temperature on them.

8. Discuss the rate of water loss during the first 60 min of operation. Do the process kinetics indicate an equilibrium stage?

9. How can the process be optimized based on the data obtained in this lab exercise?

10. Plot the normalized moisture content (NMC) and normalized solids content (NSC) with time at all temperatures. Discuss the effect of temperature on them.

11. Plot and estimate the effects of the *WL* to *SG* ratio for each temperature. How does this ratio change with temperature?

12. Estimate the overall apparent mass transfer coefficient *K* for solids during osmotic dehydration at different temperatures and discuss its effect on the process (optional).

14.5 SUGGESTED READINGS AND REFERENCES

1. H. N. Lazarides, 1994. "Osmotic preconcentration: Developments and prospects." In *Minimal Processing of Foods and Process Optimisation. An Interface*, Boca Raton, FL: CRC Press.

2. M. LeMaguer, 1989. "Osmotic dehydration: Review and future directions." In *Proceedings of the International Symposium on Progress in Food Preservation Progresses*, Vol. 1, Brussels: CERIA.

3. H. N. Lazarides, E. Katsanidis, and A. Nickolaidis. "Mass transfer kinetics during osmotic preconcentration aiming at minimal solid uptake." *J. Food Eng.* 25 : 151 (1995).

DATA SHEET 14.1

Osmotic dehydration experiment at 20°C

Time (min)	Weight of Material	Moisture (%)	Dry Solids (%)	WR (g/g DM_i)	SG (g/g DM_i)	Rate of SG (g/min g DM_i)	WL (g/min g DM_i)	Rate of WL (g/min g DM_i)
0								
20								
40								
60								
80								
100								
120								

WR = weight reduction in fruit (g/g DM_i).
SG = solid gain in the fruit (g/g DM_i).
WL = water loss in the fruit (g/g DM_i).
DM_i = initial dry matter content (g).

DATA SHEET 14.2

Osmotic dehydration experiment at 30°C

Time (min)	Weight of Material	Moisture (%)	Dry Solids (%)	WR (g/g DM_i)	SG (g/g DM_i)	Rate of SG (g/min g DM_i)	WL (g/min g DM_i)	Rate of WL (g/min g DM_i)
0								
20								
40								
60								
80								
100								
120								

WR = weight reduction in fruit (g/g DM_i).
SG = solid gain in the fruit (g/g DM_i).
WL = water loss in the fruit (g/g DM_i).
DM_i = initial dry matter content (g).

DATA SHEET 14.3

Osmotic dehydration experiment at 40°C

Time (min)	Weight of Material	Moisture (%)	Dry Solids (%)	WR (g/g DM_i)	SG (g/g DM_i)	Rate of SG (g/min g DM_i)	WL (g/min g DM_i)	Rate of WL (g/min g DM_i)
0								
20								
40								
60								
80								
100								
120								

WR = weight reduction in fruit (g/g DM_i).
SG = solid gain in the fruit (g/g DM_i).
WL = water loss in the fruit (g/g DM_i).
DM_i = initial dry matter content (g).

15

MICROWAVE HEATING OF FOODS

15.1 INTRODUCTION

In conventional cooking, heat is transferred to food by conduction, convection, and radiation. In the microwave heating of foods, on the other hand, energy in the form of electromagnetic radiation at microwaves frequencies is coupled directly to the food so that energy absorption and, hence, heating take place throughout the food. Therefore, the conduction of heat throughout the food is not the primary mode of heating and heating is quite rapid.

However, since microwave energy is absorbed as it penetrates the food, effective heating will not take place at greater depths. Therefore, thicker foods will still rely on conduction to complete the heating process. The geometry of the product is, therefore, an important consideration in designing microwavable foods. Maximum thickness for samples should be limited to two to three times the penetration depth for uniform cooking at high power. In other cases, low power must be used to heat foods more slowly and avoid overcooking of the surfaces.

Other factors to consider in microwave heating include the absence of crusting or browning of the food surfaces, as the surfaces tend to remain cool during microwave cooking. Moisture diffuses to the surface of the food, where it evaporates into the cool microwave oven cavity, resulting in cool, soggy surfaces. Venting is often used to remove any vapors that evolve during cooking in order to prevent condensate on the food surfaces.

Some of these problems can be overcome through the selection of the proper microwave package. Packaging can be either microwave passive, microwave active, or microwave reflective. Microwave passive materials are transparent to microwaves and, therefore, would be used for packaging microwave products. Microwave active materials (susceptors) absorb microwave energy and heat directly in a microwave field. The susceptor then transfers its heat to the food and also can be used to provide high surface temperatures at the food/susceptor interface to develop crispness and browning qualities.

15.2 BACKGROUND

This section reviews the basic principles of physics pertaining to microwave heating. In particular, the concepts of field strength, dipoles, capacitance, dielectric constants, and dielectric losses are considered.

15.2.1 Electric Fields

1. *Energy.* Energy is the capacity to do work, and work is defined as the product of a force acting over a distance, that is,

$$E = W = F \cdot x \tag{15.1}$$

where E = energy, W = the equivalent work, F = force that performs the work, x = distance a mass is moved by the force. Thus, if you push a sofa across a room, you have exerted a force over a distance and so do work. Energy was consumed in doing that work. Electricity is a form of energy and, therefore, must involve some force that moves something. Our first goal is to thus identify that force, determine its origin, and determine what it moves. We will then seek mathematical ways of describing the energy involved.

2. *Atomic particles.* All matter are composed of atoms. Atoms, in turn, consist of a nuclei surrounded by orbiting electrons. The nucleus consists of positively charged protons and uncharged neutrons. The surrounding electrons are negatively charged. In neutral atoms, the number of protons in the nucleus equals the number of electrons, resulting in a 0 net charge.

3. *Electrostatic forces.* If some electrons are removed from a piece of material, the protons will outnumber the electrons and the material will take on a positive charge. Similarly, if some electrons are added to a piece of material, that material will take on a negative charge. If two positively charged objects are brought near to each other, they will each feel a force pushing them apart. Similarly, if two negatively charged objects are brought together, they will each experience a force pushing them apart. On the other hand, if a negatively charged object is brought near a positively charged object, each will experience a force pulling them together. In other words,

   · Like charges repel each other.
   · Unlike charges attract each other.

4. *Coulomb's law.* If two charges of magnitude q_1 and q_2 are separated by a distance r as shown in Figure 15.1, each will feel a force of magnitude:

$$F = k \frac{q_1 q_2}{r^2} \tag{15.2}$$

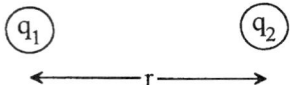

Figure 15.1 Principle of Coulomb's law.

It is clear from Eq. (15.2) that the force is proportional to the magnitude of each charge and inversely proportional to the square of the distance between them. If, for example, we double the charge on either object, the force will double. On the other hand, if we double the distance between them, the force will be reduced to $\frac{1}{4}$ of its previous value.

5. *Units and signs.* In the SI system of units, the variables in the above equation have the following units as shown in Table 15.1.

> **Example 1** *If an object with a negative charge of magnitude* $-2\mu C$ *(electrons exceed protons by* 6×10^{12}*) is placed 3 cm away from an object with a negative charge with a magnitude of 3 μC, the force will be*

$$F = k\frac{q_1 q_2}{r^2} = \left(9 \times 10^9 \ \frac{Nm^2}{C^2}\right) \frac{(-2 \times 10^{-6}\,C)(-3 \times 10^{-6}\,C)}{(3 \times 10^{-2}\,m)^2} = 60\,N$$

> *The positive sign of the force indicates that the objects are being pushed apart as you would expect with like charges.*

> **Example 2** *If an object with a negative charge with a magnitude of 3 μC is placed 6 cm away from an object with a positive charge with a magnitude of 4 μC, the force will be*

$$F = k\frac{q_1 q_2}{r^2} = \left(9 \times 10^9 \ \frac{Nm^2}{C^2}\right) \frac{(-3 \times 10^{-6}\,C)(4 \times 10^{-6}\,C)}{(6 \times 10^{-2}\,m)^2} = -30\,N$$

> *The negative sign of the force indicates that the objects are being pulled together as you would expect with unlike charges.*

Table 15.1 Units in Coulomb's law

Quantity	Units
q_1, q_2	Coulombs (C), where a coulomb $= 6 \times 10^{18}$ protons. The sign $(+$ or $-)$ of q should be that of the charge. Thus, 6×10^{18} electrons have a charge of -1 C, whereas 6×10^{18} protons have a charge of $+1$ C.
r	Meters (m), always positive.
F	Newtons (N). A $+$ sign indicates a force that will increase separation, that is, a repulsion. A $-$ sign indicates a force that will decrease separation, that is, an attraction.
k	$= 9.0 \times 10^9 \ Nm^2/C^2$

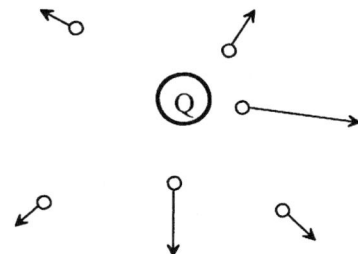

Figure 15.2 Forces around charge Q.

6. *Electric fields.* Electrostatic force is defined as "force at a distance" (Eq. 15.1). If we have a charge Q, and a test charge q is placed a distance r away from it, Q will push on q across that distance as shown in Figure 15.2. The magnitude of the push will depend on the magnitudes of Q, q, and r as given in Eq. (15.2).

   Another way to look at this is to say that Q creates a field in the space that surrounds it. At any point in that space, the field will have a strength E that depends on Q and r. If a test charge q is placed at some point in the space, the field at that point will push on it with a force that depends on the field strength E at that point and on q. To make these two explanations mathematically equivalent, we separate Eq. (15.2) into two parts; thus,

$$F = k\frac{Qq}{r^2} = \left(k\frac{Q}{r^2}\right)(q) \tag{15.3}$$

The second part is simply the charge of the second particle. The first part we call E, the field strength at distance r away from Q:

$$E = \left(k\frac{Q}{r^2}\right) \tag{15.4}$$

Now the force on q can be defined in terms of the field strength times the magnitude of q:

$$F = E \cdot q \tag{15.5}$$

7. *Lines of force.* We can map the field surrounding a charge by drawing lines parallel to the direction of the forces. By convention, we show the lines pointing away from positive charges and pointing toward negative charges. This is the direction a positive test charge is pushed by the field. A field surrounding a positive and negative charge is shown in Figure 15.3.

8. *Units of field strength.* Since field strength E times charge q equals force ($F = E \cdot q$), it follows that field strength itself must exist in units of force per unit charge. In SI units,

$$E[=]\text{N/C} \tag{15.6}$$

where [=] reads as "has units of."

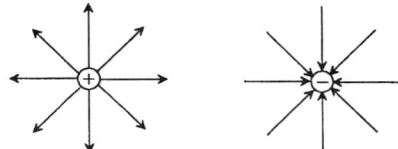

Figure 15.3   Fields around a single charge.

Example 3   An object with a charge of $-30\,\mu C$ is hung by a thread in the middle of a room. What is the field strength $5\,cm$ away from the charge? What force would a $3\text{-}\mu C$ charge experience if placed at that spot?

Solution   The field strength is

$$E = k\frac{Q}{r^2} = \left(9 \times 10^9 \; \frac{\text{Nm}^2}{\text{C}^2}\right) \frac{(-30 \times 10^{-6}\,\text{C})}{(5 \times 10^{-2}\,\text{m})^2} = -1.08 \times 10^8 \frac{\text{N}}{\text{C}}$$

The negative sign indicates a field pulling toward the charge that produces it. The force at this location on the $3\text{-}\mu C$ charge will be

$$F = E \cdot q = \left(-1.08 \times 10^8 \frac{\text{N}}{\text{C}}\right)(3 \times 10^{-6}\,\text{C}) = -324\,\text{N}$$

This is the same result that you would get from Coulomb's law.

9. *Dipoles.* When two charges of opposite sign but equal magnitude are held a fixed distance apart, we call the configuration a dipole. The field at any location around a dipole is the vector sum of the fields produced by each charge. The vector sum of fields at two points around a dipole are shown in Figure 15.4, whereas the field that results from the vector sum at all points surrounding a dipole is illustrated in Figure 15.5. Electrons in water molecules tend to concentrate around the oxygen atom, giving it a net negative charge and leaving the hydrogen atoms with a net positive charge. Because of the bond angles within water molecules, the hydrogen atoms are both on the same side of the oxygen. This results in the center of positive charge at a different place from the center of negative charge, as shown in Figure

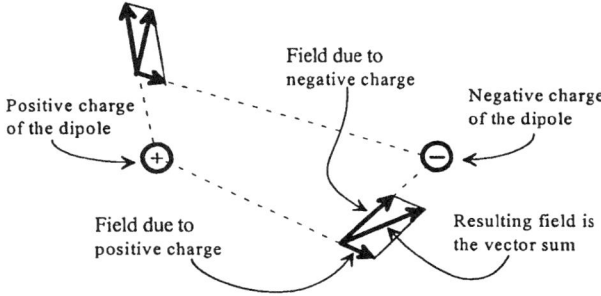

Figure 15.4   Vector sum of the field around a dipole.

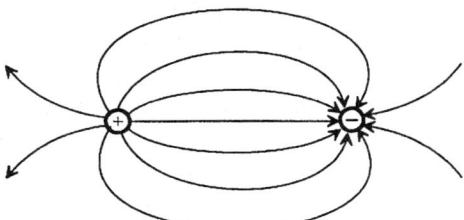

Figure 15.5 Field of a dipole.

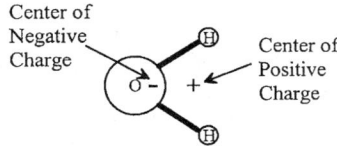

Figure 15.6 Water molecule as a dipole.

15.6. Thus, water molecules are natural dipoles and they have a profound effect on microwave cooking.

10. *Uniform fields.* It is possible, by the proper arrangement of many electric charges, to produce a field in which all lines of force are parallel and in which the field strength is the same at all locations (Fig. 15.7). Wherever you place a charged particle anywhere in this field, it will experience the same force in the same direction.

11. *Potential energy.* Just as an object loses potential energy as it falls in a gravitational field, so a charged particle loses potential energy as it "falls" through an electric field. A particle falling through a uniform field is shown in Figure 15.7. If the field has strength E and a particle with charge q moves a distance x along the field, the work done on the charge, and the potential energy change is

$$PE = -W = -F \cdot x = -E \cdot q \cdot x \qquad (15.7)$$

As shown in the figure, only the component of distance parallel to the field is considered in computing this potential energy loss (Eq. 15.7). The minus sign in this equation indicates that when a positive charge moves in the positive direction of the field, it loses potential energy, just as a falling body, moving with the gravitational field, loses potential energy. A negative charge loses potential energy when it moves against the direction of the field.

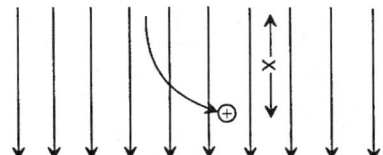

Figure 15.7 Charge moving in a uniform field.

The SI unit of potential energy is the Newton meter or Joule (J).

12. *Electrical potential.* In any given field, the potential energy lost by a particle is proportional to the charge on the particle. To express a change in potential energy in a form that is independent of the magnitude of the charge, we define potential as potential energy per unit charge:

$$\text{Potential} = V = \frac{PE}{q} = \frac{E \cdot q \cdot x}{q} = Ex \qquad (15.8)$$

The SI unit of potential is the

$$V [=] J/C$$

In other words, if a charge of 1 C moves between points A and B in an electric field and in the process loses 1 J of potential energy, then the potential difference between points A and B is 1 V. A charge of $\frac{1}{2}$ C following the same track will lose $\frac{1}{2}$ J of energy, but the potential change will be the same, namely 1 V.

> ***Example 4*** *A uniform electric field is set up between two plates with a field strength of 10×10^2 N/C. A charge of $-600\,\mu C$ is moved 3 mm parallel to the field in the direction of the lines of force, then 2 mm at right angles to the field. What is the potential energy change of the charge? What is the change in potential?*
>
> ***Solution***
>
> $$\Delta PE_{\text{Move 1}} = -\left(10 \times 10^2\,\frac{N}{C}\right)(-6 \times 10^{-4}\,C)(3 \times 10^{-3}\,m)$$
> $$= +1.80 \times 10^{-3}\,\text{Nm }(J)$$
>
> Since the potential energy change is positive, we conclude that the charge gained potential energy and, therefore, must have been moved by an outside force, not by the field. Since the second move occurs at right angles to the field, it involves no change in potential energy so the total change in potential energy is just 1.8×10^{-3} J. The change in potential is
>
> $$V = \frac{\Delta PE}{q} = \frac{1.8 \times 10^{-3}\,J}{(-6 \times 10^{-6}\,C)} = -300\,V$$

13. *Field strength and potential.* As we have pointed out, the force experienced by a particle in a field depends on the strength of the field and size of the charge. On the other hand, field strength depends only on the location in the field and so is a property of field position. Similarly, potential energy loss depends on the magnitude of the charge and the place where it moves in the field. Potential, on the other hand, depends only on the starting and ending point of the move. It is a

property of position change in the field. We defined field strength in terms of force per unit charge (Newtons/coulomb). However,

$$V = Ex,$$

so

$$E = \frac{V}{x} \qquad (15.9)$$

Thus, field strength can also be expressed in terms of the rate of potential change per unit distance. We also can verify this by showing that "Newton per coulomb" is equivalent to "volts per meter":

$$E \,[=]\, \frac{N}{C} = \frac{Newton\ meter}{coulomb\ meter} = \frac{J/C}{m} = \frac{V}{m} \qquad (15.10)$$

14. *Batteries.* A battery is a device that uses chemical reactions to move electrons from one material to another. Since it separates electrical charges, a battery creates a field and, hence, a potential difference between two locations. A standard symbol for a battery is shown in Figure 15.8. Although we frequently use more sophisticated devices to generate electrical potentials, we will represent them in diagrams as batteries.

15. *Dipole in a field.* If we place a dipole in an external field such as in a random position in a uniform field, its typical behavior is shown in Figure 15.9. The positive pole of the dipole experiences a force in the direction of the field and the negative pole experiences a force in the opposite direction. This pair of forces is called a couple and it has the affect of rotating the dipole until it is aligned with the field. Notice that when the dipoles become reoriented as shown in Figure 15.10, their positive poles are oriented toward the negative pole of the field and vice versa. But, as noted above, dipoles create their own fields. Because of their orientation, their field will be in the opposite direction to the field in which they are placed. The

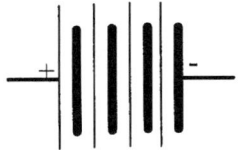

Figure 15.8 Battery symbol.

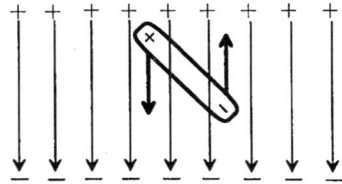

Figure 15.9 Rotational forces on a dipole in a uniform field.

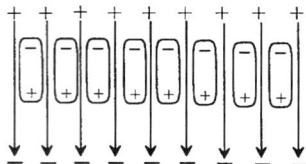

Figure 15.10   Weakening dipoles in a field.

two fields will, therefore, partially cancel each other. Thus, placing dipoles in an electric field weakens the field.

15.2.2 Capacitance

1. *Capacitors*. If we bring two metal plates close together so that they are parallel but separated by an electrical insulator as shown on the right side of Figure 15.11, we have a device called a capacitor.

2. *Charging a capacitor*. If a battery or other source of electrical potential is connected across the plates of the capacitor, electrons will be moved from one plate to the other until a potential difference exists between the plates that is equal in magnitude to the potential of the battery. This condition is illustrated in Figure 15.11.

3. *Capacitance*. The magnitude of the charge on the capacitor will depend on the applied voltage and capacitance of the capacitor:

$$Q = VC \qquad (15.11)$$

where Q = amount of charge transferred from one plate to the other (in C), V = applied potential (in V), C = capacitance of the capacitor (in F). Thus, if a potential of 1 V is applied across a 1-F capacitor, 1 C of charge will move from one plate to the other. Since a farad is a large capacitance, most capacitors have units of microfarad (μF or mF) or picofarads (pF).

Capacitance, which measures the charge-holding capacity of a capacitor, is

• Proportional to the area of the plates.

• Inversely proportional to the distance between them:

$$C = \varepsilon_0 \frac{A}{d} \qquad (15.12)$$

where C = capacitance (F), A = area of plates (m^2), d = distance between plates (m), $\varepsilon_0 = 8.85 \times 10^{-12}$ F/m or 8.85×10^{-14} F/cm. This proportionality constant is called the "permitivity of free space."

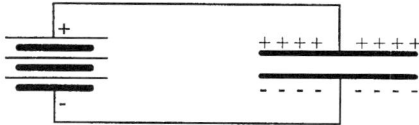

Figure 15.11   Charged capacitor.

Example 5   *Two 2 × 2 cm aluminum plates are held parallel to each other with a space of 0.5 mm between them. What is the capacitance of this configuration? If each plate is connected to a different pole of a 9-V battery, how much charge will be moved from one plate to the other? The capacitance is*

$$C = \varepsilon_0 \frac{A}{d} = \left(8.85 \times 10^{-12}\, \frac{\text{F}}{\text{m}}\right) \frac{(2 \times 10^{-2}\,\text{m})^2}{(5 \times 10^{-4}\,\text{m})} = 7.1 \times 10^{-12}\,\text{F} \quad (7.1\, pF)$$

The charge moved is

$$Q = VC = (9\,\text{V})(7.1 \times 10^{-12}\,\text{F}) = 6.4 \times 10^{-11}\,\text{C}\ (6.4 \times 10^{-5}\,\mu C)$$

4. *Field in a capacitor.* Since one plate of the capacitor bears a positive charge and the other a negative charge, a field will be set up between the plates. If the plates are parallel, this field will be uniform except near the edge of the plates.

5. *Energy storage.* Suppose you charge a capacitor to some potential and then disconnect it from the battery. If there is no connection between the plates, the charges will remain where the battery put them. This means that the capacitor will retain the potential difference even without the battery. In other words, the capacitor is storing electrical energy. This is analogous to the energy stored when you lift an object to a high shelf. If the plates of a charged capacitor are connected by a conductor, the charges will rush back to their original location, discharging the capacitor and releasing the stored energy. If some device like a flashbulb is placed in the connection, the energy can be converted to useful work. This, in fact, is exactly how the electronic flash used in photography works. Such an event is analogous to pushing the object off the high shelf.

6. *Dielectrics.* Suppose you charge a capacitor, disconnect it, and slip a slab of nonconductive material (say, rubber, mica, ceramic, etc.) between the plates. A voltmeter will now show a lower potential between the plates. Since there is no way for the charge to change, the material must have reduced the strength of the field between the plates. We call that material a dielectric. Consider another experiment. If a dielectric is inserted between the plates of a capacitor and it is then attached to a battery, the capacitor will again be charged to the potential of the battery, but this time, a larger charge will be moved before the full potential is reached. On the basis of these experiments, we conclude that a dielectric increases the capacitance of a capacitor. The amount of this increase will depend on the material used as the dielectric so we will define a new constant, called the dielectric constant that is characteristic of the material. The capacitance of a capacitor is now given by the equation

$$C = \varepsilon' \varepsilon_0 \frac{A}{d} \tag{15.13}$$

Table 15.2 Dielectric constants for some selected materials

Material	ε'
Vacuum	1.0
Air	1.0
Water	78.0
Paper	3.5
Mica	5.4
Titanium dioxide	100.0

where ε' = dielectric constant that is defined as the ratio:

$$\varepsilon' = \frac{\text{Charge per volt with the dielectric}}{\text{Charge per volt without the dielectric}} \qquad (15.14)$$

A list of dielectric constants for some materials is given in Table 15.2.

Example 6 *What would the capacitance in Example 5 be if the space between the plates was filled with titanium dioxide? What charge would the 9 V battery move? The capacitance is*

$$C = 100(7.1 \, \text{pF}) = 710 \, \text{pF}$$

The charge it takes is

$$Q = 100(6.4 \times 10^{-11} \, \text{C}) = 6.4 \times 10^{-9} \, \text{C}$$

7. *Molecules in a dielectric.* There are several types of dielectrics and they all behave somewhat differently. However, they shift charges in such a way that the field strength in the capacitor is reduced.
   - *Polar.* Compounds like water contain molecules that are natural dipoles. These dipole molecules will form couples with the electric field of the capacitor and rotate as shown in Figure 15.12A. The rotated dipoles will tend to reduce the strength of the capacitors field, allowing a greater charge per volt, and hence increase the capacitance of the capacitor.
   - *Non polar.* If a dielectric contains a non polar material such as oil, the electric field will distort the arrangement of charges within molecules and induce dipoles

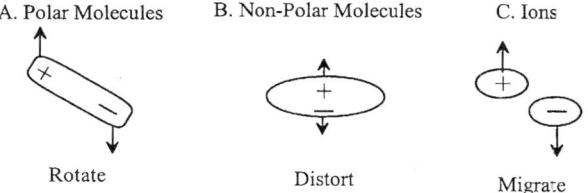

A. Polar Molecules   B. Non-Polar Molecules   C. Ions

Rotate   Distort   Migrate

Figure 15.12 Effects of fields on different types of molecules.

as shown in Figure 15.12B. The effect on the capacitor's field is the same as if already existing dipoles had been rotated.

- *Electrolyte*. If the dielectric contains electrolytes such as salts or acids, positive ions will migrate toward the negative plate of the capacitor and negative ions will migrate toward the positive plate as shown in Figure 15.12C. The field will be reduced again.

In each of these cases, there will be a restoring force trying to return the molecules or ions to their original condition, much as a stretched spring tries to return to its relaxed state. Just as the spring stores mechanical energy, the shifting of charges in a dielectric stores electrical energy. If the capacitor is discharged, this energy is released.

8. *Oscillating fields*. If a capacitor is connected to an alternating current, charges will flow first in one direction, then in the other. This will create an alternating potential between the plates and, hence, an alternating field as represented by the sine wave in Figure 15.13. This, in turn, will cause the molecules in the dielectric to rotate, distort, or migrate repeatedly one way and then the other as shown at the top of Figure 15.13. When the field is changing in one direction, energy is being stored in the dielectric. When the field changes back, that energy is released.

9. *Dielectric loss*. Circuits can be set up to cause oscillation of almost any frequency. However, there is a limit to the rate at which molecules in the dielectric can respond. In Figure 15.14, we have plotted the angle of rotation of the field versus the angle of rotation of the dipoles. Effectively, the horizontal axis represents a force and the vertical axis represents a distance so the area under any curve is proportional to the work done when a dipole is rotated. Specifically, the area under the upper curve represents the energy stored and the area under the lower curve represents the energy recovered. The area inside a closed curve is the difference

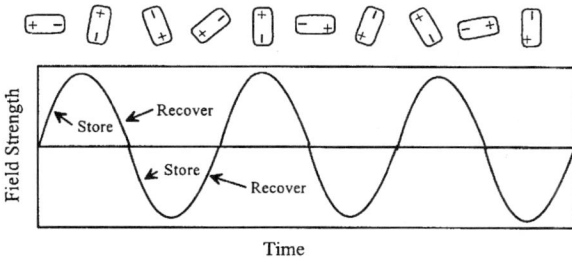

Figure 15.13 Dipoles in an oscillating field.

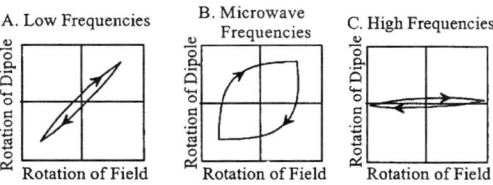

Figure 15.14 Energy loss at various frequencies.

between these two energies and, therefore, represents the energy lost as heat during one cycle. Using these graphs, we can see the relationship between frequency and energy loss.

- At very low frequencies, the molecules will have no trouble oscillating with the field as shown in Figure 15.14A. Under these circumstances, most of the energy that is stored during one half-cycle will be recovered during the next half-cycle and the area inside the curve is small.

- At very high frequencies, the molecules in the dielectric will be unable to respond to a change in the field before the change is reversed. Instead of rotating, they just quiver as shown in Figure 15.14C. Since there is less movement, little work is done on or by the dielectric and less energy is stored or released. Hence, the area inside the curve is small.

- At some intermediate frequency, the molecules will move with the field but always lag somewhat behind as illustrated in Figure 15.14B. Thus, only a portion of the work done in one half-cycle is recovered in the next half-cycle. The large area inside the curve indicates the large energy loss and this energy heats the food.

- The effects of dielectric loss on heating are given by

$$\text{Heat} = Q = 2\pi f \varepsilon_0 \varepsilon'' E^2 \tag{15.15}$$

where f = frequency of field oscillation (in Hz= cycles/s), ε_0 = permitivity of free space, ε'' = dielectric loss, E = root-mean-square (rms) field strength. Note that rms average field strength is used because the field strength is constantly changing. The rms average rather than arithmetic average is used to give correct values for power computations.

Dielectric loss is dependent on:
- Nature of the dielectric.
- Frequency of oscillations of the field.
- Temperature of the dielectric.

10. *Ohmic loss.* When the dielectric contains ions, another source of heating exists. As the field changes, the ions move through the dielectric. In effect, this is an electric current and obeys Ohm's law:

$$I = \frac{V}{R} \tag{15.16}$$

where I = electric current (flow of charges) in amperes (A) = C/s, V = potential in V, R = resistance in ohms (Ω.) But an electric current will dissipate energy at a rate of

$$P = V \cdot I \tag{15.17}$$

where P = power in W= J/s. This dissipated power will appear as heat in the same way as heat is generated when you pass an electric current through the heating elements of a toaster. We speak of this type of heating as ohmic loss or ohmic heating.

11. *Capacitive ovens.* In capacitive ovens, a conveyor belt carries food between plates of a capacitor with an oscillating field. The oscillations are timed to produce maximum hysteresis and, hence, maximum heating.

15.2.3 Electromagnetic Radiation

Ordinary mechanical energy is defined in terms of a force moving a mass through a distance. It requires that an object be moved. Electromagnetic radiation is a form of energy that propagates through a vacuum in the absence of any moving material. We observe electromagnetic radiation as light and use it as radio waves, X-rays, etc. Here, we are interested in a form of electromagnetic radiation called microwaves that can be used to heat foods. Let us see how it originates.

1. *Reversing the dipole.* If the dipole suddenly reverses itself so that the positive charge is at the bottom and the negative at the top, the field will also reverse itself (Fig. 15.15A and 15.15B). The charge placed in that field will feel a reversal of the force acting on it.

2. *Speed of light.* It turns out, however, that charge will not feel the change the instant the dipole changes. Rather, there will be a very short delay while the change propagates to that spot. The speed of this propagation will be the speed of light.

3. *Oscillation.* If the dipole continues to reverse itself over and over again, we say that the dipole is oscillating. The oscillation of the dipole will, in turn, cause the electric field to oscillate and these oscillations will propagate into space as shown in Figure 15.15C.

4. *Magnetic fields.* As the dipole reverses, charges in the dipole move from place to place and this produces a magnetic field at a right angle to the dipole. This field will also propagate into space at the speed of light, but at right angles to the electric field.

5. *Electromagnetic radiation.* The two oscillating fields, one electric and one magnetic, together form what we call electromagnetic radiation.

6. *Wavelength and frequency.* If a dipole oscillates f times per second, we say that it is oscillating at a frequency of f Hertz (Hz). The electromagnetic field produced by this dipole will also oscillate at f Hz. A charge or compass placed at a fixed location in the field will each feel a force that oscillates at f Hz. The distance in space between two oscillations is called the wavelength of the radiation. Since electromagnetic

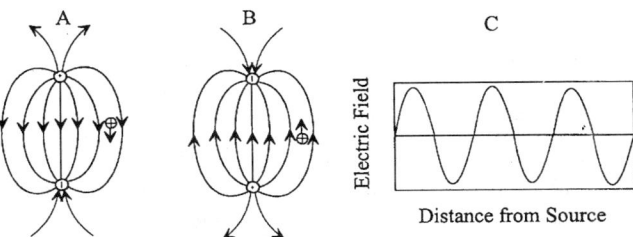

Figure 15.15 Oscillating dipole.

Table 15.3   List of various electromagnetic spectrum waves

Approximate Range		
Wavelength (m)	Frequency (Hz)	Band Name
10^8–10^3	1–10^6	Longwave radio
10^3–10^1	10^5–10^7	AM radio
10^1–1	10^7–10^8	FM radio and TV
10^2–10^{-4}	10^6–10^{12}	Shortwave radio
10^{-1}–10^{-3}	10^9–10^{11}	Microwaves
10^{-4}–4×10^{-7}	10^{12}–7×10^{14}	Infrared
4×10^{-7}–7×10^{-7}	7×10^{14}–4×10^{14}	Visible light
7×10^{-7}–10^{-9}	4×10^{14}–10^{17}	Ultraviolet
10^{-7}–10^{-12}	10^{15}–10^{20}	X-rays
10^{-10}–10^{-14}	10^{18}–10^{22}	Gamma rays

radiation propagates at the speed of light, this wavelength will equal the speed of light divided by the frequency:

$$\lambda = \frac{c}{f} \tag{15.18}$$

where λ = wavelength in m, f = the frequency in Hz (cycles/s), $c = 3 \times 10^8$ m/s = the speed of light.

Example 7   *The electromagnetic field in red light oscillates about 7.5×10^{14} times/s (Hz), whereas an AM radio wave oscillates around 10^6 Hz. What are the wavelengths?*

Solution

$$\lambda_{\text{red}} = \frac{3 \times 10^8 \text{ m/s}}{7.5 \times 10^{14}} \text{ Hz} = 4 \times 10^{-7} \text{ m} \quad (400 \text{ nm})$$

$$\lambda_{\text{AM}} = \frac{3 \times 10^8 \text{ m/s}}{10^6 \text{ Hz}} = 300 \text{ m}$$

7. *Electromagnetic spectrum*. Electromagnetic radiation can exist in a wide range of frequencies, each with its own character. This spectrum is divided into bands according to the use we make of it as shown in Table 15.3.
8. *Microwave wavelength*. Microwave lengths range from roughly 1 mm to around 30 cm or between 100 to 1000 MHz. This range is useful because it rotates dipoles in food with a slight lag and, therefore, induces heating. At lower frequencies, these dipoles would show little lag and, hence, little heating. At higher frequencies, these dipoles would not be able to respond to the changing field. For home cooking, the frequency used is 2450 MHz. This corresponds to a wavelength of 12.2 cm.

15.2.4 Microwave Oven

1. The microwave oven consists of three major parts.
   - The magnetron is the device that generates the microwaves.
   - Wave guides direct these waves to the oven cavity.
   - The oven cavity holds the food to be cooked so that microwaves can impinge on them.
2. *Magnetron.* It generates microwaves and consists of the following parts:
   - *Central cathode.* The cathode is a metal cylinder at the center of the magnetron that is coated with an electron-emitting material. In operation, the cathode is heated to a temperature high enough to cause electrons to boil off the coating.
   - *Outer anode.* There is a metal ring called an anode around the magnetron that is maintained at a large positive potential (voltage) relative to the cathode. This sets up an electrostatic field between the cathode and anode (Fig. 15.16A) that accelerates the electrons toward the anode.
3. *Magnetic field.* A strong magnetic field is placed next to the anode and cathode in such an orientation that it produces a magnetic field at right angles to the electrostatic field (Fig. 15.16B). This field has the effect of bending the path of the electrons so that, instead of rushing to the anode, they begin to circle in the space between the cathode and anode in a high-energy swarm.
4. *Resonant cavities.* They have been built into the anode. Random noise in the electron swarm causes occasional electrons to strike these cavities and set up electromagnetic oscillations. The dimensions of the cavities are such that most radiation frequencies die out. Microwave frequencies, on the other hand, bounce around the cavities and tend to grow, thus getting their energy from electrons captured from the swarm. Some of this radiation emerges from the magnetron, passes through the wave guides, and enters the oven cavity.
5. *Standing waves.* A wave is a series of peaks and valleys, frequently sinusoidal in form.
   - In a traveling wave, the peaks move through the medium (Fig. 15.17A). The waves that propagate outward when a pebble is thrown into a pond are traveling waves.

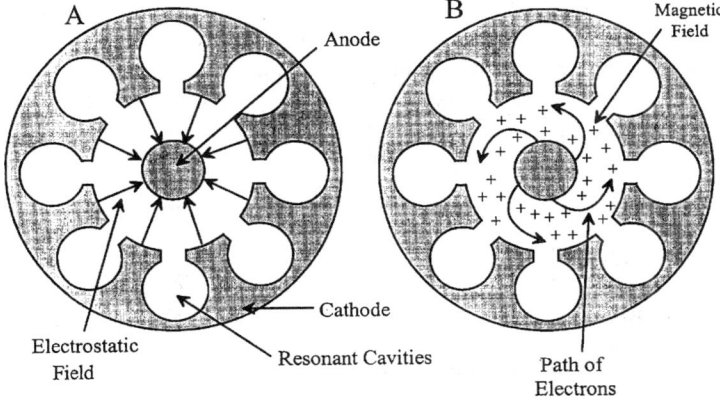

Figure 15.16 Magnetron.

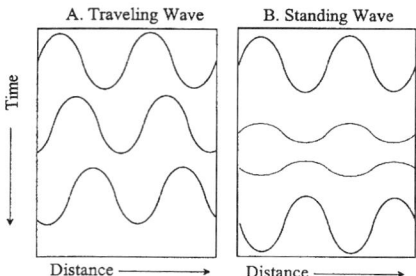

Figure 15.17 Traveling and standing waves.

- If traveling waves strike a nontransmitting location such as the shore of the pond, they are reflected back. You can easily see wave reflection by tieing a rope to a wall or other fixed object, stretching it out to nearly full length, and flipping the free end once up and down. A wave will travel down the rope from your hand to the wall, then reflect back along the rope toward your hand.

- If you move your hand continuously up and down, a series of traveling waves will move down the rope and reflect back. The reflected waves will meet the newly arrived waves and combine with them. At certain locations, the incoming and outgoing waves tend to cancel and the rope does not move. At some locations, where the incoming and outgoing waves reinforce each other, the rope moves a great deal. At some locations, the resulting waves appear to stand still, they are called standing waves (Fig. 15.17B).

- The points on the rope that remain stationary are called nodes, whereas the moving lengths in between are called internodes. All the energy that your hand is putting into the rope has become concentrated at the internodes with none in the nodes.

- Strings of guitars vibrate in this manner. If you pluck a guitar at the midpoint of a string, the string will tend to vibrate as one big standing wave. This will produce a single fundamental tone and mellow sound. If you pluck the same string near one end, many nodes will be set up that produce higher overtones and a brighter sound. Flamenco guitarists strum for these brighter tones.

6. *Wave stirrers.* In the oven cavity, microwaves bounce back and forth and form standing waves with low-energy nodes and high-energy internodes. This results in uneven heating of your food. To avoid this, a wave stirrer is frequently placed near the location where the wave guide opens into the oven cavity. This is usually an irregularly shaped piece of metal that is slowly rotated so it reflects microwaves in different directions. This moves nodes from place to place within the oven cavity to produce a more even distribution of energy.

15.3 LAB EXERCISE

15.3.1 Objectives

The main objectives of this lab exercise are to:

1. Become familiar with microwave oven components, operation, and temperature measurements.

2. Observe the effect of microwave power on the heating rate, power absorbed, and coupling efficiency.

3. Observe the effect of differences in water, oil, and salt content on the heating rate and power absorption.

4. Observe the effect of microwaving foods that are spherical in shape.

15.3.2 Materials

1. Research-grade microwave oven or any other suitable microwave oven.

2. Fiber optic probes and signal conditioning unit for temperature measurements.

3. Microwave containers and materials to be heated.

15.3.3 Operating Instructions

1. Place a suitable load (food or other material) in the oven and close the door.

2. Insert one or more temperature probes into the sample so that the tip is at the location where the temperature is to be measured.

3. Turn the power switch on.

4. Toggle the microwave selector switch.

5. Set the microwave percent power control to the desired position.

6. Process the food for the desired time.

7. Make sure that microwave power is stopped when the door is opened.

8. Toggle the turntable switch if desired.

9. Push the start button and observe the "process time" and "microwave power" indicators.

10. You can open the door before time-out if you need to temporarily stop microwave power, the turntable, and the timing, to examine oven load.

11. Push the start button to restart the "process timer" and "microwave power".

12. At time-out, a tone sounds for a few seconds and microwave power is turned off.

15.3.4 Data Logging

Temperature data is obtained by placing the tip of a probe at the location to be measured. Measure and record the temperature either manually or use a computer program to enter data in an electronic file.

15.3.5 Experimental Procedures

1. Following the operating instructions, heat 150 g of distilled water at 50 and 80% of full power. Be sure to observe safety precautions when using a microwave oven. Start from the same initial temperature, place thermocouple in the same location, and heat to at least 80°C.

2. Repeat this procedure with distilled water plus 5% NaCl.

3. Repeat the same procedure for vegetable oil.

4. Determine the density of each material heated or obtain a literature value.

5. Heat 150 g of water and 150 g of vegetable oil simultaneously (in separate containers) at 50 and 80% power.

6. Place the temperature probes in a potato, say, 2 in. in diameter near the surface and at the center. Measure the weight and approximate diameter of the potato. Heat at 50 and 80% power for about 180 s.

7. Experiment with the commercial products obtained from the grocery store. Cook with and without susceptors and note any differences.

8. Record the data in Data Sheets 15.1 through 15.3.

15.3.6 Results

1. For each heating run of water, oil, and salt solutions:
   a. Plot temperature versus time data for each run.
   b. From this plot, determine and tabulate the rate of heating as $\dfrac{\Delta T}{\Delta t}$ in °C/s.
   c. Compute the power coupled as $Q = \rho C_p \dfrac{\Delta T}{\Delta t}$ when W/cm^3.
   d. Compute the percent efficiency of power coupled as $\left(\dfrac{Q \times \text{Volume}}{\text{Oven wattage}}\right) \times 100$.

2. Plot the time-temperature data for the water and oil heated simultaneously and determine the rate of heating and power control for both power levels.

3. Estimate the electric field strength E from the experimental data as

$$E = \left[\frac{\rho C_p (dT/dt)}{2\pi f \varepsilon_0 \varepsilon''}\right]^{1/2}$$

Determine the apparent field strengths for 5% salt solutions and pure water for the 50 and 80% power conditions. It appears that the loss factor for salt solutions is fairly insensitive to temperature so we will use the following values for 2450 MHz at an average temperature from 25 to 100°C.

Salt Concentration (%)	ε''
0	5.8
1	23.6
5	71.1

4. For the potato experiment, plot the time-temperature data for the different locations on the same graph.

15.3.7 Discussion

Discuss your results including the effect of power level on the rate of heating and power coupled for water, vegetable oil, and salt solutions. Discuss the relationship between the calculated electric field strength and heating rates for the salt solutions versus water. In the case of oil and water heated simultaneously, how does this compare to the case where they are heated individually? Discuss the temperature gradients that exist in the solid potato as compared to what you would normally expect when heated by conduction. What happens to the temperature gradient over time? Compare these results to what you think happens during the oven-baking of potatoes. Discuss the quality and overall performance of the commercial microwave products that you tested.

15.4 SUGGESTED READINGS AND REFERENCES

1. A. K. Datta. "Heat and mass transfer in the microwave processing of food." *Chem. Engr. Prog.*, 86(6): 47 (1990).
2. T. Ohlsson. "Fundamentals of microwave cooking." *Microwave World* 4(2): 4, (1983).
3. C. R. Buffler, 1993. *Microwave Cooking and Processing: Engineering Fundamental for the Food Scientist*, New York: Van Nostrand Reinhold.

DATA SHEET 15.1

Time-temperature data during microwave heating of water, 5% salt solution, and vegetable oils

Time (s)	Water Temperature at 50% Power Level (°C)	Water Temperature at 80% Power Level (°C)	Temperature of 5% Salt Solution at 50% Power Level (°C)	Temperature of 5% Salt Solution at 80% Power Level (°C)	Oil temperature at 50% Power Level (°C)	Oil temperature at 80% Power Level (°C)
0						
15						
30						
45						
60						
75						
90						
105						
120						
135						
150						
165						
180						

DATA SHEET 15.2

Time-temperature data during simultaneous microwave heating of water and vegetable oil in separate containers

Time (s)	Water Temperature at 50% Power Level (°C)	Water Temperature at 80% Power Level (°C)	Oil temperature at 50% Power Level (°C)	Oil temperature at 80% Power Level (°C)
0				
15				
30				
45				
60				
75				
90				
105				
120				
135				
150				
165				
180				

DATA SHEET 15.3

Time-temperature data during microwave heating of a potato

Time (s)	Surface Temperature at 50% Power Level (°C)	Center Temperature at 50% Power Level (°C)	Surface Temperature at 80% Power Level (°C)	Center Temperature at 80% Power Level (°C)
0				
15				
30				
45				
60				
75				
90				
105				
120				
135				
150				
165				
180				

16

FRYING OF FOODS

16.1 BACKGROUND

Frying is considered to be one of the oldest cooking methods to create unique flavors and texture in processed foods. The immersion frying process, also called deep-fat frying, involves chemical and physical changes in foods including starch gelatinization, protein denaturation, water vaporization, and crust formation. Various factors such as the rate of heating, oil penetration into the food, oil–food interactions, and oil degradation affect the texture and final product quality.

16.1.1 Heat Transfer Stages in Frying

Two distinct types of heat transfer take place during the frying process: conduction and convection. Conductive heat transfer in an unsteady state occurs within the solid food, which is influenced by the thermal properties of the food including thermal diffusivity, thermal conductivity, specific heat and density. Convective heat transfer occurs between a solid food and the surrounding oil. The surface interactions between the oil and food material are influenced by the vigorous movement of water vapors escaping from the food into the oil as shown in Figure 16.1. These escaping vapor bubbles cause considerable turbulence in the oil, preventing efficient heat transfer. The amount of escaping water vapor bubbles decreases with longer frying times as the moisture remaining in the food material decreases. Also, any solutes present may elevate the boiling point above that of water. As the frying process proceeds, more water evaporates from the outer region of the food and, consequently, its temperature begins to rise above the boiling point. Based on the above observations, the frying process can be divided into the following four stages.

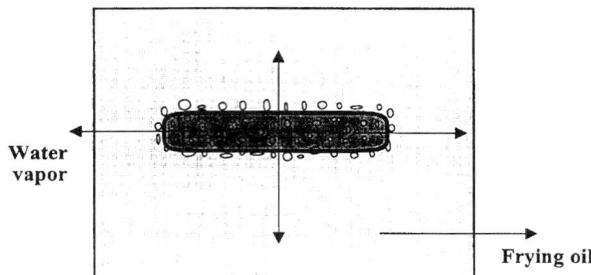

Figure 16.1 French fry during frying process. Water vapor bubbles are expelled from the food.

16.1.1.1 Initial Heating During the initial heating stage, the surface of a food submerged in oil heats to a temperature equivalent to the elevated boiling point of the oil. The mode of heat transfer between the oil and food occurs by natural convection and no vaporization of water occurs from the food surface.

16.1.1.2 Surface Boiling During this stage, the vaporization of water begins from the food surface. The mode of heat transfer changes from natural convection to forced convection due to turbulence in the oil surrounding the food.

16.1.1.3 Falling Rate In this stage of the frying process, more moisture leaves the food, and the internal core temperature rises to the boiling point. Some physico-chemical changes such as starch gelatinization and protein denaturation take place in the internal core region. The surface crust layer continues to increase in thickness, and the vapor transfer rate continues to decrease at the surface.

16.1.1.4 Bubble End Point This stage is observed if the frying is continued for a long period of time. The rate of moisture removal diminishes and no more bubbles are seen escaping from the food surface. As the frying process continues, the crust layer thickness continues to increase.

16.1.2 Mass Transfer in Frying

As the frying process proceeds, several important changes take place in the surrounding oil. The viscosity increases, surface tension decreases, fat oxidizes, and interactions take place between the oil, water, and other food materials. Mass transfer during frying is described by the movement of oil into the food and the movement of water in vapor form from the food into the oil. Various semiempirical models have been proposed to illustrate water loss and oil gain. In some models, moisture content has been found to be proportional to the square root of frying time.

   Oil content in the product is independent of the frying temperature, but closely related to moisture content, frying time, oil type, and oil quality. In the case of frying potato chips, oil uptake is proportional to the potato chip surface area and its thickness. Final oil content decreases linearly with increased thickness. During the frying process, the oil is contaminated with food components leaching into the oil, water vapor condensation into oil, thermal breakdown of oil, and oxygen absorption at the oil–air interface.

16.1.3 Factors Affecting Oil Uptake in Fried Foods

Various factors have been reported to affect the oil uptake in foods. Some of the main factors are as follows:

16.1.3.1 Oil Quality and Composition   Fat uptake and oil deterioration increase with frying time. Surface active agents produced by oxidation reduce the oil surface tension of the immiscible materials and affect the heat transfer at the oil–fat interface. These surfactants are produced to a greater degree in degraded oils than fresh oil.

16.1.3.2 Frying Temperature and Duration   Generally, it has been observed that increasing the frying temperature decreases the oil uptake. However, in many cases, increasing the frying temperature is not always beneficial, as frying time is independent of frying temperature in the range of 160 to 200°C. A higher-surface-to-mass ratio increases oil absorption. Surface roughness that increases the surface area also increases the oil uptake.

16.1.3.3 Moisture Content   Generally, oil uptake occurs as moisture is removed from the food during frying. A higher initial moisture content results in higher fat uptake such as in potato chips. In some other products, such as donuts and French fries, a higher initial moisture results in lower fat uptake. Retention of water by some food additives such as alginates and cellulose affects the oil uptake and moisture loss in food.

16.1.3.4 Composition   High initial fat content in some foods increases fat uptake. However, it is not the case in some meat and fish products. The addition of soy protein to cake donut reduces fat uptake. Pretreatment of potatoes in hot solutions of sodium chloride or calcium chloride has been reported to reduce oil uptake.

16.1.3.5 Prefrying Treatment   Blanching or lowering the moisture level of food prior to frying decreases oil absorption. Prewashing the product with oil containing emulsifier reduces oil uptake. Freezing prior to frying seems to reduce fat uptake in French frying. The predrying of a product reduces the fat uptake, whereas freeze drying increases the fat uptake.

16.1.3.6 Surface Treatments and Interfacial Tension   Surface coating with a hydrocolloid inhibits oil uptake during frying. Batter and breading influences fat absorption. Batter-coated fishsticks absorb less oil than uncoated fishsticks. The batter coating apparently reduces water loss during frying, which in turn reduces fat uptake.

Oil uptake increases markedly when the initial surface tension is lower, suggesting that a hydrophobic surface would increase oil uptake during deep-fat frying. Similarly, treatments or additives that increase interfacial tensions should reduce oil uptake.

16.1.3.7 Gel Strength and Crust   In structured foods, gel strength is important to control oil uptake. Oil uptake is reduced with increased gel strength, perhaps by the creation of a barrier and hindrance to water movement to the surface by evaporation. Crust is formed during the deep-fat frying process and its development influences heat transfer and oil uptake. Oil uptake during deep-fat frying is localized in the crust. Crust thickness is higher when gel strength is low compared with the case when gel strength is high. When crust is fully developed and quite dry, other factors such as porosity, tortuosity, and permeability play a significant role in oil uptake.

16.1.3.8 Porosity Initial product porosity is directly related to the oil uptake level in deep-fat fried foods. Although porosity increases with frying time, oil uptake is significant only during the initial stages of frying. Product porosity can be controlled by changing product formulation such as the addition of natural ingredients that form a film around the product surface.

16.2 LAB EXERCISE

16.2.1 Objectives

The main objectives of this lab exercise are to:

1. Evaluate the effect of frying temperature and time on the surface and center product temperature.
2. Estimate the moisture loss and fat absorption during the process.
3. Study the changes in physical, textural, and sensory quality of the fried product.

16.2.2 Materials

1. A selected variety of potatoes suitable for French fries such as Russet burbank, Russet rural, Sebago, Kennebec, and red potatoes. Typical fresh potatoes consist of 75 to 80% moisture, 0 to 1% fat, 13 to 18% starch, 1 to 4% sugar, and 1 to 2% protein.
2. Four to five thermocouples.
3. An infrared thermometer or surface thermocouple to measure the product surface temperature.
4. A data logger to record temperature at preselected time intervals.
5. A Hunter-Lab colorimeter to measure color parameters L, a, and b of the fried product.
6. A penetrometer or other means to measure product texture.
7. An electronic weighing balance.
8. Hot air oven.
9. Aluminum dishes.
10. An oil bath or French fryer.
11. A wire-mesh-type basket to fry the product.

16.2.3 Procedure

1. Wash and peel the potatoes. Cut them into a suitable size of approximately $10\,mm \times 10\,mm \times 45\,mm$.

2. Pour the vegetable oil or other frying oil into the French fryer and raise its temperature to the desired levels.

3. Select three oil bath temperatures, for example, 180, 190, and 200°C, to study the effect of oil temperature on French fry quality.

4. Insert thermocouples into the center of four to five potato pieces for each cooking load of 30 to 40 pieces.

5. Measure the center temperature at 60 s intervals and report the results in Data Sheets 16.1 through 16.3 at different oil temperatures. Take the average temperature of four to five samples of French fries.

6. Withdraw the French fry sample at an interval of every 60 s to determine their color, texture, and total moisture content.

7. Record the center temperature, percent moisture loss in Data Sheets 16.1 through 16.3.

8. If chemical analysis facilities are available, the percentage of fat uptake data may also be obtained. However, it is optional.

16.2.3.1 *Moisture Content*

1. Weigh predried empty aluminum dishes.

2. Periodically collect the French fry sample. Grind or homogenize or mesh the sample thoroughly for uniform distribution in aluminum dishes.

3. Place the ground sample on the dish.

4. Weigh them to note the initial sample weight.

5. Put the aluminum dishes in a forced air dry oven or conventional dry oven at 75°C for 24 h.

6. Put the dishes in a desiccator and weigh them again until there is no change in weight.

7. Determine the percentage of moisture content gravimetrically by subtracting the weight of total dried materials from the initial sample weight.

16.2.3.2 *Fat Content*

1. Take about 5 g of ground sample for fat analysis.

2. Place the sample in a Soxhelt flask extraction thimble.

3. Extract the sample with 125 to 150 mL of petroleum ether, for 6 to 8 h.

4. Cool and evaporate the flask content.

5. Dry the flask at 75 to 80°C until a constant weight is reached.

6. Determine the fat content gravimetrically.

7. Assume 0 fat content in the raw potato samples before frying.

16.2.3.3   Color Measurement

1. Turn on the lab calorimeter or Macbeth color eye.
2. Calibrate the instruments per the manufacturer's instruction manual.
3. Place the French fry sample wrapped in transparent plastic film at the view port and close the sample holder.
4. Follow the instructions per the instrument's manual and record the L, a, and b values of the color parameters.

16.2.4   Report

1. Plot the product sample center temperature versus time for each oil temperature. Discuss how it remains essentially constant irrespective of different oil bath temperatures.
2. Discuss why the center temperature of the French fry samples does not exceed more than approximately 103°C.
3. Plot the percentages of moisture loss and fat uptake with time at different oil bath temperatures.
4. Discuss the results and interpret the mechanism of the percentage of fat uptake. How does it relate to the percentage of moisture loss?
5. Plot surface color parameters values L, a, and b with time. How do they differ at different oil temperatures?
6. Discuss the product texture parameter with frying time at different oil temperatures. Does it change with respect to different oil temperatures?
7. How would the percentage of fat uptake vary if three types of oil with high, medium, and low viscosity are used for French fries? (See the references given below)

16.3   SUGGESTED READINGS AND REFERENCES

1. P. Fellows, 1988. "Frying." In *Food Process Technology, Principles and Practice*, Chichester, UK: Ellis Horwood Ltd.
2. R. P. Singh. "Heat and mass transfer in foods during deep fat frying." *Food Technol.* 49: 134 (1995).
3. M. M. Blumenthal. "A new look at the chemistry and physics of deep fat frying." *Food Technol.* 45: 68 (1991).

DATA SHEET 16.1

Type of vegetable oil: _____

Oil temperature: 180°C

Dry solids in potato: _____

Time (s)	Average Center Temperature	Total Moisture Content % (wet basis)	Fat Uptake % (wet basis)	Color Parameter (L value)	Texture (penetrometer reading)
0					
60					
120					
180					
240					
300					
360					
420					

DATA SHEET 16.2

Type of vegetable oil: _____

Oil temperature: 190°C

Dry solids in potato: _____

Time (s)	Average Center Temperature	Total Moisture Content % (wet basis)	Fat Uptake % (wet basis)	Color Parameter (*L* value)	Texture (penetrometer reading)
0					
60					
120					
180					
240					
300					
360					
420					

DATA SHEET 16.3

Type of vegetable oil: _____

Oil temperature: 200°C

Dry solids in potato: _____

Time (s)	Average Center Temperature	Total Moisture Content % (wet basis)	Fat Uptake % (wet basis)	Color Parameter (L value)	Texture (penetrometer reading)
0					
60					
120					
180					
240					
300					
360					
420					

17

EXTRUSION COOKING
OF FOODS

17.1 BACKGROUND

Extrusion is one of the most versatile operations available to the food industry for transforming (either cooking or forming) ingredients into intermediate or finished products. Extrusion cooking is generally regarded as a high-temperature short-time (HTST) process capable of generating high temperatures up to $180°C$ or so, high pressures up to 2000 psi, and relatively high shear rates of 10 to 200 s^{-1}. The residence times in extruders are of the order of 5 s to 3 min, which are very short as compared with conventional cooking processes.

Extrusion cooking is a process by which starchy and/or proteinaceous biopolymer materials are plasticized with added water and cooked with a fairly high degree of mechanical shear. A cooking extruder embodies a number of unit operations, for example, microbial destruction, denaturation of protein and/or enzymes, gelatinization of starch, polymerization or depolymerization of proteins, and ultimately texturization and shaping of the end product into a desirable form. A wide range of thermomechanical treatments are possible because of the control of important process variables like screw speed, screw profile, temperature, moisture, feed rate, and die size/shape. Extrusion cooking is widely used in the food industry for the processing of cereals and grains into snacks and ready-to-eat breakfast cereals, whereas sugars or sugar/cereal blends are also extruded to produce confectionery products. Basically, any screw extrusion process involves feeding and conveying of the raw materials through a particular screw configuration that rotates in a tightly fitting cylindrical barrel. Extrusion processing causes changes in the physical and chemical properties of the raw materials, which in turn determines the quality attributes of the food product.

17.1.1 Single- and Twin-screw Extruders

There are basically two types of continuous screw extruders used in the food and petfood business; single-screw and twin-screw. A schematic cutway view of both the single- and twin-screw extruder is shown in Figure 17.1.

17.1.1.1 Single-screw Extruders In a single-screw extruder (SSE) the only force that keeps the material rotating with the screw and advancing it ahead is its friction against the inner barrel surface. This fact tends to limit the formulations that can be extruded with a SSE. High-moisture and high-fat formulations may be difficult to extrude with a SSE.

The flow in a SSE is a combination of drag and pressure flow. Drag flow results from viscous drag and is proportional to screw speed. Pressure flow in the reverse direction is caused by the higher pressure at the die end of the extruder. The mixing of ingredients within the single-screw channel is also limited because laminar flow conditions generally exist. These limitations can be improved by the use of grooved inner barrel surfaces and a cut-flight-type screw design. Increased back pressure behind the die can also help to improve mixing performance. SSEs seem to be the extruder of choice for producing fish feeds for aquaculture, petfoods, and pasta. This is probably largely a matter of economics.

17.1.1.2 Twin-screw Extruders Twin-screw extruders, as the name suggest, consist of two intermeshing screws. They can be divided into co-rotating and counter-rotating types, depending on the the relative direction of the screw rotation. Generally, counterrotating screws act like a positive displacement pump due to a closed C-shaped chamber formed by the two screws, which progresses from the feed to die end and prevents material leakage from one screw to another. However, this also reduces the extent of

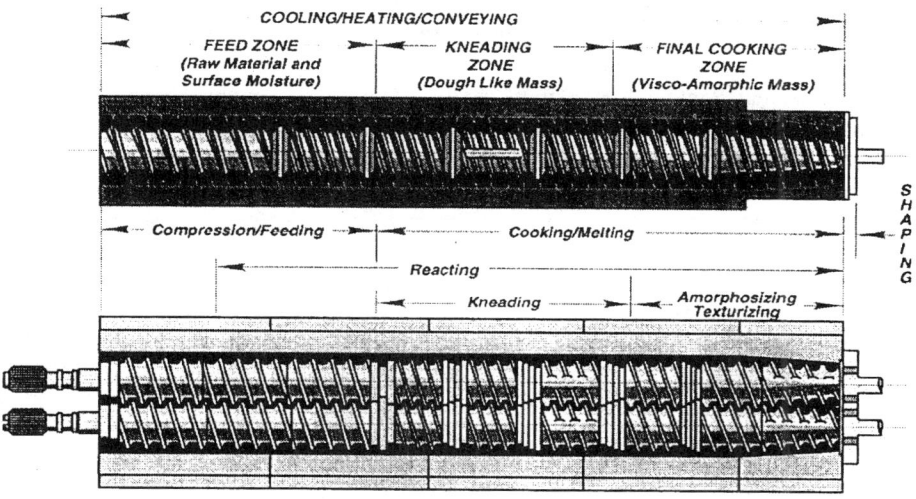

Figure 17.1 Cutway views of a single- and twin-screw extruder (courtesy of Wenger Manufacturing, Inc., Sabetha, KS).

mixing and narrows the residence time distribution so that it approaches plug flow. The intermeshing of screws and their pressure profile also dictate fairly low screw speeds and correspondingly low shear rates. Therefore, these extruders are used for thermally sensitive materials that require uniform processing at low overall shear rates and with a narrow residence time distribution. They are especially useful for running low-viscosity materials, slurries, or rapidly solubilizing sugars and gums in which high pressure is useful. They are often cited as the machine of choice for "bioreactor" types of processes and are used for the production of licorice.

In co-rotating twin-screw extruders, closed chambers are not formed and the combined screw flights produce passages that allow material to move from one screw channel to another. Hence, tangential pressure does not build up, and when the pressure is high for one screw, it is low for the other. Thus, co-rotating screws form axially open channels and allow material exchange lengthwise in the machine. There are no localized pressure points and small clearances between the screws give them a self-cleaning action. They have a lower degree of positive conveying action, but better mixing capabilities. The co-rotating screws can be operated at higher speeds compared to counterrotating screw extruders and are therefore suitable for high shear extrusion cooking processes. Overall, co-rotating extruders are considered to be more useful for manufacturers producing a variety of products. Their good mixing characteristics, high screw speeds, and acceptable throughput rates have made twin-screw extruders a popular choice for both the snack and breakfast food industries.

17.1.2   Description of Extrusion Process

An extruder system, whether single- or twin-screw, is made up of several subcomponents as shown in Figure 17.2. A bin provides a buffer for raw material at the inlet so that the extruder can operate continuously and without interruption. A variable-speed feeding screw is used to continuously and uniformly discharge material from the bin and feed it to the extruder. A preconditioning cylinder is sometimes used to preblend steam and/or water with the raw feed. Ideally, the retention time here is sufficient so that each cereal particle achieves temperature and moisture equilibrium. The extruder barrel itself consists of jacketed heads that contain the rotating screws. The heads can be heated by electric cartridge, steam, hot water, or thermal oil or can be cooled by water or cooling media.

The various operations of heating, cooling, conveying, feeding, compressing, reacting, mixing, homogenizing, melting, cooking, texturing, and shaping are carried out in the different processing zones of an extruder. The main components of an extruder are as follows.

17.1.2.1   Feeding Zone   In this area, low-bulk-density raw materials are introduced into the extruder barrel. The overall feed rate is limited by the ability of the screws in this section to transport the dry feed. Water is typically injected downstream of the feeding zone to hydrate biopolymers and possibly to enhance conductive heat transfer, if barrel heating is being used.

17.1.2.2   Kneading Zone   In the kneading zone, compression continues, and the screws of the extruder start to achieve a higher degree to fill as their screw pitch decreases. The raw material loses its granular identity texture, and its density begins to increase as does the pressure inside the barrel. Shearing begins to play a role as the screws fill. Stream

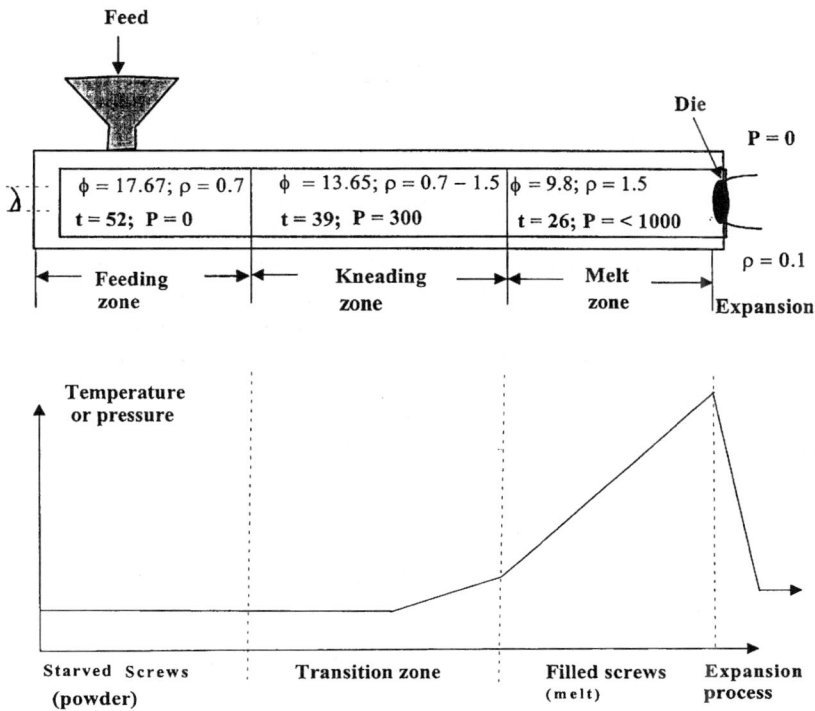

Figure 17.2 Representative direct expansion process in an extruder (helix angle ϕ in degrees, material density ρ in g/cm^3; screw lead t in mm; barrel pressure P in psig).

may also be injected to help with cooking. The kneading zone is basically a transition zone between raw particulate material and the homogenous viscoelastic material found in the melt zone.

17.1.2.3 Final Cooking Zone

In this area, temperature and pressure typically increase most rapidly due to the presence of the die and small pitch of the screw. The final transformation of the raw material also occurs here, which largely influences the density, color, and functional properties of the final product.

17.1.2.4 Die

The product is actually formed (shaped) by passing through a die and possibly a cutting mechanism attached directly to the extruder. Additional postextrusion devices can be used to establish the final product length and shape. The die also plays an important role in puffed products by promoting a sudden flash-off of moisture as steam. Important product properties of puffed products are bulk and particle density, texture, and porosity. This type of puffing process will be the subject of the following lab exercise.

17.2 LAB EXERCISE

One of the earliest cooking extrusion processes was for the expanded corn snack. This process remains commercially important today and will be the subject of this lab exercise. An interesting aspect of extrusion cooking of expanded snacks is that the corn meal is converted into a homogenous viscoelastic material (often referred to as a melt, indicative of a polymeric material in a fluid state) within the extruder at moisture content of only around 16 to 17%. The energy for this low-moisture cooking process is essentially the viscous dissipation of the motor power used to turn the screws.

A typical extrusion process is described schematically in Figure 17.2. Corn meal (10%) moisture is fed into the extruder and mixed with added water so that the in-barrel moisture content is about 16 to 17%. The bulk density of the feed is about 0.7 g/cm^3. As the material is transported down the extruder, the volume of the screws decreases so that the moistened corn meal is compressed and eventually gives rise to the so-called filled section behind the die as shown in Figure 17.2. The material density increases to about 1.5 g/cm^3 in this filled section. The entrance to the die represents the highest temperature and pressure in the process. A typical melt temperature at this point would be about 180°C, whereas the pressure depends on the die geometry, throughput rate, and rheological properties of the melt. However, a typical value might be between 300 to 600 psi for a small extruder. As discussed earlier, much of the temperature increase comes from dissipation of the frictional forces due to the shear stresses between the inner barrel surface and the melt within the filled section. The geometry of the screws and inner barrel surface of the figure-8-shaped twin-screw extruders makes the calculation of the actual power requirements for a particular process difficult. However, for a SSE, Crawford (1987) gives the power requirement as shown in Eq. (17.1):

$$P(W) = \tau \cdot N \cdot L \cdot \pi^2 \cdot D^2 \qquad (17.1)$$

where P = motor power (W), τ = shear stress (N/m^2), N = screw speed (1/s), L = length of filled section (m), D = screw diameter (m).

Thus for a particular machine and if we assume constant material properties, power is proportional to the screw speed, shear stress, and length of the filled section. Given the complex nature of extrusion cooking, a general system analysis approach is often used in place of explicit mathematical models to describe and understand the process. This will be demonstrated here.

17.2.1 System Analysis

There are several independent process variables in a typical cooking extrusion process. Generally, they are total throughput, screw speed, barrel temperature profile, and in-barrel moisture content. For a typical expanded snack food extrusion process, the moisture content will be around 16% and the screw speed about 350 rpm. The throughput rate will generally depend on the size of the machine, whereas the barrel temperature profile plays a secondary role in this type of process, since essentially all of the temperature increase is due to viscous dissipation.

Given this scenario, one can begin to appreciate the important role of moisture content in such an extrusion process.

17.2.1.1 Role of Moisture Content All other factors being equal, the moisture content has a profound effect on the shear stresses associated with a particular process and hence the motor power dissipated, which in turn directly determines the melt temperature at the die. The melt temperature then determines the driving force for expansion, as the melt is depressurized on its way through the die. It is generally assumed that the driving force for expansion is $(T_M - 100°C)$, where T_M is the melt temperature. In other words, the amount of moisture that will "flash off" to steam is that required to lower the melt temperature to 100°C, the boiling point of water at 1-atm pressure. The volume increase is due to the expansion of liquid water at T_M to steam at 100°C, which causes the melt to expand or "puff."

Another critical factor also determining final product density and texture is the material properties of the cereal melt itself, which are a function of both temperature and moisture content. For a biopolymer, such as starch, the material properties are related to an underlying glass transition temperature. Thus, while the role of the extruder can be viewed as providing a viscoelastic cereal melt to the die at a particular temperature, an understanding of the postextrusion puffing process requires that the thermodynamic driving force for expansion be considered together with the underlying material properties. This can be accomplished with a state diagram such as shown in Figure 17.3.

Figure 17.3 depicts how the glass transition temperature T_g of starch, in general, varies with the moisture content in the range relevant to the production of puffed snacks. It is clear that T_g decreases as the moisture content increases, and in particular drops below 100°C at about 11% moisture content. Since T_g represents the transition between a brittle solid (in this case, a corn puff) and a flowable cereal melt, clearly the expansion process itself can only occur for temperatures greater than T_g. An ideal extrusion puffing process might be considered one in which the thermodynamic driving force for expansion $(T_M - 100°C)$ lies above the underlying T_g versus moisture content curve, but not so far above that the puffed extrudate collapses and densifies upon cooling to give an overly hard texture. This apparent zone of solidification is identified as a critical material stiffness zone in Figure 17.3. Such an optimal process might be identified by overlaying experimental process data on the underlying materials property data conducted at different moisture contents and endpoints of the puffing process. These endpoints will be denoted

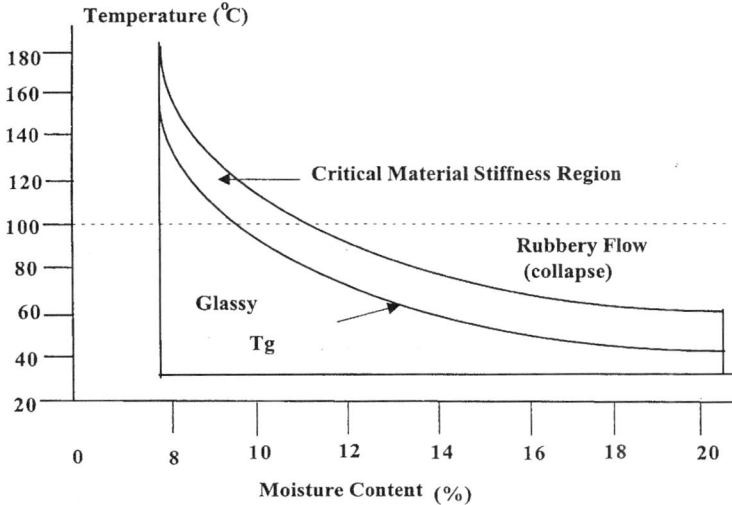

Figure 17.3 Schematic state diagram showing the expected relationship between a glass transition temperature and critical moisture stiffness to prevent the "collapse" phenomenon in puffed products.

by the final moisture content (at the end of puffing) and a temperature of 100°C. The final moisture content can be estimated as in Eq. (17.2):

$$M_s = \frac{C_p(T_M - 100°\text{C})}{\Delta H_{fg}} \tag{17.2}$$

where M_s = kg water lost/kg extrudate, C_p = specific heat of extrudate (kJ/kg °C), T_M = melt temperature at die, ΔH_{fg} = latent heat of vaporization of water at 100°C (2257 kJ/kg). C_p can be determined for each initial moisture content as

$$C_p = C_{pw} M + C_{ps}(1 - M) \tag{17.3}$$

where C_{pw} = specific heat of water (4.2 kJ/kg°C), C_{ps} = specific heat of cereal solids (1.2 kJ/kg °C), M = moisture content (weight fraction).

Example Calculate the C_p of a 16% moisture content cereal melt.

If we substitute the values in Eq. (17.3) it becomes

$$C_p = 4.2 \text{ kJ/kg°C}(0.16) + 1.2 \text{ kJ/kg°C}(1 - 0.16) = 1.68 \text{ kJ/kg°C}$$

If the melt temperature of this material was 180°C, then the estimated amount of water flashed off to steam would be

$$M_S = \text{kg water lost/kg extrudate}$$

Substituting values of C_p and latent heat of fusion into the following equation, we get

$$\begin{aligned} M_s &= \frac{C_p(T_M - 100°\text{C})}{\Delta H_{fg}} \\ &= [1.68 \text{ kJ/kg °C}(180 - 100°\text{C})]/2257 \text{ kJ/kg} \\ &= 0.0595 \text{ kg water lost/kg extrudate} \end{aligned} \tag{17.4}$$

Thus, the final moisture content at 100°C on a 1 kg of initial extrudate basis would be

$$\begin{aligned} M_f &= \left(\frac{\text{Initial moisture} - \text{Moisture lost}}{\text{Final weight}}\right) \times 100 \\ &= \left(\frac{0.16 \text{ kg} - 0.06 \text{ kg}}{0.94 \text{ kg}}\right) \times 100 = 10.7\% \end{aligned} \tag{17.5}$$

It is generally known that an in-barrel moisture content of around 16% is optimal for this process. Therefore, you will evaluate extrusion processes at 14, 17, and 20% moisture contents to determine why this is the case.

17.2.2 Objectives

The main objectives of this lab exercise are to:

1. Determine the effect of moisture content on specific mechanical energy (SME), product temperature T_m, and pressure for a typical expanded snack extrusion process.
2. Evaluate the correlation between moisture content and SME, and other selected product attributes.
3. Evaluate the interaction between the thermodynamics of expansion by moisture flashoff and the effect of moisture content on the material properties of the extruded products.

17.2.3 Materials and Methods

17.2.3.1 Corn Meal A typical proximate analysis for corn meal is shown in Table 17.1. This is the usual base material for expanded corn snacks. Cornstarch is a typical cereal starch consisting of roughly 75% amylopectin and 25% amylose.

17.2.3.2 Typical Twin-Screw Extruder

The extruder used may be single- or twin-screw. Determine its L/D ratio as the length divided by the screw diameter. A L/D of about 15 is typical for this type of process. Note the screw profile used. It should be designed to provide a high-pressure/temperature environment just behind the die. Note the dimensions of the circular dies used to shape the product.

Corn meal will be "starved-fed" via the feed section for a twin-screw extruder. Additional moisture is added via an inlet port located on the top of the barrel through a variable-speed, variable-stroke, positive displacement pump, which is calibrated for mass flow rate before use.

17.2.3.3 Instrumentation Screw speed, feed screw rpm or feed rate, and barrel temperature profile are manually set at the control panel. Barrel temperatures and product temperatures are measured with melt thermocouples. However, an immersible thermocouple probe is needed for the accurate measurement of melt temperature at the die. Die pressure is measured with a strain-gauge pressure transducer and may be indicated on a

Table 17.1 Typical proximate analysis of corn meal

Components	% (w/w)
Moisture	11.2
Protein	7
Fat	0.7
Fiber	0.5
Ash	0.4
Carbohydrate	80.2

digital display. Appropriately conditioned signals may also be logged to a data logger microcomputer, depending on the particular installation.

17.2.3.4 *Operating Procedure*   This will depend on the actual extruder used, but in any case the extruder will be brought to the initial steady-state condition and changes will be made to the moisture content (MC) by adding water to the product. Record the steady-state process values for the die pressure and product temperature at the die and motor power in Data Sheets 17.1 and 17.2. Make additional step changes in MC. Collect the samples for moisture and bulk density analysis approximately 15 min after an input change. Generally, moisture contents of 14, 17, and 20% should give a range of products with varying physical properties and sensory quality.

17.2.4   Evaluation Procedure

17.2.4.1 *Bulk Density (BD)*   Bulk density (g/cm³) is determined by filling a container of known volume with a sample and weighing it. This is important for obtaining the proper fill weight for the package volume.

17.2.4.2 *Piece Density (PD)*   Determine the apparent density of individual pieces by using a sand displacement method. This method is indicative of the porosity and texture of individual pieces. Alternatively, a cylindrical sample can be weighed and its volume estimated as a cylinder.

1. Weigh a vial. Put about 70 g of sand into the vial.
2. Randomly collect approximately 5 g of the sample. Weigh the sample and put it into the vial.
3. Fill the funnel with the sand (with the stopper at the lower opening) until it reaches the mark point. Then hang the funnel on the stand as shown in Figure 17.4.
4. Place the vial (with the sample) under the funnel and remove the stopper to fill the vial with sand.
5. Remove excess sand from the vial, using a metal bar to sweep the opening to obtain the exact amount of sand in the vial. Do not pack or shake the vial, as this will affect the accuracy of the measurement.
6. Weigh the swept contents of the vial.

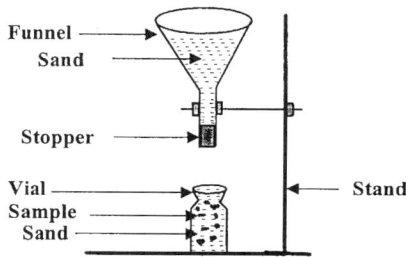

Figure 17.4   Schematic diagram showing the experimental setup to determine particle density.

7. Repeat steps 2 to 6, 3 times for the sample. Record average related data in Data Sheet 17.3.

8. Calculate the piece density of the sample as follows:

$$PD = \frac{W_p}{V_v - \dfrac{W_s}{D_s}}$$

(17.6)

where PD = piece density (g/cm^3); W_p = sample weight (g); V_v = vial volume (cm^3); W_s = sand weight (g); and D_s = sand density (1.6 g/cm^3).

17.2.4.3 *Particle Length and Diameter*

Determine the length and diameter for each treatment as the average of 10 pieces using digital calipers. It is important to note how the piece geometry changes with moisture content. Keep the cutter speed constant. If a cutter is not used, obtain a sample by facecutting with a putty knife and using the same time between cuts. You should obtain different-length samples for each moisture level. Why does it change?

17.2.4.4 *Porosity*

Determine the porosity (%) of the product sample by using the following formula:

$$\text{Porosity}\left[P = \left(1 - \frac{PD}{SD}\right)\right] \times 100$$

(17.7)

where PD = particle density and SD = true density of the solids only. Use a value of 1.4 g/cm^3 for SD.

17.2.4.5 *Specific Mechanical Energy (SME)*

Estimate the SME (kJ/kg) of the process using the following equation:

$$\text{SME} = \left(\frac{\% \text{ load}}{100}\right) \times \left(\frac{SS}{SS_{\text{rated}}}\right) \times \frac{\text{Motor power (kW)}}{\text{Throughput (kg/s)}}$$

(17.8)

where SS is given by the manufacturer of twin-screw extrudates. Record the motor power and throughput (kg/s). If power is measured directly, then the SME is

$$\text{SME} = \frac{\text{Power (kW)}}{\text{Throughput (kg/s)}}$$

(17.9)

17.2.4.6 *Calculation of Starch T_g using the Couchman–Karasz Equation*

The following equation has been used to estimate T_g for a starch melt:

$$T_g = \frac{X_W \Delta C_{pW} T_{gW} + X_s \Delta C_{ps} T_{gs}}{X_W \Delta C_{pW} + X_S \Delta C_{ps}}$$

(17.10)

where x = mass fraction, ΔC_p = change in heat capacity at T_g, T_g = glass transition temperature, subscript w = water; s = starch.

In the case of plasticized-amylopectin melts, good results have been found with the following values:

$$\Delta C_{pW} = 1.94 \text{ J/g} \cdot \text{K}; T_{gW} = 134°\text{K}$$
$$\Delta C_{pS} = 0.47 \text{ J/g} \cdot \text{K}; T_{gS} = 500°\text{K}$$

17.2.5 Report

Write a report with an introduction, procedures, results, discussion, and conclusion. Include the following:

1. Plot the melt temperature and SME versus in-barrel moisture content on the same graph.
2. Plot pressure versus moisture content.
3. Plot the BD and PD versus moisture content.
4. Taste sample products for their texture and "tooth-packing" properties.

17.2.5.1 Discussion

1. Discuss the results in terms of how varying moisture content affects the process variables (temperature, pressure, and SME) and product attributes such as densities, L/D, and porosity.
2. Discuss the differences in the shape, surface appearance, and/or internal cell structure as a function of moisture content (a light microscope would be useful for looking at a cell: slice through a piece and dab it on an inkpad to provide contrast).
3. Qualitatively evaluate the sensory attributes of each treatment such as texture, color, flavor, and teeth packing, etc. Do you detect any "burnt" flavors?
4. It is important to note that certain variables are nearly a linear function of moisture content, for example, SME and temperature, whereas density is a nonlinear function of moisture content.
5. Observe carefully the expected "collapse" phenomenon at 20% moisture content, which gives a hard, dense product.
6. Develop a state diagram showing T_g versus MC using Eq. (17.10) for moisture content between 5 and 25% in 5% intervals. Overlay your process data vectors (melt temperature corresponding to the initial moisture content point connected to the 100°C, final moisture content point for each moisture content) to help explain this observation.

17.3 SUGGESTED READINGS AND REFERENCES

1. J. L. Brent, S. S. Mulvaney, C. Cohen, and J. A. Bartsch. "Thermomechanical glass transition of extruded cereal melts." *J. Cereal Sci.* 26:301–312 (1997).

2. R. C. Miller. "Continuous cooking of breakfast cereals." *Amer. Assoc. Cereal Chemist.* 33:284–291 (1988).

3. L. N. Bailey, B. W. Hauck, E. S. Sevatson, and R. E. Singer, 1991. "Systems for manufacture of ready-to-eat breakfast cereals using twin screw extrusion." *Amer. Assoc. Cereal Chemist* 36 (10):863–869 (1991).

4. J. Fan, J. R. Mitchell, and J. M. V. Blanshard. "A computer simulation of the dynamics of bubble growth and shrinkage during extrudate expansion." *J. Food Eng.* 23:317–356 (1994).

5. R. J. Crawford, 1987. "Processing of plastics." In *Plastic Engineering*, 2nd ed., Elmsford, NY: Pergamon Press, Chap. 4.

DATA SHEET 17.1

Operating conditions: _____

Dry feed rate: _____

Screw speed: _____

Rate of added water: _____

Barrel control temperatures at different sections

Moisture Content (%)	#1	#2	#3	#4	#5	#6
	—	—	80°C	120°C	120°C	—
14						
17						
20						

Die geometry: _____

Cutter speed: _____

Other notes: _____

DATA SHEET 17.2

Moisture Content (%)	Pressure (psi/kPa)	Melt Temperature, T_m ($^\circ$C)^a	Load or Torque (%)	SME (kJ/kg)	T_g ($^\circ$C)	$T_M - T_g$ ($^\circ$C)
14						
17						
20						

^{a} Note that an immersible thermocouple must be used to obtain an accurate melt temperature at the die.

DATA SHEET 17.3

Moisture Content (%)	Length L (mm)	Diameter D (mm)	BD (g/cm^3)	PD (g/cm^3)	L/D	Porosity (%)
14						
17						
20						

18

PACKAGING OF FOODS

18.1 BACKGROUND

Packaging is an integral part of food processing. It plays two important roles in the food industry: first to protect the shelf life of foods to a predetermined degree and second to advertise food at the point of sale. In general, the main factors responsible for the deterioration of foods during transportation, distribution, and storage include the following:

1. Mechanical forces such as impact, vibration, compression, or abrasion.
2. Environmental factors such as UV light, moisture, oxygen, and temperature that can cause physical or chemical changes.
3. Postcontamination of foods by microorganisms, insects, or soil.
4. Pilferage, tampering, or adulteration.
5. Migration of toxic compounds from the package to the food.

The package should be aesthetically pleasing, have a functional size and shape, retain the food in a convenient form, be suitable for easy disposal or reuse. Package design should also meet any legislative requirements for the labeling of foods. Therefore, packaging provides a barrier between the food itself and the environment. It controls light transmission and the transfer of heat, moisture, and gases.

Light transmission is required to display package contents, but is restricted when foods are susceptible to deterioration by light such as the oxidation of lipids, destruction of riboflavin, and loss of color. Selecting the insulating properties of packaging materials controls heat transfer to foods. The rates of moisture or oxygen migration control the shelf life of dehydrated foods and are the most important properties of the packaging materials.

Water vapor or gases such as O_2, N_2, or CO_2 can permeate the packaging materials through microscopic pores or activated diffusion due to concentration gradients. The gas or vapor permeability for one-dimensional diffusion can be calculated by using Fick's law of diffusion (Eq. 18.1):

$$J = -D_g A \frac{dc}{dx} \tag{18.1}$$

where J = rate of diffusion (mol/s), A = surface area (m^2), D_g = gas diffusivity (m^2/s), c = gas concentration (mol/m^3), x = distance in the direction of diffusion (m). According to Henry's law,

$$c = S \cdot P$$

where S = gas solubility (mol/Pa $\cdot$ m^3) and P = gas partial pressure (Pa).

Therefore, we can rewrite Eq. (18.1) as

$$J = -D_g S A \frac{dP}{dx} \tag{18.2}$$

The permeability coefficient or simply permeability is defined as a product of the diffusivity coefficient and solubility and can be written as

$$B = D_g S \quad \text{(mole-thick mm/m}^2 \text{ s Pa or [mol/s} \cdot \text{Pa} \cdot \text{m])}$$

It defines the amount of gas permeated through a unit film thickness per unit time per unit packaging surface area and per unit pressure difference between the environment and packaged material. Therefore, Eq. (18.2) can be simplified to read:

$$J = -BA \frac{\Delta P}{\Delta x} \tag{18.3}$$

Intact packaging materials are a barrier to microorganisms, but seals are a potential source of contamination. Sometimes, contaminated air or water may be drawn through pinholes in hermetically sealed containers as the head space vacuum forms. The ability of packages to protect foods from mechanical damage is measured by tensile strength, Young's modulus of elasticity, and yield strength.

18.2 TYPES OF PACKAGING MATERIALS

There are two main types of packaging materials: (1) shipping containers that contain and protect the contents of a package during transport and distribution such as wooden, metal, fiberboard cases, barrels, drums, and sacks, etc. (2) Retail containers that protect and advertise the packaged food in convenient quantities for retail sale and home storage such as metal cans, glass bottles, jars, collapsible tubes, flexible plastic bags, sachets, and overwraps, etc.

Plastic has been used widely in the packaging industry since its invention in the 1960s. It is not a single plastic material, but rather an array of plastic materials that can be used for food packaging applications. Several common examples include polyethylene terephthalate (PET), high- or low-density polyethylene (HDPE, LDPE), oriented polypropylene (OPP), polyvinyl chloride (PVC), polypropylene, polyvinylidene chloride (PVDC), ethylene vinyl alcohol (EVOH), and nylon.

In the past, the shortcomings of some of the plastic materials had limited their usefulness in food packaging due to poor shelf life, heat resistance, and their permeability to water and oxygen. However, with improvements in barrier polymers, these plastics provide good opportunities as food packaging material. For example, high nitrile resins (HNR) are used extensively in packaging for their gas barrier and chemical resistance properties. A few products currently packaged in HNRs include processed meats, cheeses, bakery products, sauces, peanut butter, cooking oil, fresh pasta, herbs, and spices. HNRs have the ability to retain carbon dioxide and are so used in modified atmosphere packaging.

Nylon barrier materials provide clarity and toughness over a broad range of temperatures, and resistance to fats and aromas. Ethylene vinyl alcohol copolymer (EVOH) offers a superior barrier to gases, odors, and aromas. EVOH was the first material used to package orange juice in 64-oz plastic containers. It provided a structure that was a barrier to oxygen and prevented flavor scalping. PVDC is the oldest material to be used as an excellent barrier to gases and liquids. It is marketed under the brand name Saran^{TM} by Dow Chemical Company. Saran^{TM} in the form of coating on a polyester, cellophane, and polypropylene is used to package snack foods, processed meat, and cheese.

Metalizing is a process to improve the barrier properties of clear film. Films are specially treated to enhance metal adhesion and are then plated with a thin coating of metal, usually aluminum. The layer of aluminum is generally 30 nm thick and provides barrier properties that are difficult to obtain by other methods. Metalized films greatly enhance moisture, gas barrier properties, and also keep out light, which causes rancidity in most snack products. Metalized oriented polypropylene is being used widely in snack food packaging such as that for popcorn.

Aseptic packaging is a part of aseptic processing. Aseptic processing involves the continuous sterilization of a food product, sterilization of the container, and then filling and sealing of the container in a sterile environment. In most cases, aseptic packaging involves the use of specially treated paperboard layered with plastic film and aluminum foil. The Tetra Brik container from Tetra Pak consists of five to seven plies including polyethylene, paper, and aluminum foil. Hydrogen peroxide is used as the sterilizing medium.

18.2.1 Shelf Life of Packaged Foods

The shelf life of packaged food is controlled by the properties of foods, including water activity, pH, susceptibility to enzymatic or microbiological deterioration, and the barrier properties of the package to oxygen, light, moisture, and carbon dioxide. Moisture loss or uptake is one of the most important factors controlling the shelf life of foods. The microclimate within a package is controlled by the vapor pressure of moisture in the food at the temperature of storage. The changes in moisture content depend on the water vapor transmission rate of the package. Control of moisture exchange is also necessary to prevent condensation inside the package, which may result in mold growth or prevent freezer burn in frozen foods. To control the food moisture content inside a package, the water vapor permeability of the packaging material, its surface area and thickness should be selected based on the required storage or shelf life time.

Some foods are susceptible to oxidation and, therefore, should be stored in a package with low oxygen permeability. This also reduces vitamin C losses in foods. Sometimes, desiccants or oxygen scavengers can be placed inside a package to control water vapor and oxygen. Packaging should also retain desirable odors or prevent odor pickup from plasticizers, printing inks, adhesives, or solvents used in the manufacture of the packaging material. Most foods deteriorate more rapidly at higher temperatures, and storage conditions should be controlled to minimize temperature fluctuations.

18.2.2 Measurement of Water Vapor Transmission Rate

Water vapor transmission rate, water vapor permeance, and water vapor permeability describe the water vapor transmission characteristics of film. The water vapor transmission rate (WVTR) is defined as the gram of water vapor transmitted from $1\,m^2$ of film area in 24 h (Eq. 18.4):

$$\text{WVTR} = \frac{24 \cdot m_v}{t \cdot A} \tag{18.4}$$

where m_v = mass gain or loss (g), t = time (h), A = film surface area (m^2).

Water vapor permeance is defined (Eq. 18.5) as the gram of water vapor transmitted through $1\,m^2$ of film area in 24 h when a vapor pressure difference of $1\,mm\,Hg$ is maintained:

$$\text{Water vapor permeance} = \frac{\text{WVTR}}{\Delta P} = \frac{\text{WVTR}}{(P_1 - P_2)} = \frac{\text{WVTR}}{P_s(RH_1 - RH_2)} \tag{18.5}$$

where ΔP = vapor pressure difference, $RH_1 = P_1/P_s$ and $RH_2 = P_2/P_s$ are relative humidity on each side of the film specimen, and P_s = saturation vapor pressure (mm Hg).

Water vapor permeability is defined (Eq. 18.6) as a gram of water vapor permeated through $1\,m^2$ of film area in 24 h when the vapor pressure difference is 1 mm Hg and film thickness is 1 cm:

$$\text{Water vapor permeability} = \frac{\text{Water vapor permeance}}{\text{Film thickness (cm)}} \tag{18.6}$$

An Arrhenius-type equation can be used to measure the effect of temperature on the permeability coefficient B as given by Eq. (18.6):

$$B = B_0 e^{-E_a/RT} \tag{18.7}$$

where B_0 = constant, E_a = activation energy (kJ/kg $\cdot$ mol), R = gas constant (8.314 kJ/kg $\cdot$ mol $\cdot^\circ$ K), T = absolute temperature ($^\circ K$).

18.2.3 Quality of Packaged Foods

The crispness of dry-cereal-based foods such as crackers, fried snacks, and potato chips is a function of water activity a_w. The crispness of these foods decreases as water activity increases. It has been observed that at an a_w of 0.4 to 0.45, crackers and potato chips lose their crispness. Therefore, the moisture sorption isotherm can be used to predict the critical moisture content m_c or a_w above which the product will lose its crispness. Loss of crispness is probably caused by changes in the rate at which starch molecules can slip past each other and by the decrease in velocity and sound intensity as moisture content increases.

18.2.3.1 *Prediction of Packaging Time*   Based on the moisture sorption isotherm, simple equations can be used to estimate gain or loss of moisture held in a semipermeable membrane. Labuza and Contreas Medellin (1981) derived the following equation to predict a change in weight in packaged dried foods:

$$\ln \Gamma = \ln\left(\frac{m_e - m_i}{m_e - m}\right) = \frac{B}{x}\frac{A}{W_s}\frac{P_0}{b} \cdot t \tag{18.8}$$

where Γ = unaccomplished moisture ratio (sometimes called the gamma); m_e = moisture content of the isotherm based on a straight-line approximation, one that is in equilibrium with the external temperature and humidity (g water/g solids); m_i = initial moisture content, on a dry basis; m = moisture content at time t (h or day); B/x = film permeability (g moisture/m$^2 \cdot$ 24 h $\cdot$ mm Hg); A = film packaging area (m^2); W_s = weight of dry solids in the package (g); P_0 = vapor pressure of pure water at temperature T (mm Hg); b = isotherm slope (g water/g solids per a_w unit).

A typical sorption isotherm for potato chips at 22°C is shown in Figure 18.1. The m_e and slope b obtained from the moisture sorption isotherm can be used in Eq. (18.8) to predict shelf life in packaged bags before the product loses its crispness. A straight-dashed-line equation for the working isotherm can be written as

$$m = b \cdot a_w + Y \tag{18.9}$$

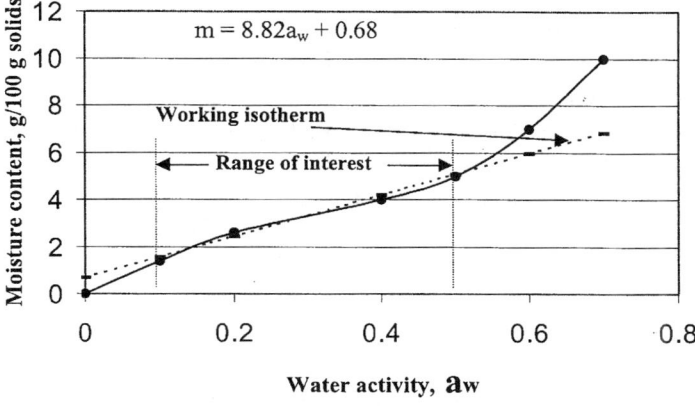

Figure 18.1   Typical moisture sorption isotherm of potato chips at 22°C.

where $b =$ isotherm slope (g water/g solids per a_w unit), $Y =$ intercept, $a_w =$ water activity.

If moisture content is plotted as g water/100 g solids, the b value will be multiplied by a factor of 100. A typical working isotherm equation for potato chips is shown in Eq. (18.10):

$$m = 8.82 \cdot a_w + 0.68 \tag{18.10}$$

where $b = 8.82$ g water/100 g solids per a_w unit.

The above equation can be used to estimate the initial moisture content m_i, critical moisture content m_c, and equilibrium moisture content m_e of potato chips at any given a_w. For example, to estimate m_e at 70% RH, it will give us the following value: $m_e = 8.82 \times 0.7 + 0.68 = 6.85$ g/100 g solids. Similarly at 100% RH, m_e will be 9.5 g/ 100 g solids.

18.3 LAB EXERCISE

18.3.1 Objectives

The main objectives of this lab exercise are to:

1. Become familiar with the characteristics of packaging films such as water permeability.
2. Predict the shelf life of a snack food such as potato chips packaged in three packaging films with high-, medium-, and low-moisture permeability.
3. Verify predicted values by carrying out an actual experiment.

18.3.2 Materials

1. Relative-humidity-measuring device such as a water activity meter.
2. Temperature-measuring device and psychrometer to estimate the dry bulb and wet bulb temperatures.
3. Weighing balance.
4. Controlled environment chamber.
5. Packaging bag made of three different films with high-, medium-, and low-water permeability characteristics. For example, a parchment paper bag (7.5 mil thick) with high water permeability, waxed paper bag (7.5 mil thick) with medium water permeability, and Saran film bag (1 mil thick) with low water permeability.
6. Fresh potato chips, about 0.5 kg worth.
7. Packaging bag sealer.

18.3.3 Procedure

1. Determine the moisture permeability B/x in g moisture per day $\cdot$ m^2 $\cdot$ mm Hg of three selected films using either the water method or desiccator method as described in Rizvi and Mittal (1992) or use the values supplied by the film manufacturers.

2. Estimate the total surface area A of both sides of the packaging bag.

3. Estimate the initial moisture content m_i using Eq. (18.10) if the initial a_w is known or assume it as $1 \, g/100$ solids.

4. Determine the dry solids W_s of potato chips to be packaged in the bag.

5. Estimate the m_c corresponding to a_w 0.45 and m_e corresponding to the desired a_w, say, 0.7 (70% RH) at 22°C.

6. Estimate the saturated vapor pressure of pure water P_0 at room temperature, say, at 22°C (20.05 mm).

7. Substitute all the values in Eq. (18.8) and estimate the time t to predict the shelf life of the potato chips.

8. The same equation can also be used to estimate the film characteristics, such as what the water permeability would be of packaging film capable of yielding a shelf life of 100 days when potato chips are stored at room temperature at 70% RH.

18.3.4 Experimental Verification

1. Procure about 0.5 kg of fresh-quality potato chips, which are crisp and crunchy.

2. Estimate the initial water activity a_w of the potato chips.

3. Prepare two packaging bags of each selected film, which can hold about 50 g of potato chips.

4. Seal the bags carefully after packaging approximately 50 g of chips.

5. Place the packaged and sealed bags in a controlled environment chamber at 70% RH at room temperature.

6. Draw the sample bags after the predicted periods of time and estimate their a_w by using the water activity meter. Evaluate the crispness of the samples by breaking and snapping them into pieces. Record the measurements in Data Sheet 18.1.

18.3.5 Results and Discussion

1. Calculate the predicted shelf life of potato chips using moisture permeability values. Evaluate the a_w and crispness of potato chips after storing the product under the same conditions. Do these experimental values agree with those used in the predicted calculations?

2. What film properties would you recommend to have about 3 months of shelf life under abusive conditions of 100% RH at room temperature.

3. Which film is most suitable for packaging potato chips? Discuss your results with reference to the properties of packaging films.

SUGGESTED READING AND REFERENCES

1. T. P. Labuza, and R. Contreas Medellin. "Prediction of moisture protection requirements for foods." *Cereal Food World* 26:335 (1981).

2. S. S. H. Rizvi, and G. S. Mittal, 1992. "Water vapor transmission of food packaging films." In *Experimental Methods in Food Engineering*, New York: Van Nostrand Reinhold.

DATA SHEET 18.1

Length × width of packaging bag: _____

Film surface area of both sides (A): _____

Temperature (room): _____

Saturated vapor pressure of pure water: _____

Relative humidity of controlled chamber: _____

Moisture permeability of film

1. Parchment paper: _____ g H_2O/day m^2 mm Hg

2. Wax paper: _____ g H_2O/day m^2 mm Hg

3. Saran film: _____ g H_2O/day m^2 mm Hg

Estimation of predicted shelf life and experimental a_w

Type of Films	Predicted Days (shelf life) to Attain 0.45 a_w at 70% RH and 22°C	Experimental a_w after Storage of Predicted Shelf Life Time at 70% RH and 22°C
Parchment paper		
Waxed paper		
Saran film		

19

PROCESS CONTROLS IN FOOD MANUFACTURING

19.1 BACKGROUND

Process controls are important to improve product uniformity and production efficiency and to reduce process costs. In the food industry, they are used for detailed production planning, scheduling of materials and resources, monitoring the flow of a product through the process, managing orders and recipes, and evaluating the process and product. In general, two main types of control systems are used: (1) local equipment controllers that are an integral part of the process plant and (2) a process computer that monitors and controls the process in a specified zone.

19.1.1 Sensors

Sensors are the primary component of a process control system and measure process variables such as temperature, pressure, weight, color, etc. Various types of sensors are used in the food industry. For example, strain gauges and load cells are used for weight, thermocouples or resistance thermometers for temperature, digital pressure or Bourdon gauges for pressure and vacuum, tachometers for pump and motor speeds, absorbance meters for turbidity, capacitance gauges for conductivity measurement, etc.

Solid-state electronic sensors have largely replaced mechanical sensors in recent years. They have greater reliability, accuracy, and precision, and faster response times. However, they also have the limitation of measuring the more complex and important quality attributes of foods, such as color, texture, surface appearance, and shape. Nonetheless, the application of microprocessor controls along with computers enables large amounts of information about a food to be stored, measured, and compared with specifications. This type of control is used in sorting the colors and shapes of food items such as potato chips. Microprocessors have the ability to analyze data from sensors rapidly and analyzed results can be directly used in automatic control process.

19.1.2 Controllers

Sensors measure a process variable and convey the information to the controller, where it is compared with a set point. When the input deviates from the set point, the controller alters an actuator such as a motor or solenoid valve to correct the deviation. In a closed-loop system, there is a continuous flow of information around the loop. A simple closed-loop control is a feedback control, which can be used to control temperature in a heat exchanger as shown in Figure 19.1. It consists of a temperature sensor or temperature-measuring device (temperature transmitter); a controlling device, usually a proportional integral device or programmable logic controller (PLC); and a manipulating device (control valve). When the temperature falls from the desired set point, the sensor gives a signal to the controller, which actuates a pneumatically operated steam valve to allow more steam (manipulated variable) into the heat exchanger and, hence, temperature increases. Basically, the feedback controller determines the correct steam valve opening for the desired set point. This is fairly useful when we consider the alternative, which is to conduct a mass and energy balance for each possible process condition.

19.1.2.1 Pneumatic Control Valves
In the food industry, steam control valves are operated pneumatically as shown in Figure 19.2. These valves operate as follows:

1. When the valve disk is pressed against the valve seat, the flow of fluid through the valve stops. As the disk moves away from the seat, the flow increases. The position of the valve disk is controlled by a sliding valve stem that runs the length of the valve.

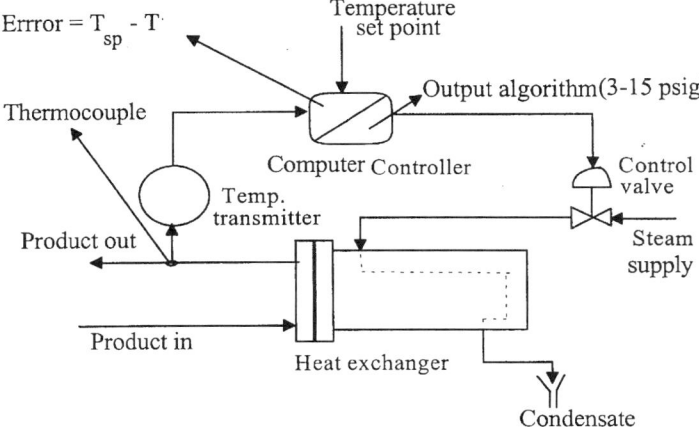

Figure 19.1 Schematic diagram of a feedback control system in a heat exchanger. (Note that additional signal conditioning is needed to convert thermocouple readings to a voltage or current proportional to temperature and the controller digital output to a 3 to 15 psig pneumatic signal; T_{sp} = set point temperature; T = product temperature.)

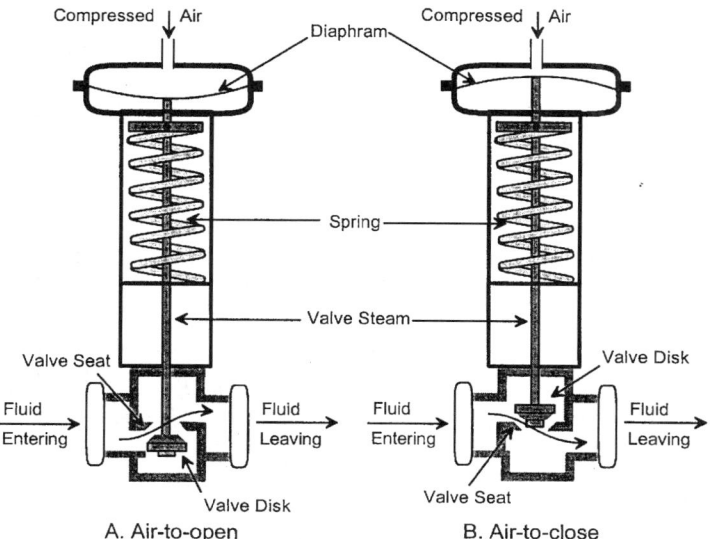

Figure 19.2 Schematic diagrams of pneumatic control valve (*A*, air-to-open type; *B*, air-to-close type).

2. The valve stem is pushed up by a spring and down by a diaphragm.
3. The diaphragm is pushed by compressed air. The position of the diaphragm and, hence the stem and valve disk, is determined by the balance between the spring force and air pressure. This means that the position of the valve and, hence, the fluid flow can be controlled by varying the magnitude of the air pressure. Since force is a product of pressure times area, a large diaphragm is used to magnify the force exerted by the air.
4. Pneumatic control valves are of two types: Air-to-open valves (Fig. 19.2A) are configured to be closed by the spring and opened by increasing the air pressure. Air-to-close valves (Fig. 19.2B) are configured to be opened by the spring and closed by increasing air pressure. The choice of valve depends on the answer to the question: "If the air pressure fails, is it safer to have the valve closed or open?" For the steam lines' feeding retorts, it is clearly safer to have the valve closed if a failure occurs and, therefore, air-to-open valves are always used in this situation.

19.1.2.2 *Pneumatic Controllers* Pneumatic control valves are often used to control temperature and pressure in retorts. Since these valves are manipulated by air pressure, a suitable process controller must be available to regulate air pressure in response to changes in the temperature and pressure of the retort. For manual operation, a set of knobs are generally used to adjust the desired pressure (say, 15 psig) and temperature (around 250°F). The controllers then regulate the supply air that in turn regulates the steam. A typical schematic diagram of a pneumatic temperature controller is shown in Figure 19.3. It works as follows:

1. A reservoir supplies air at a constant pressure of 15 psig.
2. A temperature bulb is placed at the location, where the temperature is to be controlled. Changes in temperature at this location cause changes in pressure

within the bulb. This change in pressure is transmitted through a tube to a Bourdon tube. (For pressure regulators, the space to be controlled is connected directly to the Bourdon tube.)

3. When pressure increases within the Bourdon tube, the tube attempts to straighten out, pulling on the link as it does so.

4. The link rotates the pen arm about a pivot. The pen arm can act as either a pointer in a dial or move a pen on a recorder.

5. When the temperature increases, the control rod moves to the right, restricting the movement of a flapper. When the temperature decreases, the control rod moves to the left, increasing the room for the flapper to move.

6. Air pressure within the nozzle pushes the flapper valve out more when the temperature is below the desired temperature and less (or not at all) when it is above.

7. When the temperature rises above the set point and the flapper is forced toward the nozzle, it closes so that air passing through the reducing tube increases the pressure behind the nozzle. This pressure expands the diaphragm that moves the plunger into the right-hand seat of the pilot valve. This shuts off the compressed air supply to that valve and allows air to escape through the open left-hand seat, so that pressure within the pilot valve decreases. This reduced pressure is transmitted to the control valve, allowing the spring to close the valve (Fig. 19.2A) and reducing the flow of steam. The result is a drop in temperature.

8. When the temperature falls below the set point and the flapper is allowed to move away from the nozzle, air escapes through the nozzle and the pressure behind the nozzle drops (friction of the reducing tube allows a large pressure drop to exist between the supply and nozzle). The reduced pressure contracts the diaphragm, which moves the plunger toward the right-hand seat of the pilot valve. Compressed air can now reenter the pilot valve while the escape of air is reduced. This results in increased pressure in the pilot valve that is transmitted to the control valve (Fig. 19.2A), forcing it to open and increasing the flow of steam. This results in a rise in temperature.

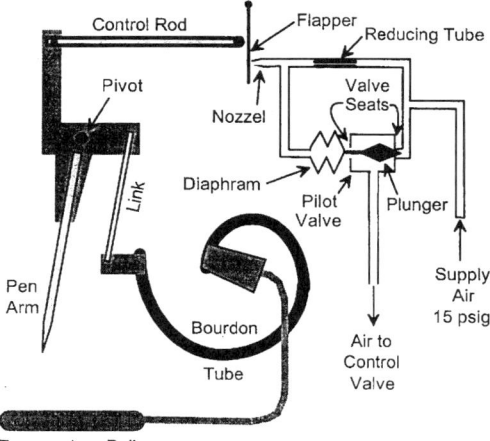

Figure 19.3 Schematic diagram of a temperature control system.

19.1.3 Programmable Logic Controllers

Programmable logic controllers (PLCs) are based on microprocessors that have a fixed program stored in two modes. In one mode, instructions are programmed into the computer memory via a random access memory keyboard, and in the second mode, the program is executed automatically in response to data received from sensors. This is carried out with the use of software building blocks termed algorithms. Each algorithm carries out a specific function and sequences of algorithms are programmed to operate a particular application.

A display monitor provides information on the progress status of the control. If a process parameter exceeds a preset limit, it may either be used to activate a warning signal for the operator or may automatically correct the deviation from specification. PLCs are very reliable, easy to install and use. They can be easily programmed and thus allow greater flexibility modifying the process conditions or changing product formulations. PLCs are also useful to collect and process data such as length of process stoppage and energy consumption, which can then be used to design maintenance programs more effectively.

19.1.4 Applications

19.1.4.1 Product Formulation PLCs are useful to produce smaller batches of products with more frequent changes to processing conditions and product formulations. The equipment has a microprocessor that stores information about the types and weights of ingredients for all products made on the production line. The microprocessor controls the flow of ingredients from storage tanks, through automatic weighers, to mixing vessels. These types of PLCs are used in snack foods, baked foods, and ice cream manufacture. A similar system can be used for the automatic control of flow blending, for example, to adjust the alcohol content and strength of beer to preset specifications. In a fermentation tank, high-strength beer is mixed with deaerated water to the correct proportions based on their set specific gravity. PLCs are also used to determine the least-cost formulation needed to produce the required specification from different combinations of raw materials.

19.1.4.2 Weighing PLCs are used to control the weights of packaged foods by avoiding overfilling and underfilling the containers. A checkweigher is programmed with the target weight for all products produced on a particular production line. It weighs each package as it passes on a conveyor belt and continuously updates the mean weight and standard deviation of the pack weights. If the average weight falls below a preset level, a warning is given to the operator or process computer, and the checkweigher automatically rejects any individual packages.

19.2 LAB EXERCISE

19.2.1 Objectives

The main objectives of this lab exercise are to:

1. Become familiar with the various components and working principles of a process control system such as temperature control or pressure control feedback type.
2. Observe the dynamics between set point changes and the response of the process variable.

19.2.2 Example

This example is for a heat exchanger with a product temperature controller. Other unit operations could also be used such as a retort, extruder, etc.

19.2.3 Materials

1. Milk, apple juice, or another available fluid to heat or cool.
2. Temperature recorder.

19.2.4 Procedure

1. First circulate water throughout the heat exchanger. Once all the air and gas are out of the system, start circulating the desired food product, which is at room temperature initially.
2. Set the temperature controller to 60°C and record the product temperature with time. Also record the average actual temperature of the product at steady state. Report the data in Data Sheet 19.1. Estimate the maximum rate of temperature change and record the results in Data Sheet 19.2.
3. Increase the set point temperature to 70°C and repeat step 2.
4. Repeat step 2 at temperatures of 80 and 90°C.
5. While at steady state, increase or decrease the product flow while maintaining the same set point. This is known as a process deviation.

19.2.5 Report

1. Draw a schematic diagram of the system showing the temperature sensor, process controller, steam or hot water control valve.
2. Plot the set point temperature and actual steady-state temperature. Is the relationship linear? Is there any offset between the set point and actual temperature?
3. Plot the dynamic temperature response for each temperature change. Are they similar? Do they show any overshoot of the desired set point?
4. Discuss how these results may differ with a change in product properties such as viscosity or a process variation such as a change in the pilot plant steam supply pressure. How would these variations be handled if there was no closed-loop controller to monitor the situation?

19.3 SUGGESTED READINGS AND REFERENCES

1. P. Harriott, 1983. *Process Control*, Malabar, FL: Robert E. Krieger Publishing Co.
2. T. H. Tsai, J. W. Lane, and C. S. Lin, 1986. *Modern Control Techniques for the Processing Industries*, New York: Marcel Dekker.
3. P. Fellows, 1988. "Material handling and process control." In *Food Processing Technology*, Chichester, UK: Ellis Horwood Ltd.

DATA SHEET 19.1   Dynamic temperature response for various set point changes

Time (s)	$25 \rightarrow 60°C$	$60 \rightarrow 70°C$	$70 \rightarrow 80°C$	$80 \rightarrow 90°C$
0				
5				
10				
15				
20				
30				
40				
50				
60				

DATA SHEET 19.2 Set point temperature and actual steady-state temperature

Set Point at Controller (°C)	Steady-state Actual Product Temperature (°C)	Temperature Difference (set point − actual) if any (°C)	Maximum Rate of Temperature Change, dT/dt (°C/s)[a]

[a] Determine as tangent to process response curve at maximum rate from Data Sheet 19.1.

APPENDIX A

VALUES OF log (g), f_h/U AT $Z = 18\,F$ AND $j = 1 \sim 1.2$ USED IN THERMAL PROCESS DESIGN

Table A.1 Values of log(g) for given values of f_h/U

f_h/U	log (g)	f_h/U	log (g)	f_h/U	log (g)	f_h/U	log (g)
0.35	−2.147	0.59	−0.985	0.83	−0.494	1.175	−0.11
0.36	−2.068	0.60	−0.949	0.84	−0.479	1.2	−0.09
0.37	−1.993	0.61	−0.928	0.85	−0.463	1.225	−0.072
0.38	−1.922	0.62	−0.907	0.86	−0.448	1.25	−0.054
0.39	−1.854	0.63	−0.886	0.87	−0.434	1.275	−0.036
0.40	−1.790	0.64	−0.864	0.88	−0.420	1.30	−0.019
0.41	−1.729	0.65	−0.843	0.89	−0.406	1.325	−0.003
0.42	−1.671	0.66	−0.821	0.90	−0.392	1.35	0.013
0.43	−1.616	0.67	−0.80	0.91	−0.379	1.375	0.028
0.44	−1.563	0.68	−0.778	0.92	−0.366	1.4	0.042
0.45	−1.512	0.69	−0.757	0.93	−0.353	1.425	0.057
0.46	−1.464	0.70	−0.736	0.94	−0.341	1.45	0.07
0.47	−1.418	0.71	−0.715	0.95	−0.328	1.475	0.084
0.48	−1.373	0.72	−0.694	0.96	−0.317	1.50	0.097
0.49	−1.331	0.73	−0.674	0.97	−0.305	1.55	0.122
0.50	−1.290	0.74	−0.654	0.98	−0.293	1.60	0.146
0.51	−1.251	0.75	−0.635	0.99	−0.282	1.65	0.168
0.52	−1.213	0.76	−0.616	1.00	−0.271	1.70	0.189
0.53	−1.177	0.77	−0.597	1.025	−0.245	1.75	0.210
0.54	−1.142	0.78	−0.579	1.05	−0.220	1.80	0.229
0.55	−1.108	0.79	−0.561	1.075	−0.196	1.85	0.248
0.56	−1.076	0.80	−0.544	1.10	−0.173	1.90	0.265
0.57	−1.044	0.81	−0.527	1.125	−0.151	1.95	0.282
0.58	−1.014	0.82	−0.511	1.15	−0.130	2.0	0.289

(continued)

Table A.1 (*continued*)

f_h/U	log (g)	f_h/U	log (g)	f_h/U	log (g)	f_h/U	log (g)
2.05	0.314	3.25	0.564	4.45	0.698	8.25	0.903
2.1	0.329	3.3	0.571	4.5	0.702	8.5	0.911
2.15	0.343	3.35	0.578	4.55	0.707	8.75	0.919
2.2	0.357	3.4	0.585	4.6	0.711	9	0.927
2.25	0.37	3.45	0.591	4.65	0.715	10	0.955
2.3	0.383	3.5	0.598	4.7	0.719	20	1.112
2.35	0.396	3.55	0.604	4.75	0.723	30	1.187
2.4	0.408	3.6	0.61	4.8	0.727	40	1.235
2.45	0.419	3.65	0.616	4.85	0.731	50	1.270
2.5	0.43	3.7	0.622	4.9	0.734	60	1.296
2.55	0.441	3.75	0.628	4.95	0.738	70	1.318
2.6	0.452	3.8	0.634	5	0.742	80	1.336
2.65	0.462	3.85	0.639	5.25	0.759	90	1.352
2.7	0.472	3.9	0.645	5.5	0.776	100	1.365
2.75	0.481	3.95	0.65	5.75	0.791	110	1.377
2.8	0.491	4	0.655	6	0.805	120	1.388
2.85	0.5	4.05	0.66	6.25	0.819	130	1.397
2.9	0.508	4.1	0.665	6.5	0.831	140	1.406
2.95	0.517	4.15	0.67	6.75	0.843	150	1.414
3	0.525	4.2	0.675	7	0.854	160	1.422
3.05	0.533	4.25	0.68	7.25	0.865	170	1.429
3.1	0.541	4.3	0.684	7.5	0.875	180	1.435
3.15	0.549	4.35	0.689	7.75	0.885	190	1.441
3.2	0.556	4.4	0.694	8	0.894	200	1.447

Table A.2 f_h/U values for given log(g)

log(g)	0	0.002	0.004	0.006	0.008
−1.95	0.3759	0.3757	0.3754	0.3751	0.3748
−1.90	0.3831	0.3828	0.3826	0.3823	0.382
−1.85	0.3906	0.3903	0.39	0.3897	0.3894
−1.8	0.3984	0.3981	0.3978	0.3975	0.3971
−1.75	0.4065	0.4062	0.4058	0.4055	0.4052
−1.7	0.4149	0.4146	0.4143	0.4139	0.4136
−1.65	0.4237	0.4234	0.423	0.4227	0.4223
−1.6	0.4329	0.4325	0.4322	0.4318	0.4314
−1.55	0.4425	0.4421	0.4417	0.4413	0.4409
−1.5	0.4525	0.4521	0.4517	0.4513	0.4509
−1.45	0.463	0.4625	0.4621	0.4617	0.4613
−1.40	0.4739	0.4735	0.473	0.4726	0.4721
−1.35	0.4854	0.485	0.4845	0.484	0.4836
−1.3	0.4975	0.497	0.4965	0.496	0.4955
−1.25	0.5102	0.5097	0.5092	0.5086	0.5081
−1.2	0.5236	0.523	0.5225	0.5219	0.5214
−1.15	0.5376	0.5371	0.5365	0.5359	0.5353

Table A.2 (*continued*)

log(g)	0	0.002	0.004	0.006	0.008
−1.1	0.5525	0.5519	0.5513	0.5507	0.5501
−1.05	0.5682	0.5675	0.5669	0.5663	0.5659
−1	0.5848	0.5841	0.5834	0.5828	0.5821
−0.9	0.6232	0.6223	0.6213	0.6204	0.6194
−0.8	0.6698	0.6689	0.6679	0.667	0.6661
−0.7	0.7173	0.7163	0.7153	0.7144	0.7134
−0.6	0.7686	0.7675	0.7664	0.7654	0.7643
−0.5	0.8266	0.8253	0.8241	0.8228	0.8216
−0.4	0.8941	0.8926	0.8911	0.8897	0.8882
−0.3	0.9742	0.9725	0.9708	0.969	0.9673
−0.2	1.0706	1.0685	1.0664	1.0643	1.0622
−0.1	1.1873	1.1848	1.1822	1.1796	1.1771
0	1.33	1.3331	1.3363	1.3395	1.3427
0.1	1.5059	1.5099	1.5138	1.5178	1.5218
0.2	1.7258	1.7308	1.7357	1.7407	1.7457
0.3	2.0052	2.0116	2.0179	2.0243	2.0308
0.4	2.368	2.3763	2.3847	2.3931	2.4016
0.5	2.8518	2.8631	2.8744	2.8859	2.8974
0.6	3.5186	3.5344	3.5504	3.5664	3.5826
0.7	4.4739	4.497	4.5203	4.5438	4.5675
0.8	5.9057	5.9431	5.9771	6.0133	6.0498
0.9	8.1682	8.2259	8.2841	8.343	8.4025
1	11.969	12.069	12.17	12.272	12.376
1.05	14.865	15.001	15.138	15.277	15.417
1.1	18.828	19.015	19.205	19.397	19.592
1.15	24.367	24.632	24.9	25.172	25.448
1.2	32.295	32.678	33.067	33.462	33.864
1.25	43.935	44.505	45.084	45.673	46.272
1.3	61.505	62.378	63.266	64.171	65.091
1.4	132.8	135.05	137.35	139.7	142.1
1.45	206.03	206.03	206.03	206.03	206.03

Adapted from American Can Company, Barrington, IL. For detail refer to *CRC Handbook of Lethality Guides for Low Acid Canned Foods*, 1983, by C. R. Stumbo, K. S. Purohit, T. V. Ramakrishnan; D. A. Evans, and F. J. Francis. CRC Press, Inc. Boca Raton, FL.

APPENDIX B

CATECHOL TEST AND THERMOPHYSICAL PROPERTIES

1. Cut the apple fruit slices in half perpendicular to the longitudinal axis.
2. Dip the fruit slices into a 1% catechol solution for 1 min and place them on a paper towel.
3. Keep the sample for 30 to 45 min and observe for color development and location.
4. Illustrate the results in your laboratory report.

Table B.1 Thermal properties of some food products

Food	Water (% wet basis)	Specific heat[a], kJ/kg · K (BTU/lb · °F)	Specific heat[b], kJ/kg K (BTU/lb · °F)	Average Freezing Point °C (°F)
Apples	84	3.60 (0.86)	1.84 (0.44)	−2.0 (28.4)
Apple juice	87	3.85 (0.92)	1.93 (0.46)	−1.4 (29.4)
Orange juice	86	3.89 (0.93)	1.93 (0.46)	−1.2 (29.8)
Carrots	88	3.77 (0.90)	1.93 (0.46)	−1.4 (29.5)
Beef, lean	68	3.50 (0.84)	2.05 (0.49)	−1.70 (28.9)
Frankfurter	60	3.60 (0.86)	2.34 (0.56)	−1.7 (28.9)

[a]Above freezing.
[b]Below freezing.

Table B.2   Heat transfer coefficients at various conditions

Freezing Condition	$W/m^2 \cdot K$	$BTU/h \cdot ft^2 {}^\circ F$
Still air freezing (no radiation)	5.7	1
Air blast freezing (500 ft/min)	17.0	3
Air blast freezing (1000 ft/min)	28.4	5
Liquid nitrogen immersion freezing	170	30.0
Slowly circulating brine freezing	56	9.88
Rapidly circulating brine freezing	85	15.0
Boiling water heating	568.0	100

Table B.3.   Thermal conductivity of some materials

Material	k, $BTU/(h \cdot ft {}^\circ F)$	k, $W/(m \cdot K)$
Cardboard	0.040	0.06
Plastic	0.154	0.26
Tin	35.240	61.0
Steel, 1% C	26.0	45.0
Stainless steel, 304	8.08	14
Stainless steel, 308	8.66	15
Aluminum	118.5	203
Air (1 atm), $-45^\circ C$ ($-50^\circ F$)	1.184	2.04
Air (1 atm), $-18^\circ C$ ($-0^\circ F$)	1.311	2.27
Air (1 atm), $-10^\circ C$ ($-50^\circ F$)	1.436	2.49

Table   B.4   Enthaply   of   pure water above and below the freezing state

Temperature ($^\circ C$)	H(kJ/kg)
10	453.47
0 (liquid)	411.00
0 (ice)	78.29
-8.88	59.86
-20.55	36.58
-28.88	20.58
-40.0	0.0

Table B.5 Properties of ice as a function of temperature

Temperature ($^\circ$C)	Thermal Conductivity (W/m $\cdot$ K)	Specific Heat (kJ/kg $\cdot$ K)	Density (kg/m^3)
0.0	2.22	2.05	924.2
-7	2.27	2.02	922.6
-12	2.32	1.98	919.4
-18	2.37	1.95	919.4
-23	2.41	1.92	919.4
-45.5	2.72	1.78	917.8
-73	3.08	1.58	916.2

APPENDIX C

ADDITIONAL READING MATERIAL ON EXPERIMENTAL DESIGN

C.1 OTHER FACTORIAL DESIGNS

In addition to one-variable and two-factor designs, there are many other treatment designs. We mention a few of them here for your information.

- *More than two factors.* Factorial experiments can have any number of factors, each with a different number of levels.

- *Optimization experiments.* A number of designs have been developed to determine the best combination of conditions, say, the temperature, time, and sugar combination that produces the highest yield. Optimization experiments typically use three to five levels of each factor, but may not use every combination of levels. Some classes of optimization designs include "central composite" designs, "mixture" designs, "steepest ascent" designs, and "simplex" designs. See a statistician for help in selecting the best design for your purposes.

- 2^k *factorial.* When we want to screen many variables in order to find out which ones are worth further study, an experiment in which each factor occurs at just two levels is most efficient. If an experiment is to study k different factors, it will require 2^k-different treatment combination, for example, a four factor experiment would require 16 treatments. The two levels of each factor can represent:

    The presence and absence of something such as an ice cream mix with and without stabilizers. We would pick this as a factor if we wanted to know whether stabilizers had an effect.

305

Two different types of something such as two different sweeteners. We would pick this as a factor if we knew that sweeteners were required, but wanted to know whether the type made a difference.

Two quantitative levels such as 10% sugar and 15% sugar to see if quantity had an effect.

- *Fractional factorial designs.* As the number of factors increases, the number of treatments goes up rapidly. There are designs based on the 2^k-design that do not require using every combination. One-half, one-quarter, and even smaller fractions of the available combinations can be used. For example, a 2^k-experiment with 7 factors requires 128 treatments. There is a design based on the 2^7 fractional factorial design that only requires 8 treatments and still yields useful information about all 7 factors. Note, however, that the 8 treatments must be very carefully selected and the results carefully interpreted, so consult a statistician or a book on experimental design before attempting this design.

- Both 2^k and fractional factorial experiments are useful when we want to screen many variables to select the ones that most affect our product. In screening experiments, the two levels of quantitative variables are set fairly far apart so that important variables are likely to show a difference.

- Both 2^k and the fractional factorial designs can also be used in a sensitivity study to determine which variables must be most carefully controlled to maintain quality. In these experiments, one level of each factor is set at the optimal level. The other level is set a small distance from the optimum so that sensitivity to small changes will be detected.

C.2 LEVELS OF QUANTITATIVE VARIABLES

When we are designing an experiment with a quantitative experimental variable, we need to decide how many levels to use and what those levels should be. Our decision will depend in part on how much we know before performing the experiment.

Our goal with quantitative variables is frequently to determine trends rather than compare specific treatments. We should, therefore, select enough treatments to show the trend clearly and place them where they will best define the trend. Do not, however, select more levels than we need as this is wasteful. The box below shows some guidelines.

SELECTING QUANTITATIVE TREATMENTS

- If possible, determine the range of values that are likely to be practical and select one level at either end of this range. This may be dictated by the processing equipment, past experiments, etc. If we are sure that the relationship between response and experimental variable is linear, these two treatments are all we need (Fig. C.1A).

- Unless we are sure that the relationship is linear, select at least one other treatment between these two to check for curvature. In most cases, a point half-way between is most efficient (Fig. C.1B).

- If we expect sharper curvature in one place than another, put the third treatment near the sharper curvature (Fig. C.1C).
- If we think the relationship is not a simple curve, say an experimental curve or S-shaped curve, or if we are not sure, use more than three levels (Fig. C.1D).

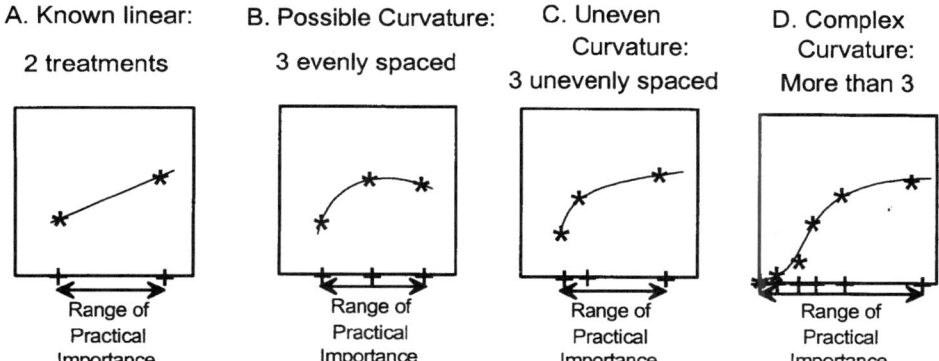

Figure C.1 Some trends of quantitative treatments.

C.3 NUMBER OF REPLICATES

The question now arises, "How many replicates should we perform?" Frequently, there is no answer to this question until after we have done the experiment. However, if we have the following advance information, it is possible to roughly estimate the number of replicates we need:

1. *Required difference.* We must be able to specify the smallest difference in the response variable that we want to detect. Let us call this d and design the experiment so that, if the "true" difference in response between two treatments is d or greater, the experiment will tell us there is a difference. For example, if increasing cheese yield from 92 to 92.4% or more is of practical importance, then specify that $d = 0.4\%$. The smaller d is, the more replicates will be needed.

2. *One- or two-sided.* Sometimes, the goal of an experiment is to determine whether two treatments differ without specifying which treatment will, on average, be higher. This leads to a two-sided test since we test for differences in two directions. In other experiments, we are only interested in determining whether treatment A is greater than treatment B. This leads to a one-sided test since differences in only one direction are tested for. In general, two-sided tests require more replicates.

3. *Expected variation.* We must be able to specify the standard deviation expected among replicates of the same treatment. This information can come from the literature, previous similar experiments, a pilot experiment, or the experience of an expert. Let us say that s stands for the expected standard deviation within treatments. It can be obtained in several ways.

4. If we have a set of comparable data $(X_1, X_2, X_s, \ldots, X_n)$ from n units that were all treated alike, and the average of the n units is $\bar{x}$, estimate s by

$$s = \sqrt{\frac{\sum(X_i - \bar{x})^2}{n - 1}} \qquad \text{(C.1)}$$

5. If we have an analysis of variance from a previous comparable experiment, estimate s by

$$s = \sqrt{\text{Error mean square}} \qquad \text{(C.2)}$$

6. If we have prior experience, try specifying the largest and smallest value we expect for units receiving the same treatment. Estimate s by

$$s = \frac{\text{Largest} - \text{smallest}}{4} \qquad \text{(C.3)}$$

The larger s is, the more replicates we will need.

C.3.1 *Risks of error.* Experimental results can lead to two types of errors.

C.3.1.1 *Type I error.* The results can lead us to say that treatment means differ when they really do not. This is called a type I error and the risk of such an error is usually symbolized by the Greek alpha (α). α is frequently chosen to be 0.05 but can be smaller (say, 0.025 or 0.01) if we decide that type I errors are serious or larger (say, 0.10) if we decide they are less serious.

C.3.1.2 *Type II error.* The results can lead us to say that the treatments do not differ when, in fact, they differ by an amount $= d$. This is called a type II error and the risk of such an error is usually symbolized by the Greek beta (β). Choose a small value (say, 0.05 or 0.01) if we decide that type II errors are serious and a larger value (say, 0.10 or 0.20) if we decide they are less serious. The smaller the risks we are willing to take, the more replicates we will need.

ESTIMATING THE NUMBER OF REPLICATES

- Let $d =$ the smallest difference in the response to be detected, $s =$ the expected standard deviation within treatments, $\alpha =$ the acceptable risk of a type I error, $\beta =$ the acceptable risk of a type II error and the number of sides (one or two) to the test.
- In a table of the normal distribution (Table C.1), find a value of z for a risk of α if this is to be a one-sided test and $\alpha/2$ if this is to be a two-sided test.
- In the same table, find the z value for a risk of β.
- Compute n, the approximate number of replicates per treatment with Eq. (C.4).
- It is usually best to treat this as a lower limit for n.

$$r = \left[\frac{s}{d} (z_\alpha + z_\beta) \right]^2 \qquad \text{(C.4)}$$

Table C.1 List of z values corresponding to risk values (α,β)

Risk	0.20	0.10	0.05	0.025	0.02	0.01	0.005
z	0.842	1.282	1.645	1.960	2.055	2.333	2.575

Example 1   *We wish to know if a treatment increases cheese yield by 0.4%. From a previous experiment, we obtain an error mean square (EMS) of 0.09. We decide to accept a 5% risk of finding an increased yield when there is not one ($\alpha = 0.05$) and a 10% risk of saying there is not an increase when a difference as large as 0.4% exists ($\beta = 0.10$).*

Solution   Since we are only looking for increase in yield, not decrease, this is a one-sided test. Checking a normal table, we find for $\alpha = 0.05$, $z_a = 1.645$. For $\beta = 0.10$, $z_b = 1.282$. We have specified that $d = 0.4\%$ and from the EMS of 0.09, we estimate that $s = 0.3\%$, so

$$n = \left[\frac{0.3}{0.4}(1.645 + 1.28) \right]^2 = 4.8$$

Thus, we should replicate each treatment at least five times. One or more would not hurt if it is economically feasible. Since the experiment will have two treatments, we will need a total of 20 experimental units (batches).

C.4 SETTING UP DESIGNS

Most experiments require randomization in assigning experimental units to treatments. The boxes below present two techniques for generating the necessary random numbers.

RANDOM PERMUTATION FROM A RANDOM NUMBER TABLE

- Start at some randomly selected place on the table and select n two-digit numbers.
- If any numbers duplicate previous ones, do not add the new occurrence to the list.
- Find the smallest number in your list and place a 1 next to it. Find the next smallest and place a 2 next to it. Continue until all n numbers have been ranked.
- The ranks will be the numbers from 1 to n in random order.

Example 2 (Completely Random Design or CRD)   *An experiment is to be run in which loaves are the experimental unit with 3 treatments and 4 replicates of each treatment. Before running the experiment, select 12 loaves of bread and number them 1 to 12. They should all come from the same batch or each from a different* *batch. In Minitab, give the following commands (or use the menus):*

```
MTB >    random 12 in C1;
SUBC >   uniform.
MTC >    set C2
DATA >   1:12
DATA >   end;
MTB >    sort C1 C2 into C3 C4;
SUBC>    by C1.
MTB >    print C4
         9  8  1  2  3  4  6  7  10  11  5  12
```

Loaves 9, 8, 1, and 2 go to treatment 1, loaves 3, 4, 6, and 7 go to treatment 2, and loaves 10, 11, 5, and 12 go to treatment 3.

RANDOM PERMUTATION WITH MINITAB PROGRAM

· For a permutation of 12 integers, for example, generate a set of 12 random numbers with the following commands (or select Calc|Random Data from the menus):

```
MTB > RANDOM 12 in C1;
SUBC > UNIFORM.
```

· Generate a sequence of integers, with the command (or Calc|SetPatterned Data):

```
MTB > SET in C2
DATA > 1:12
```

· Randomize the sequence while unrandomizing the random numbers with the following commands (or select Manipulated|Sort):

```
MTB >    SORT C1 C2 into C3 C4;
SUBC >   BY C1.
```

· To see the permutation, use the command

```
MTB >    PRINT C4
         9  2  5  4  11  1  7  12  8  3  10  6
```

Example 3 (Randomized Complete Block or RCB) *An experiment compares 4 starter cultures on the effect of cheese yield. Three replicates are to be performed on each culture. It is not practical to make more than 5 vats of cheese per day, but it is thought that the variation in milk from day to day may affect the results so it is decided to treat days as a block and prepare 4 vats per day, one for each treatment.*

The order of the treatments is to be randomized each day:

```
MTB >    Random 4 in C1-C3;
SUBC >   Uniform.
MTB >    Sec c4
DATA >   1:4
DATA >   end;
MTB >    Sort C1 C4 to C5 C6;
SUBC >   By C1.
MTB >    Print C6
         4  1  3  2
MTB >    Sort C2 C4 to C5 C6;
SUBC >   By C2.
MTB >    Print C6
         2  1  4  3
MTB >    Sort C3 C4 to C5 C6;
SUBC >   By C3.
MTB >    Print C6
         1  4  3  2
```

On the first day, the treatments are performed in the order 4, 1, 3, 2. On the second day in the order 2, 1, 4, 3 and on the third day in the order 1, 4, 3, 2.

C.5   ANALYSIS

The following analyses are done with the aid of Minitab. Most other statistical packages can be used, but the procedure would be different.

C.5.1   Evaluate Data

EVALUATING DATA IN MINITAB

- Enter all the data for each response variable. Use one column for each response variable and one row for each experimental unit.
- In an empty column, place a number in each row, indicating which treatment the data in that row came from, numbering the treatments from 1 to t.
- If data is in C1 and treatment numbers in C2, plot the data versus treatment number with the command:

  ```
  MTB > Plot C1*C2
  ```

- The graph will consist of a column of points for each treatment, one point for each replicate.

- On this plot, do most treatments show approximately the same amount of variation among replicates? If not, the analysis given below will not be correct.
- Are there any outliers that are widely separated from the rest of the replicates of the same treatment? Check that they are not copying or typing errors? If not, are the outliers associated with anything abnormal in the conduct of the experiment?

Example 4 (One-variable, CRD) In order to compare the effect of 3 baking temperatures on loaf volume, a batch of dough was divided into 12 loaves. Four loaves were baked at each of 3 temperatures. The resulting loaf volumes are as given in Table C.2. These data are entered into column C1 of the Minitab worksheet and treatment numbers are entered into column C2 as shown in Table C.3.

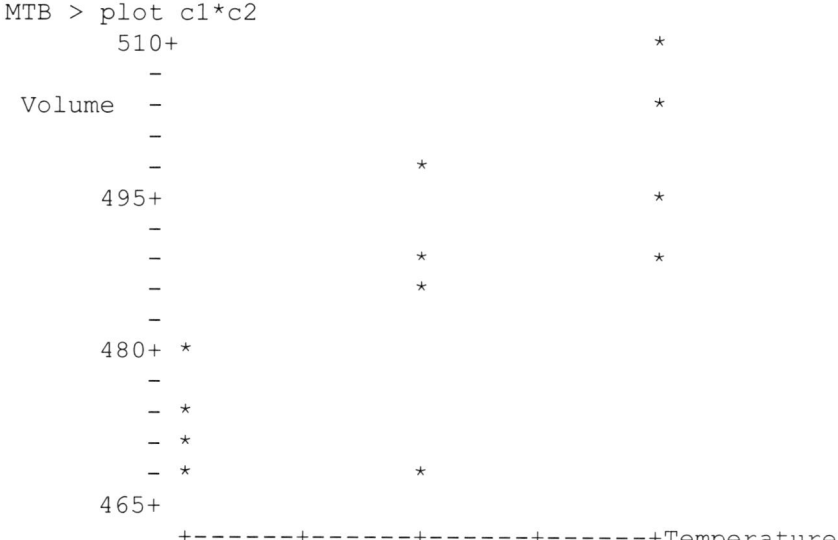

```
MTB > plot c1*c2
        510+                                      *
           -
Volume     -                                      *
           -
           -                       *
        495+                                      *
           -
           -               *              *
           -               *
           -
        480+  *
           -
           -  *
           -  *
           -  *              *
        465+
          +------+------+------+------+Temperature
```

The three treatments show roughly the same spread among replicates and there are no obvious outliers although you might doublecheck the low value in treatment 2. Thus, there are no problems with the data and we can proceed with further analysis.

Table C.2 Data on the effect of baking temperature on bread volume, CRD

1. (340°F)	2. (360°F)	3. (380°F)
460	490	495
480	475	510
475	488	488
468	498	503

Table C.3   Minitab worksheet

	C1	C2
	Volume	Treatment
1	460	1
2	480	1
3	475	1
4	468	1
5	490	2
6	475	2
7	488	2
8	498	2
9	495	3
10	510	3
11	488	3
12	503	3

ANALYSIS OF ONE-VARIABLE EXPERIMENT USING CRD

- Enter and evaluate your data as described in "Evaluating Data in Minitab."
- If the data are in column C1 and treatment numbers in column C2, use the command:

```
MTB > ONEWAY C1 C2
```

(or Stat|ANOVA|Oneway from the menus) to obtain an analysis of variance (ANOVA) table and a table of means with confidence intervals for each response variable.

- On the right-hand side of each ANOVA table, find a p value. If this is less than 0.05, then it is quite likely that at least two of the treatment means are really different from each other. We say that there is a significant difference between the treatments. The smaller the p value, the more convincing the evidence.
- If the p value is much larger than 0.05, then this experiment has failed to separate treatment effects (if any) from those of interfering variables. We say that the difference between treatments is not significant.
- If the p value indicates significant differences, compare the means in the accompanying table.
- The interpretation of confidence intervals will be given later.
- The square root of the error mean square estimates the standard deviation within treatments (pooled standard deviation), a measure of random variation in the experiment.

Example 5 (One-variable, RCB)   *Using the data from Example 4, we obtain*

```
MTB > oneway c1 c2
```

```
ANALYSIS OF VARIANCE on Volume
SOURCE    DF        SS        MS        F         p
Treat.    2      1618.2     809.1     9.41     0.006
ERROR     9       773.5      85.9
TOTAL    11      2391.7
                           INDIVIDUAL 95% CI'S FOR MEAN
                           BASED ON POOLED STDEV
LEVEL     N      MEAN      STDEV --+------+------+------+--
  1       4     470.75     8.64  (----*----)
  2       4     487.75     9.54          (------*------)
  3       4     499.00     9.56                  (------*------)
                                  --+------+------+------+----
    POOLED STDEV =    9.71     465    480    495    510
```

The p value of 0.006 is quite small, telling us the evidence is quite convincing that temperature affects loaf volume. The table of means show a steady upward trend that seems to form a straight line.

ANALYSIS OF ONE-VARIABLE EXPERIMENTS, BLOCKED

- Enter and evaluate your data as described in "Evaluating Data in Minitab."
- If data are in column C1, treatment numbers in column C2, and block numbers in column C3, use the command:

  MTB > ANOVA C1 = C2 C3

  (or Stat|ANOVA|Balance ANOVA from the menus) to obtain an analysis of variance (ANOVA) table and a table of means with confidence intervals for each response variable.

- If the p value for treatments is less than 0.05, then it is quite likely that at least two of the treatment means are really different from each other. We say that there is a significant difference between the treatments. The smaller the p value, the more convincing the evidence.

- If the p value for blocks is less than 0.10, the blocking you did was worth doing. If it is larger, you may want to skip blocking by this variable in subsequent experiments.

- If the p value for treatments is much larger than 0.05, then this experiment has failed to separate treatment effects (if any) from those interfering variables. We say that the difference between treatments is not significant.

- If the p value indicates significant differences, compute a table of means as follows (be sure to include the semicolon and period as shown):

  MTB > TABLE C2;
  SUB > MEANS C1.

- The square root of the error mean square estimates the standard deviation within blocks and treatments, a measure of random variation in the experiment.

Table C.4   Effect of baking temperature on bread volume (data with blocks)

Block (Batch)	340°F	360°F	380°F
1	460	475	495
2	480	490	510
3	475	498	503
4	468	488	488

Example 6 (Two factors, CRD)   *In a repeat of the experiment in Examples 4 and 5, 4 batches of dough are prepared. Three loaves are taken from each batch and randomly assigned to treatments. In Table C.4, columns represent different treatments and rows represent different blocks. (For the purpose of comparison, exactly the same data as in Example 4 are used here, but arranged in blocks. The data in blocks 2 and 3 are uniformly higher than those in blocks 1 and 4. Table C.5 shows how these data look in Minitab.*

To determine whether these data show an effect of temperature on volume, we use the ANOVA command, placing the response variable to the left of the equal sign and the experimental and blocking variables to the right, in either order:

```
MTB > anova c1=c2 c3

    Factor    Type    Levels           Values
    Temp.     fixed      3          1    2    3
    Blocks    fixed      4          1    2    3    4
```

Table C.5   RCB data in Minitab worksheet

	C1	C2	C3
	Volume	Treatment	Block
1	460	1	1
2	480	1	2
3	475	1	3
4	468	1	4
5	475	2	1
6	490	2	2
7	498	2	3
8	488	2	4
9	495	3	1
10	510	3	2
11	503	3	3
12	488	3	4

```
             Analysis of Variance for Volume

      Source    DF        SS         MS        F        P
      Temp.      2    1618.17     809.08    27.30    0.001
      Blocks     3     595.67     198.56     6.70    0.024
      Error      6     177.83      29.64
      Total     11    2391.67
```

The p value for temperature (0.001) is extremely small so the evidence for a treatment effect is very convincing.

The p value for blocks is also quite small (0.024), indicating that batch-to-batch variation had a significant effect in volume and we did well to use blocks to eliminate its effect on loaf volume and should plan to use this blocking design in future similar experiments.

Notice that the error mean square, a measure of random variation, has changed from 85.9 in the previous example to 29.64 in this one. This reduction in errors is a result of blocking. This, in turn, causes the P value for treatments to change from 0.006 to 0.001, increasing the confidence in the results.

The ANOVA command does not produce confidence intervals so we use the "two-way" command:

```
MTB > twoway c1 c2 c3;
      SUBC> means c2 c3.

      ANALYSIS OF VARIANCE   Volume

      SOURCE    DF        SS         MS
      Temp.      2     1618.2      809.1
      Blocks     3      595.7      198.6
      ERROR      6      177.8       29.6
      TOTAL     11     2391.7


             Individual 95% CI
   Temp.     Mean      ---+---------+---------+---------+--
      1     470.8      (------*------)
      2     487.7                     (------*------)
      3     499.0                               (------*------)
                       ---+---------+---------+---------+--
                       470.0     480.0     490.0     500.0
             Individual 95% CI
   Blocks    Mean      ---+---------+---------+---------+--
      1     476.7      (------*------)
      2     493.3                        (------*------)
      3     492.0                       (------*------)
      4     481.3           (------*------)
                       ---+---------+---------+---------+--
                       472.0     480.0     488.0     496.0
```

The first table of means shows the treatment means with confidence limits. It indicates a definite increase in volume with increasing temperature.

The second table shows block means and confidence intervals. This is included here just to show the nature of the block-to-block variation. It shows that two batches produced higher than average volumes and two produced lower than average. By blocking the experimental units, we made sure that this variation did not interfere with the effects of the treatments that really interested us.

ANALYSIS OF TWO-FACTOR EXPERIMENTS, UNBLOCKED

- Enter and evaluate your data as described in "Evaluating Data in Minitab."
- In each row of two blank columns, enter numbers identifying the levels of the factor applied to the experimental unit in the row.
- If the data are in C1, factor 1 levels in C2, and factor 2 levels in C3, use the command:

  ```
  MTB > ANOVA C1 = C2 C3 C2*C3
  ```

  (or Stat|ANOVA|Balance ANOVA from the menus) to obtain an analysis of variance table and test whether the treatments are having a significant effect.
- If the p value for the interaction is less than 0.05, then it is quite likely that an interaction exists between the factors. Make an interaction plot (like Fig. 11.2A or B in Chap 11).
- If the p value for interaction is larger than 0.05 but the p value for main effects is less than 0.05, plot that main effect.
- The square root of the error mean square estimates the standard deviation within treatments after block-to-block variation is removed, a measure of random variation in the experiment.

Example 7 (Two factors, RCB) *An experiment is conducted to compare the effect of aluminum versus steel pans on loaf volume over a range of temperatures. Three temperatures were used, so this experiment requires 6 treatments. Each treatment is replicated twice, a total of 12 loaves of bread (experimental units) will be used. Each loaf comes from a different batch so completely random design (no blocking) can be used. Table C.6 lists the resulting data. The same data are entered into Minitab as shown in Table C.7.*

Table C.6   Effect of metal on loaf volume (2 × 3 factorial, CRD)

Metal	240°F	260°F	280°F
1. (Fe)	460	483	496
	472	502	512
2. (Al)	476	486	485
	484	498	479

Table C.7 Minitab worksheet, Example 7

	C1	C2	C3
	Volume	Metal	Temperature
1	460	1	1
2	472	1	1
3	476	2	1
4	484	2	1
5	483	1	2
6	502	1	2
7	486	2	2
8	498	2	2
9	496	1	3
10	512	1	3
11	485	2	3
12	479	2	3

```
MTB > anova c1=c2 c3 c2*c3

Factor    Type    Levels    Values
Metal     fixed      2          1    2
Temp      fixed      3          1    2    3

Analysis of Variance for Loaf Volume

Source        DF         SS        MS        F        P
Metal          1      24.08     24.08     0.29    0.611
Temp           2    1028.17    514.08     6.14    0.035
Metal*Temp     2     656.17    328.08     3.92    0.082
Error          6     502.50     83.75
Total         11    2210.92
```

Always check the interaction first. Since the p value (0.082) is greater than 0.05, we conclude that this experiment was unable to find an interaction between the factors. We can go on and check the main effects.

The main effect for metal is definitely not significant ($p = 0.611$) so this experiment has failed to show any difference between the two metals. On the other hand, the main effect for temperature ($p = 0.035$) is significant, confirming that temperature has an effect on loaf volume. To see the means for treatments and main effects, we compute

```
MTB > table c2 c3;
SUBC> means c1.
```

```
ROWS:    Metal      COLUMNS:  Temp

                1           2           3         ALL

    1       466.00      492.50      504.00      487.50
    2       480.00      492.00      482.08      484.67

ALL       473.00      492.25      493.00      486.08

    CELL CONTENTS --
            Volume:MEAN
```

Since the main effects for temperature are the only ones that are significant, the three means at the bottom of this table (473.00, 492.25, and 493.08) are those to examine and interpret. They appear to show a rise in volume between 240 and 260°, followed by a leveling off.

ANALYSIS OF TWO-FACTOR EXPERIMENTS, BLOCKED

- Enter and evaluate your data as described in "Evaluating Data in Minitab."
- In one blank column, enter in each row a number identifying the block to which the experimental unit belongs.
- In two blank columns, enter in each row numbers identifying the levels of the factor applied to that experimental unit.
- If data are in column C1, block numbers in column C2, factor 1 levels in column C3, and factor 2 levels in column C4, use the command:

```
MTB> ANOVA C1 = C2  C3  C4  C3*C4
```

(or Stat|ANOVA|Balance ANOVA from the menus) to obtain an analysis of variance table and test whether the treatments are having a significant effect.
- If the *p* value for blocks is less than 0.10, the blocking you did was worth doing. If it is larger, you may want to skip blocking by this variable in subsequent experiments.
- If the *p* value for the interaction is less than 0.05, then it is quite likely that there is an interaction.
- If the *p* value for interaction is larger than 0.05 but the *p* value for main effects is less than 0.05, plot that main effect.
- The square root of the error mean square estimates the standard deviation within blocks and treatments, a measure of random variation in the experiment.
- The interpretation of confidence intervals will be given later. As a rule of thumb, if two intervals overlap by less than 50% (see below), then the two treatments are very probably different.

```
(---------*---------)
        (---------*---------)
```

Example 8 *Example 7 is repeated except that only 2 batches of dough are made. Each batch is considered a block and divided into 6 loaves that are randomly assigned to the 6 treatments. The data are shown in Table C8. These same data are entered into a Minitab as shown in Table C.9:*

```
MTB > anova c1=c2 c3 c4 c3*c4

Factor      Type      Levels    Values
Block       fixed        2          1    2
Metal       fixed        2          1    2
Temp        fixed        3          1    2    3

Analysis of Variance for Volume

Source       DF          SS         MS         F         P
Block         1       310.08     310.08      8.06     0.036
Metal         1        24.08      24.08      0.63     0.465
Temp          2      1028.17     514.08     13.36     0.010
Metal*Temp    2       656.17     328.08      8.53     0.024
Error         5       192.42      38.48
Total        11      2210.92
```

Table C.8 2 × 3 factorial, RCB data

Metal	Block	240°F	260°F	280°F
1. (Fe)	1	460	483	496
	2	472	502	512
2. (Al)	1	476	486	485
	2	484	498	479

Table C.9 Example 8 data in Minitab

	C1	C2	C3	C4
	Volume	Blocks	Metal	Temperature
1	460	1	1	1
2	472	2	1	1
3	476	1	2	1
4	484	2	2	1
5	483	1	1	2
6	502	2	1	2
7	486	1	2	2
8	498	2	2	2
9	496	1	1	3
10	512	2	1	3
11	485	1	2	3
12	479	2	2	3

Blocking has reduced experimental error (error mean square of 38.48 rather than 83.75). This has made the interaction significant ($p = 0.024$) so you should plot and interpret the six means in the body of the means table:

```
MTB > table c3 c4;
SUBC> means c1.
                  1           2           3          ALL

     1         466.00      492.50      504.00      487.50
     2         480.00      492.00      482.00      484.67

    ALL        473.00      492.25      493.00      486.08
```

It appears from these data that loaf volume increases steadily with temperature in the Fe pan while it changes less in the aluminum pan and shows a possible maximum around 260°F. Confidence limits must be examined before accepting these observations.

C.5.2 Plots and Confidence Intervals

The analysis of variance tests whether the treatments are affecting the response variable. It does not tell which treatments differ. One way to find out is to plot the treatment means and draw confidence intervals around them.

The means computed from experimental data are rarely equal to the "true" means for a particular treatment. By "true" mean we refer to the mean we would obtain if we replicated a treatment an infinite number of times. Confidence intervals indicate the range of values in which the true mean will most likely be found. A wide confidence interval indicates much uncertainty. The width of an interval and, hence, the uncertainty depend on three factors.

- Uncertainty increases with increasing random variation.
- Uncertainty decreases when more replicates are averaged together.
- Uncertainty increases as the level of confidence increases.

Confidence intervals are computed with the equation:

$$CI = \bar{x} \pm t \frac{s}{\sqrt{n}} \tag{C.5}$$

COMPUTING CONFIDENCE INTERVALS

- Compute the standard deviation s within treatment as the square root of the error mean square (EMS) in the analysis of variance table.
- Determine the number of experimental units n averaged in the particular mean to which the interval will refer.

- Look up Student's t value for error degrees of freedom (Table C.10).
- The confidence interval is computed using Eq. (C.5).
- For qualitative treatments, use the following rule of thumb. If two confidence intervals overlap by 50% (or less) as shown here, then the treatments are significantly different.

```
(---------+---------)
              (---------*---------)
```

Table C.10   Values of Student's t for 95% confidence intervals

df	1	2	3	4	5	6	7	8	9	10	15	20	60	Inf
t	12.7	4.3	2.18	2.78	2.57	2.45	2.36	2.31	2.26	2.23	2.13	2.09	2.00	1.96

Example 9   *In Example 7, the main effect for temperature was significant and the means for temperature were 473.00, 492.25, and 493.00. Each mean is the average of four data values and the error mean square is 83.75. Error df = 6. The confidence limits of these means are computed as follows:*

$$s = \sqrt{83.75} = 9.15 \qquad \text{with 6 df and 95\% confidence, } t = 2.45$$

$$CI = 473.00 \pm 2.45\frac{9.15}{\sqrt{4}} = 473.00 \pm 11.2 = \left\{\begin{matrix} 484 \\ 462 \end{matrix}\right\}$$

$$= 492.25 \pm 2.45\frac{9.15}{\sqrt{4}} = 492.25 \pm 11.2 = \left\{\begin{matrix} 503 \\ 481 \end{matrix}\right\}$$

$$= 493.00 \pm 2.45\frac{9.15}{\sqrt{4}} = 493.00 \pm 11.2 = \left\{\begin{matrix} 504 \\ 482 \end{matrix}\right\}$$

The three means with their limits are plotted in Figure C.2.

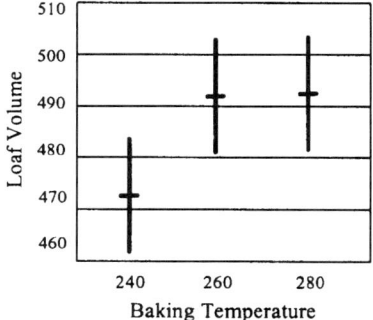

Figure C.2   Confidence intervals of loaf volume as a function of baking temperature.

Example 10 In Example 8, the interaction between metal and temperature was significant. The means for the steel pans were 466.0, 492.5, and 504.0. The means for the aluminum pans were 480.0, 492.0, and 482. Each mean is the average of two data values and the error mean square is 38.48. Error df = 5. The confidence limits of these means are computed as follows:

$$s = \sqrt{38.48} = 6.20 \qquad \text{with 5 df and 95\% confidence, } t = 2.57$$

$$CI = 466.0 \pm 2.57 \frac{6.20}{\sqrt{2}} = 466.0 \pm 10.4 = \begin{Bmatrix} 476 \\ 456 \end{Bmatrix}$$

$$= 492.5 \pm 2.57 \frac{6.20}{\sqrt{2}} = 492.5 \pm 10.4 = \begin{Bmatrix} 503 \\ 482 \end{Bmatrix}$$

$$= 504.0 \pm 2.57 \frac{6.20}{\sqrt{2}} = 504.0 \pm 10.4 = \begin{Bmatrix} 524 \\ 494 \end{Bmatrix}$$

$$= 480.0 \pm 2.57 \frac{6.20}{\sqrt{2}} = 480.0 \pm 10.4 = \begin{Bmatrix} 490 \\ 470 \end{Bmatrix}$$

$$= 492.0 \pm 2.57 \frac{6.20}{\sqrt{2}} = 492.0 \pm 10.4 = \begin{Bmatrix} 502 \\ 482 \end{Bmatrix}$$

$$= 482.0 \pm 2.57 \frac{6.20}{\sqrt{2}} = 482.0 \pm 10.4 = \begin{Bmatrix} 492 \\ 472 \end{Bmatrix}$$

These means and confidence intervals have been plotted in Figure C.3. From this, we see that the only significant difference is between the metals at 280°. At other temperatures, the two metals do not show a significant difference.

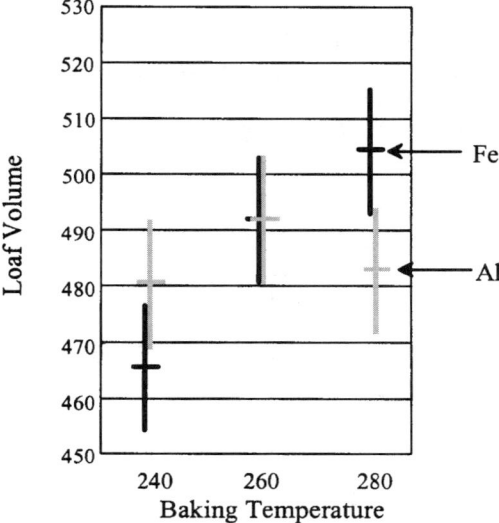

Figure C.3 Confidence intervals of loaf volume as a function of baking temperature and metal pans.

C.5.3 INTERPRETING CONFIDENCE INTERVALS

See the box below for rules of thumb to follow in interpreting confidence intervals. More precise statistical tests are available, but in most cases, these are good enough to interpret the data. The following examples illustrate the rules of thumb for the interpretation of confidence limits under several different circumstances.

RULES OF THUMB FOR INTERPRETING CONFIDENCE LIMITS

- *For qualitative treatments.* If confidence limits for two treatments overlap by 50% or less, then the experiment has succeeded in showing a significant difference between those treatments (*A* and *B*, e.g.). If the overlap is greater than 50%, the experiment has failed to show a significant difference (*B* and *C*, e.g.):

```
A.   (---------*---------)
B.                 (---------*---------)
C.                    (---------*---------).
```

- *For quantitative treatments.*

  If we can draw a horizontal line through all intervals, then the experiment has failed to show a relationship between the experimental and response variables.

  If we can draw a straight but not horizontal line through all intervals, then the experiment shows a linear relationship between the variables. If the line cannot move much without going outside the intervals, then the slope can be estimated fairly closely.

  Otherwise, the slope is poorly estimated from the data.

  In that case, a nonlinear relationship probably exists between the variables.

Example 11 *In Figure C.4A, qualitative treatments A and B overlap by more than 50% and, therefore, do not differ significantly. More data or a more precise experiment would be needed to show that these treatments really differ. Also, C and D do not differ significantly. But A and B both differ significantly from C and D.*

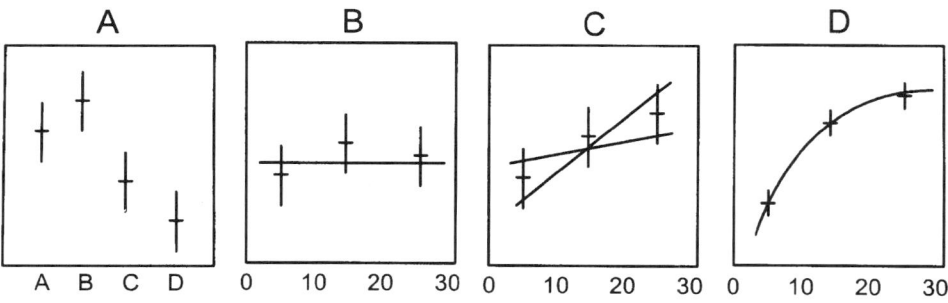

Figure C.4 Trends of different confidence intervals.

Example 12 *In Figure C.4B, quantitative treatments are compared. Here, a horizontal line will easily fit within all intervals so the data give no evidence of a relationship between the variables.*

Example 13 *In Figure C.4C, the response variable shows a linear relationship between the variables. However, because of the wide intervals, the slope of the relationship cannot be closely estimated from the data. Two possible slopes are shown.*

Example 14 *In Figure C.4D, the response variable shows a nonlinear relationship between the variables. Because of the narrow intervals, the position and shape of the line can be closely estimated.*

APPENDIX D

PROCESSING PARAMETERS IN SPRAY DRYING

Table D.1   Typical experimental data from SPRAY drying of skim milk powder

Inlet Temperature (°C)	Outlet Temperature °(C)	Moisture % (wet basis)	Bulk Density (g/mL)	Water Activity
150	90	4	0.56	0.09
180	113	1.2	0.54	0.07
210	135	1	0.51	0.06

APPENDIX E

PERMEABILITY OF SOME PACKAGING FILMS

Table E.1 Permeability ($cm^3 \cdot mil \cdot m^3$/day/atm pressure) of some plastic films at room temperature

Packaging Films	Permeant, Nitrogen	Permeant, Oxygen	Permeant, Carbon dioxide
Natural rubber	20,000	60,000	
Silicone rubber	—	10^6	6×10^6
Mylar (polyester)	20	80	260
Trithene or Kel-F	40	150	1,000
High-density polyethylene	700	2,000	10,000
Low-density polyethylene	3,500	12,000	70,000
Nylon-6	25	100	400
Saran	3	13	75

Table E.2 Permeability of various packaging materials to water vapors at 100°F and 95% versus 0% relative humidity

Packaging Materials	Permeability Range (g mil/24 h/100 in^2)
Aluminum foil, 35 mil thick	0.1–1.0
Aluminum foil, 14 mil thick	< 0.1
Polypropylene	0.2–0.4
Polyethylene, conventional	0.8–1.0
Polyethylene, low-pressure	0.3–0.5
Plain cellophane	20–100
Vinyl chloride-based films	0.5–0.8
Nitrocellulose-coated cellophane	0.2–2.0
Polytrifluorochloroethylene	0.01–0.1
Plastic paper foil laminations	< 0.1
Waxed paper	0.2–15.0
Coated papers	0.2–5.0
Mylar	0.8–1.5
Saran	0.1–0.5
Silicone rubber	> 200

APPENDIX F

FUNDAMENTAL CONSTANTS AND CONVERSION FACTORS

1. *Gas law constant (R)*

8314.34	$J/kg \cdot mol \cdot °K$
8314.35	$kg \cdot m^2/s^2 \cdot kg\ mol \cdot °K$
8314.36	$m^3 \cdot Pa/kg\ mol \cdot K$
1.9872	$g\ cal/g\ mol \cdot K$
1.9872	$Btu/lb\ mol \cdot °R$

2. *Acceleration of gravity*

   $g = 9.80665\ m/s^2 = 980\ cm/s^2 = 32.174\ ft/s^2$

   g_c (gravitational constant) $= 32.174\ lb_m \cdot s/lb_f \cdot s^2 = 980.665\ g_m \cdot cm/g_f \cdot s^2$

3. *Density*

   $1\ g/cm^3 = 62.43\ lb_m/ft^3 = 1000\ kg/m^3$

   $1\ g/cm^3 = 8.345\ lb_m/U.S.\ gallon$

   $1\ lb_m/ft^3 = 16.018\ kg/m^3$

4. *Force*

   $1\ kg \cdot m/s^2 = 1\ Newton\ (N)$

   $1\ lbf = 4.4482\ N$

   $1\ g \cdot cm/s^2\ (dyne) = 10^{-5}\ kg \cdot m/s^2 = 10^{-5}\ N = 2.2481 \times 10^{-6}\ lb_f$

   $1\ kg\ force = 9.806\ N$

5. *Heat transfer coefficient*

$1\,kcal/h \cdot m^2 \cdot °F = 0.2048\,Btu/h \cdot ft^2 \cdot °F$
$1\,Btu/h \cdot ft^2 \cdot °F = 5.6783 \times 10^{-4}\,W/cm^2 \cdot °C$
$1\,Btu/h \cdot ft^2 \cdot °F = 5.6783\,W/m^2 \cdot °C$

6. *Heat capacity, enthalpy*

$1\,Btu/lb_m \cdot °F = 4.1868\,kJ/kg \cdot °K = 1.0\,cal/g \cdot °C$
$1\,Btu/lb_m = 2326.0\,J/kg$
$1\,ft \cdot lbf/lb_m = 2.989\,J/kg$
$1\,kcal/g\,mol = 4.184 \times 10^3\,kJ/kg\,mol$

7. *Heat flux*

$1\,Btu/h \cdot ft^2 = 3.1546\,W/m^2$
$1\,Btu/h = 0.29307\,W$
$1\,cal/h = 1.1622 \times 10^{-3}\,W$

8. *Length*

$1\,in. = 2.54\,cm = 25.4\,mm$
$100\,cm = 1\,m$
$1\,\mu m = 10^{-6}\,m$
$1\,m = 3.2808\,ft = 39.37\,in.$
$1\,mile = 1760\,yd = 5280\,ft$
$1\,ft = 0.3048\,m$

9. *Mass*

$1\,lb_m = 453.59\,g = 16\,oz$
$1\,kg = 1000\,g = 2.2046\,lb_m$
$1\,metric\,ton = 1000\,kg$
$1\,short\,ton = 2000\,lb_m$
$1\,long\,ton = 2240\,lb_m$
$1\,oz = 28.349\,g$

10. *Mass flux and molar flux*

$1\,g/cm^2 \cdot s = 7.3734 \times 10^3\,lb_m/h \cdot ft^2$
$1\,g\,mol/cm^2 \cdot s = 7.3734 \times 10^3\,lb\,mol/h \cdot ft^2 = 10\,kg\,mol/m^2 \cdot s = 1 \times 10^4\,g\,mol/m^2 \cdot s$
$1\,lb\,mol/h \cdot ft^2 = 1.3562 \times 10^{-3}\,kg\,mol/m^2 \cdot s$

11. *Power*

$1\,hp = 745.7\,(Watts) = 0.7457\,kW = 550\,ft \cdot lbf/s = 0.7068\,Btu/s$
$1\,W = 14.34\,cal/min = 1\,J\,(Joule)/s$

12. *Pressure*

$1\,bar = 1 \times 10^5\,Pa = 1 \times 10^5\,N/m^2$
$1\,atm = 14.696\,psia = 101.325\,kPa = 1.01325\,bar = 760\,mm\,Hg = 29.921\,in.\,Hg$
$1\,atm = 33.90\,ft\,water\,at\,4°C$

1 psia $= 2.0360$ in. Hg $= 51.715$ mm Hg $= 6.89476$ dyne/cm^2 $= 6.89476$ kPa
1 mm Hg or 1 torr (0°C) $= 133.322$ Pa $= 0.13332$ kPa
1 Mpa $= 145.038$ psi

13. *Thermal conductivity*
    1 Btu/h · ft · °F $= 1.730735$ W/m · °K
    1 Btu/h · ft · °F $= 4.1365 \times 10^{-3}$ cal/s · cm · °C

14. *Viscosity, dynamic*
    1 cP $= 10^{-2}$ g/cm · s (poise) $= 2.4191$ lb$_m$/ft · h $= 6.7197 \times 10^{-4}$ lb$_m$/ft · s
    1 cP $= 2.0886 \times$ 10-5 lbf · s/ft^2
    1 cP $= 10^{-3}$ Pa · s $= 10^{-3}$ kg/m · s $= 10^{-3}$ N · s/m^2
    1 Pa · s $= 1$ N · s/m^2 $= 1$ kg/m · s $= 1000$ cp

15. *Viscosity, kinematic*
    1 Stokes $=$ cm^2/s $= 100$ mm^2/s $= 10^{-4}$ m^2/s $= 3.875$ ft^2/h
    1 cS $= 10^{-2}$ cm^2/s

16. *Volume*
    1 L (liter) $= 1000$ cm^3
    1 in.^{3} $= 16.387$ cm^3
    1 ft^3 $= 28.317$ L $= 0.028317$ m^3 $= 7.481$ U.S. gallons
    1 m^3 $= 1000$ L $= 264.17$ U.S. gallons
    1 U.S. gallon $= 4$ qt $= 3.7854$ L
    1 British gallon $= 1.20094$ U.S. gallons

17. *Work, energy, and heat*
    1 J $= 1$ N · m $= 1$ kg · m^2/s^2 $= 10^7$ g · cm^2/s^2 (erg)
    1 Btu $= 1055.06$ J $= 1.05506$ kJ $= 252.16$ cal $= 778.17$ ft · lb$_f$
    1 cal $= 4.1868$ J
    1 hp · h $= 0.7457$ kW · h
    1 Btu/ft^3 $= 37.25895$ kJ/m^3
    1 ft.lb$_f$/lb$_m$ $= 2.989$ J/kg

APPENDIX G

STEAM TABLES

Table G.1 Saturated steam–temperature

Temperature (°C) T	Pressure (kPa) P	Specific Volume (m³/kg)		Internal Energy (kJ/kg)			Enthalpy (kJ/kg)			Entropy (kJ/kg)		
		Liquid v_f	Saturated Vapor v_g	Saturated Liquid u_f	Saturated Evaporation u_{fg}	Vapor u_g	Saturated Liquid h_f	Saturated Evaporation h_{fg}	Vapor h_g	Saturated Liquid s_f	Saturated Evaporation s_{fg}	Saturated Vapor s_g
0.01	0.61	0.001000	206.14	00.00	2375.3	2375.3	00.01	2501.3	2501.4	0.0000	9.1562	9.1562
5	0.87	0.001000	147.12	20.97	2361.3	2382.3	20.98	2489.6	2510.6	0.0761	8.9496	9.0257
10	1.23	0.001000	106.38	42.00	2347.2	2389.2	42.01	2477.7	2519.8	0.1510	8.7498	8.9008
15	1.70	0.001001	77.93	62.99	2333.1	2396.1	62.99	2465.9	2528.9	0.2245	8.5569	8.7814
20	2.34	0.001002	57.79	83.95	2319.0	2402.9	83.96	2454.1	2538.1	0.2966	8.3706	8.6672
25	3.17	0.001003	43.36	104.88	2304.9	2409.8	104.89	2442.3	2547.2	0.3674	8.1905	8.5580
30	4.25	0.001004	32.89	125.78	2290.8	2416.6	125.79	2430.5	2556.3	0.4369	8.0164	8.4533
35	5.63	0.001006	25.22	146.67	2276.7	2423.4	146.68	2418.6	2565.3	0.5053	7.8478	8.3531
40	7.38	0.001008	19.52	167.56	2262.6	2430.1	167.57	2406.7	2574.3	0.5725	7.6845	8.2570
45	9.59	0.001010	15.26	188.44	2248.4	2436.8	188.45	2394.8	2583.2	0.6387	7.5261	8.1648
50	12.35	0.001012	12.03	209.32	2234.2	2443.5	209.33	2382.7	2592.1	0.7038	7.3725	8.0763
55	15.76	0.001015	9.568	230.21	2219.9	2450.1	230.23	2370.7	2600.9	0.7679	7.2234	7.9913
60	19.94	0.001017	7.671	251.11	2205.5	2456.6	251.13	2358.5	2609.6	0.8312	7.0784	7.9096
65	25.03	0.001020	6.197	272.02	2191.1	2463.1	272.06	2346.2	2618.3	0.8935	6.9375	7.8310
70	31.19	0.001023	5.042	292.95	2176.6	2469.6	292.98	2333.8	2626.8	0.9549	6.8004	7.7553
75	38.58	0.001026	4.131	313.90	2162.0	2475.9	313.93	2321.4	2635.3	1.0155	6.6669	7.6824
80	47.39	0.001029	3.407	334.86	2147.4	2482.2	334.91	2308.8	2643.7	1.0753	6.5369	7.6122
85	57.83	0.001033	2.828	355.84	2132.6	2488.4	355.90	2296.0	2651.9	1.1343	6.4102	7.5445
90	70.14	0.001036	2.361	376.85	2117.7	2494.5	376.92	2283.2	2660.1	1.1925	6.2866	7.4791
95	84.55	0.001040	1.982	397.88	2102.7	2500.6	397.96	2270.2	2668.1	1.2500	6.1659	7.4159
100	101.32	0.001044	1.6729	418.94	2087.6	2506.5	419.04	2257.0	2676.1	1.3069	6.0480	7.3549
105	120.82	0.001048	1.4194	440.02	2072.3	2512.4	440.15	2243.7	2683.8	1.3630	5.9328	7.2958
110	143.27	0.001052	1.2102	461.14	2057.0	2518.1	461.30	2230.2	2691.5	1.4185	5.8202	7.2387
115	169.06	0.001056	1.0366	482.30	2041.4	2523.7	482.48	2216.5	2699.0	1.4734	5.7100	7.1833

120	198.53	0.001060	0.8919	503.50	2025.8	2529.3	503.71	2202.6	2706.3	1.5276	5.6020	7.1296
125	232.1	0.001065	0.7706	524.74	2009.9	2534.6	524.99	2188.5	2713.5	1.5813	5.4962	7.0775
130	270.1	0.001070	0.6685	546.02	1993.9	2539.9	546.31	2174.2	2720.5	1.6344	5.3925	7.0269
135	313.0	0.001075	0.5822	567.35	1977.7	2545.0	567.69	2159.6	2727.3	1.6870	5.2907	6.9777
140	361.3	0.001080	0.5089	588.74	1961.3	2550.0	589.13	2144.7	2733.9	1.7391	5.1908	6.9299
145	415.4	0.001085	0.4463	610.18	1944.7	2554.9	610.63	2129.6	2740.3	1.7907	5.0926	6.8833
150	475.8	0.001091	0.3928	631.68	1927.9	2559.5	632.20	2114.3	2746.5	1.8418	4.9960	6.8379
155	543.1	0.001096	0.3468	653.24	1910.8	2564.1	653.84	2098.6	2752.4	1.8925	4.9010	6.7935
160	617.8	0.001102	0.3071	674.87	1893.5	2568.4	675.55	2082.6	2758.1	1.9427	4.8075	6.7502
165	700.5	0.001108	0.2727	696.56	1876.0	2572.5	697.34	2066.2	2763.5	1.9925	4.7153	6.7078
170	791.7	0.001114	0.2428	718.33	1858.1	2576.5	719.21	2049.5	2768.7	2.0419	4.6244	6.6663
175	892.0	0.001121	0.2168	740.17	1840.0	2580.2	741.17	2032.4	2773.6	2.0909	4.5347	6.6256
180	1002.1	0.001127	0.19405	762.09	1821.6	2583.7	763.22	2015.0	2778.2	2.1396	4.4461	6.5857
185	1122.7	0.001134	0.17409	784.10	1802.9	2587.0	785.37	1997.1	2782.4	2.1879	4.3586	6.5465
190	1254.4	0.001141	0.15654	806.19	1783.8	2590.0	807.62	1978.8	2786.4	2.2359	4.2720	6.5079
195	1397.8	0.001149	0.14105	828.37	1764.4	2592.8	829.98	1960.0	2790.0	2.2835	4.1863	6.4698
200	1553.8	0.001157	0.12736	850.65	1744.7	2595.3	852.45	1940.7	2793.2	2.3309	4.1014	6.4323
205	1723.0	0.001164	0.11521	873.04	1724.5	2597.5	875.04	1921.0	2796.0	2.3780	4.0172	6.3952
210	1906.2	0.001173	0.10441	895.53	1703.9	2599.5	897.76	1900.7	2798.5	2.4248	3.9337	6.3585
215	2104.0	0.001181	0.09479	918.14	1682.9	2601.1	920.62	1879.9	2800.5	2.4714	3.8507	6.3221
220	2318.0	0.001190	0.08619	940.87	1661.5	2602.4	943.62	1858.5	2802.1	2.5178	3.7683	6.2861
225	2548.0	0.001199	0.07849	963.73	1639.6	2603.3	966.78	1836.5	2803.3	2.5639	3.6863	6.2503
230	2795.0	0.001209	0.07158	986.74	1617.2	2603.9	990.12	1813.8	2804.0	2.6099	3.6047	6.2146
235	3060.0	0.001219	0.06537	1009.89	1594.2	2604.1	1013.62	1790.5	2804.2	2.6558	3.5233	6.1791
240	3344.0	0.001229	0.05976	1033.21	1570.8	2604.0	1037.32	1766.5	2803.8	2.7015	3.4422	6.1437
245	3648.0	0.001240	0.05471	1056.71	1546.7	2603.4	1061.23	1741.7	2803.0	2.7472	3.3612	6.1083
250	3973.0	0.001251	0.05013	1080.39	1522.0	2602.4	1085.36	1716.2	2801.5	2.7927	3.2802	6.0730
255	4319.0	0.001263	0.04598	1104.28	1496.7	2600.9	1109.73	1689.8	2799.5	2.8383	3.1992	6.0375
260	4688.0	0.001276	0.04221	1128.39	1470.6	2599.0	1134.37	1662.5	2796.9	2.8838	3.1181	6.0019
265	5081.0	0.001289	0.03877	1152.74	1443.9	2596.6	1159.28	1634.4	2793.6	2.9294	3.0368	5.9662
270	5499.0	0.001302	0.03564	1177.36	1416.3	2593.7	1184.51	1605.2	2789.7	2.9751	2.9551	5.9301

(*continued*)

Table G1. *(continued)*

Temperature (°C) T	Pressure (MPa) P	Specific Volume (m³/kg)		Internal Energy (kJ/kg)			Enthalpy (kJ/kg)			Entropy (kJ/kg)		
		Liquid v_f	Saturated Vapor v_g	Saturated Liquid u_f	Saturated Evaporation u_{fg}	Vapor u_g	Saturated Liquid h_f	Saturated Evaporation h_{fg}	Vapor h_g	Saturated Liquid s_f	Saturated Evaporation s_{fg}	Saturated Vapor s_g
275	5.942	0.001317	0.03279	1202.25	1387.9	2590.2	1210.07	1574.9	2785.0	3.0208	2.8730	5.8938
280	6.412	0.001332	0.03017	1227.46	1358.7	2586.1	1235.99	1543.6	2779.6	3.0668	2.7903	5.8571
285	6.909	0.001348	0.02777	1253.00	1328.4	2581.4	1262.31	1511.0	2773.3	3.1130	2.7070	5.8199
290	7.436	0.001366	0.02557	1278.92	1297.1	2576.0	1289.07	1477.1	2766.2	3.1594	2.6227	5.7821
295	7.993	0.001384	0.02354	1305.20	1264.7	2569.9	1316.30	1441.8	2758.1	3.2062	2.5375	5.7437
300	8.581	0.001404	0.02167	1332.00	1231.0	2563.0	1344.00	1404.9	2749.0	3.2534	2.4511	5.7045
305	9.202	0.001425	0.019948	1359.30	1195.9	2555.2	1372.40	1366.4	2738.7	3.3010	2.3633	5.6643
310	9.856	0.001447	0.018350	1387.10	1159.4	2546.4	1401.30	1326.0	2727.3	3.3493	2.2737	5.6230
315	10.547	0.001472	0.016867	1415.50	1121.1	2536.6	1431.00	1283.5	2714.5	3.3982	2.1821	5.5804
320	11.274	0.001499	0.015488	1444.60	1080.9	2525.5	1461.50	1238.6	2700.1	3.4480	2.0882	5.5362
330	12.845	0.001561	0.012996	1505.30	993.7	2498.9	1525.30	1140.6	2665.9	3.5507	1.8909	5.4417
340	14.586	0.001638	0.010797	1570.30	894.3	2464.6	1594.20	1027.9	2622.0	3.6594	1.6763	5.3357
350	16.513	0.001740	0.008813	1641.90	776.6	2418.4	1670.60	893.4	2563.9	3.7777	1.4335	5.2112
360	18.651	0.001893	0.006945	1725.20	626.3	2351.5	1760.50	720.5	2481.0	3.9147	1.1379	5.0526
370	21.03	0.002213	0.004925	1844.00	384.5	2228.5	1890.50	441.6	2332.1	4.1106	0.6865	4.7971
374.14	22.09	0.003155	0.003155	2029.60	0.0	2029.6	2099.30	0.0	2099.3	4.4298	0.0000	4.4298

Source: From G.J. Van Wylen, R.E. Sonntag, and C. Borganakke, 1994. *Fundamentals of Classical Thermodynamics.* New York: John Wiley & Sons.

Table G.2 Superheated vapor

T		P = 0.010 MPa (Sat. 45.81°C)				P = 0.050 MPa (Sat. 81.33°C)				P = 0.10 MPa (Sat. 99.63°C)		
	v	u	h	s	v	u	h	s	v	u	h	s
Sat.	14.674	2437.9	2584.7	8.1502	3.240	2483.9	2645.9	7.5939	1.6940	2506.1	2675.5	7.3594
50	14.869	2443.9	2592.6	8.1749								
100	17.196	2515.5	2687.5	8.4479	3.418	2511.6	2682.5	7.6947	1.6958	2506.7	2676.2	7.3614
150	19.512	2587.9	2783.0	8.6882	3.889	2585.6	2780.1	7.9401	1.9364	2582.8	2776.4	7.6134
200	21.825	2661.3	2879.5	8.9038	4.356	2659.9	2877.7	8.1580	2.172	2658.1	2875.3	7.8343
250	24.136	2736.0	2977.3	9.1002	4.820	2735.0	2976.0	8.3556	2.406	2733.7	2974.3	8.0333
300	26.445	2812.1	3076.5	9.2813	5.284	2811.3	3075.5	8.5373	2.639	2810.4	3074.3	8.2158
400	31.063	2968.9	3279.6	9.6077	6.209	2968.5	3278.9	8.8642	3.103	2967.9	3278.2	8.5435
500	35.679	3132.3	3489.1	9.8978	7.134	3132.0	3488.7	9.1546	3.565	3131.6	3488.1	8.8342
600	40.295	3302.5	3705.4	10.1608	8.057	3302.2	3705.1	9.4178	4.028	3301.9	3704.7	9.0976
700	44.911	3479.6	3928.7	10.4028	8.981	3479.4	3928.5	9.6599	4.490	3479.2	3928.2	9.3398
800	49.526	3663.8	4159.0	10.6281	9.904	3663.6	4158.9	9.8852	4.952	3663.5	4158.6	9.5652
900	54.141	3855.0	4396.4	10.8396	10.828	3854.9	4396.3	10.0967	5.414	3854.8	4396.1	9.7767
1000	58.757	4053.0	4640.6	11.0393	11.751	4052.9	4640.5	10.2964	5.875	4052.8	4640.3	9.9764
1100	63.372	4257.5	4891.2	11.2287	12.674	4257.4	4891.1	10.4859	6.337	4257.3	4891.0	10.1659
1200	67.987	4467.9	5147.8	11.4091	13.597	4467.8	5147.7	10.6662	6.799	4467.7	5147.6	10.3463
1300	72.602	4683.7	5409.7	11.5811	14.521	4683.6	5409.6	10.8382	7.260	4683.5	5409.5	10.5183

T		P = 0.20 MPa (Sat. 120.23°C)				P = 0.30 MPa (Sat. 133.55°C)				P = 0.40 MPa (Sat. 143.63°C)		
	v	u	h	s	v	u	h	s	v	u	h	s
Sat.	0.8857	2529.5	2706.7	7.1272	0.6058	2543.6	2725.3	6.9919	0.4625	2553.6	2738.6	6.8959
150	0.9596	2576.9	2768.8	7.2795	0.6339	2570.8	2761.0	7.0778	0.4708	2564.5	2752.8	6.9299
200	1.0803	2654.4	2870.5	7.5066	0.7163	2650.7	2865.6	7.3115	0.5342	2646.8	2860.5	7.1706
250	1.1988	2731.2	2971.0	7.7086	0.7964	2728.7	2967.6	7.5166	0.5951	2726.1	2964.2	7.3789
300	1.3162	2808.6	3071.8	7.8926	0.8753	2806.7	3069.3	7.7022	0.6548	2804.8	3066.8	7.5662

(continued)

Table G.2 (*continued*)

T	P = 0.20 MPa (Sat. 120.23°C)				P = 0.30 MPa (Sat. 133.55°C)				P = 0.40 MPa (Sat. 143.63°C)			
	v	u	h	s	v	u	h	s	v	u	h	s
400	1.5493	2966.7	3276.6	8.2218	1.0315	2965.6	3275.0	8.0330	0.7726	2964.4	3273.4	7.8985
500	1.7814	3130.8	3487.1	8.5133	1.1867	3130.0	3486.0	8.3251	0.8893	3129.2	3484.9	8.1913
600	2.013	3301.4	3704.0	8.7770	1.3414	3300.8	3703.2	8.5892	1.0055	3300.2	3702.4	8.4558
700	2.244	3478.8	3927.6	9.0194	1.4957	3478.4	3927.1	8.8319	1.1215	3477.9	3926.5	8.6987
800	2.475	3663.1	4158.2	9.2449	1.6499	3662.9	4157.8	9.0576	1.2372	3662.4	4157.3	8.9244
900	2.706	3854.5	4395.8	9.4566	1.8041	3854.2	4395.4	9.2692	1.3529	3853.9	4395.1	9.1362
1000	2.937	4052.5	4640.0	9.6563	1.9581	4052.3	4639.7	9.4690	1.4685	4052.0	4639.4	9.3360
1100	3.168	4257.0	4890.7	9.8458	2.1121	4256.8	4890.4	9.6585	1.5840	4256.5	4890.2	9.5256
1200	3.399	4467.5	5147.3	10.0262	2.2661	4467.2	5147.1	9.8389	1.6996	4467.0	5146.8	9.7060
1300	3.630	4683.2	5409.3	10.1982	2.4201	4683.0	5409.0	10.0110	1.8151	4682.8	5408.8	9.8780

T	P = 0.50 MPa (Sat. 151.86°C)				P = 0.60 MPa (Sat. 158.85°C)				P = 0.80 MPa (Sat. 170.43°C)			
	v	u	h	s	v	u	h	s	v	u	h	s
Sat.	0.3749	2561.2	2748.7	6.8213	0.3157	2567.4	2756.8	6.7600	0.2404	2576.8	2769.1	6.6628
200	0.4249	2642.9	2855.4	7.0592	0.3520	2638.9	2850.1	6.9665	0.2608	2630.6	2839.3	6.8158
250	0.4744	2723.5	2960.7	7.2709	0.3938	2720.9	2957.2	7.1816	0.2931	2715.5	2950.0	7.0384
300	0.5226	2802.9	3064.2	7.4599	0.4344	2801.0	3061.6	7.3724	0.3241	2797.2	3056.5	7.2328
350	0.5701	2882.6	3167.7	7.6329	0.4742	2881.2	3165.7	7.5464	0.3544	2878.2	3161.7	7.4089
400	0.6173	2963.2	3271.9	7.7938	0.5137	2962.1	3270.3	7.7079	0.3843	2959.7	3267.1	7.5716
500	0.7109	3128.4	3483.9	8.0873	0.5920	3127.6	3482.8	8.0021	0.4433	3126.0	3480.6	7.8673
600	0.8041	3299.6	3701.7	7.3522	0.6697	3299.1	3700.9	8.2674	0.5018	3297.9	3699.4	8.1333

Table G.2 *(continued)*

T	P = 0.50 MPa (Sat. 151.86°C)				P = 0.60 MPa (Sat. 158.85°C)				P = 0.80 MPa (Sat. 170.43°C)			
	v	u	h	s	v	u	h	s	v	u	h	s
700	0.8969	3477.5	3925.9	8.5952	0.7472	3477.0	3925.3	8.5107	0.5601	3476.2	3924.2	8.3770
800	0.9896	3662.1	4156.9	8.8211	0.8245	3661.8	4156.5	8.7367	0.6181	3661.1	4155.6	8.6033
900	1.0822	3853.6	4394.7	9.0329	0.9017	3853.4	4394.4	8.9486	0.6761	3852.8	4393.7	8.8153
1000	1.1747	4051.8	4639.1	9.2328	0.9788	4051.5	4638.8	9.1485	0.7340	4051.0	4638.2	9.0153
1100	1.2672	4256.3	4889.9	9.4224	1.0559	4256.1	4889.6	9.3381	0.7919	4255.6	4889.1	9.2050
1200	1.3596	4466.8	5146.6	9.6029	1.1330	4466.5	5146.3	9.5185	0.8497	4466.1	5145.9	9.3855
1300	1.4521	4682.5	5408.6	9.7749	1.2101	4682.3	5408.3	9.6906	0.9076	4681.8	5407.9	9.5575

T	P = 1.00 MPa (Sat. 179.91°C)				P = 1.20 MPa (Sat. 187.99°C)				P = 1.40 MPa (Sat. 195.07°C)			
	v	u	h	s	v	u	h	s	v	u	h	s
Sat.	0.19444	2583.6	2778.1	6.5865	0.16333	2588.8	2784.8	6.5233	0.14084	2592.8	2790.0	6.4693
200	0.2060	2621.9	2827.9	6.6940	0.16930	2612.8	2815.9	6.5898	0.14302	2603.1	2803.3	6.4975
250	0.2327	2709.9	2942.6	6.9247	0.19234	2704.2	2935.0	6.8294	0.16350	2698.3	2927.2	6.7467
300	0.2579	2793.2	3051.2	7.1229	0.2138	2789.2	3045.8	7.0317	0.18228	2785.2	3040.4	6.9534
350	0.2825	2875.2	3157.7	7.3011	0.2345	2872.2	3153.6	7.2121	0.2003	2869.2	3149.5	7.1360
400	0.3066	2957.3	3263.9	7.4651	0.2548	2954.9	3260.7	7.3774	0.2178	2952.5	3257.5	7.3026
500	0.3541	3124.4	3478.5	7.7622	0.2946	3122.8	3476.3	7.6759	0.2521	3121.1	3474.1	7.6027
600	0.4011	3296.8	3697.9	8.0290	0.3339	3295.6	3696.3	7.9435	0.2860	3294.4	3694.8	7.8710
700	0.4478	3475.3	3923.1	8.2731	0.3729	3474.4	3922.0	8.1881	0.3195	3473.6	3920.8	8.1160
800	0.4943	3660.4	4154.7	8.4996	0.4118	3659.7	4153.8	8.4148	0.3528	3659.0	4153.0	8.3431
900	0.5407	3852.2	4392.9	8.7118	0.4505	3851.6	4392.2	8.6272	0.3861	3851.1	4391.5	8.5556
1000	0.5871	4050.5	4637.6	8.9119	0.4892	4050.0	4637.0	8.8274	0.4192	4049.5	4636.4	8.7559
1100	0.6335	4255.1	4888.6	9.1017	0.5278	4254.6	4888.0	9.0172	0.4524	4254.1	4887.5	8.9457
1200	0.6798	4465.6	5145.4	9.2822	0.5665	4465.1	5144.9	9.1977	0.4855	4464.7	5144.4	9.1262
1300	0.7261	4681.3	5407.4	9.4543	0.6051	4680.9	5407.0	9.3698	0.5186	4680.4	5406.5	9.2984

(continued)

Table G.2 (*continued*)

T	P = 1.60 MPa (Sat. 201.41°C)				P = 1.80 MPa (Sat. 207.15°C)				P = 2.00 MPa (Sat. 212.42°C)			
	v	u	h	s	v	u	h	s	v	u	h	s
Sat.	0.12380	2596.0	2794.0	6.4218	0.11042	2598.4	2797.1	6.3794	0.09963	2600.3	2799.5	6.3409
225	0.13287	2644.7	2857.3	6.5518	0.11673	2636.6	2846.7	6.4808	0.10377	2628.3	2835.8	6.4147
250	0.14184	2692.3	2919.2	6.6732	0.12497	2686.0	2911.0	6.6066	0.11144	2679.6	2902.5	6.5453
300	0.15862	2781.1	3034.8	6.8844	0.14021	2776.9	3029.2	6.8226	0.12547	2772.6	3023.5	6.7664
350	0.17456	2866.1	3145.4	7.0694	0.15457	2863.0	3141.2	7.0100	0.13857	2859.8	3137.0	6.9563
400	0.19005	2950.1	3254.2	7.2374	0.16847	2947.7	3250.9	7.1794	0.15120	2945.2	3247.6	7.1271
500	0.2203	3119.5	3472.0	7.5390	0.19550	3117.9	3469.8	7.4825	0.17568	3116.2	3467.6	7.4317
600	0.2500	3293.3	3693.2	7.8080	0.2220	3292.1	3691.7	7.7523	0.19960	3290.9	3690.1	7.7024
700	0.2794	3472.7	3919.7	8.0535	0.2482	3471.8	3918.5	7.9983	0.2232	3470.9	3917.4	7.9487
800	0.3086	3658.3	4152.1	8.2808	0.2742	3657.6	4151.2	8.2258	0.2467	3657.0	4150.3	8.1765
900	0.3377	3850.5	4390.8	8.4935	0.3001	3849.9	4390.1	8.4386	0.2700	3849.3	4389.4	8.3895
1000	0.3668	4049.0	4635.8	8.6938	0.3260	4048.5	4635.2	8.6391	0.2933	4048.0	4634.6	8.5901
1100	0.3958	4253.7	4887.0	8.8837	0.3518	4253.2	4886.4	8.8290	0.3166	4252.7	4885.9	8.7800
1200	0.4248	4464.2	5143.9	9.0643	0.3776	4463.7	5143.4	9.0096	0.3398	4463.3	5142.9	8.9607
1300	0.4538	4679.9	5406.0	9.2364	0.4034	4679.5	5405.6	9.1818	0.3631	4679.0	5405.1	9.1329

T	P = 2.50 MPa (Sat. 223.99°C)				P = 3.00 MPa (Sat. 233.90°C)				P = 3.50 MPa (Sat. 242.60°C)			
	v	u	h	s	v	u	h	s	v	u	h	s
Sat.	0.07998	2603.1	2803.1	6.2575	0.06668	2604.1	2804.2	6.1869	0.05707	2603.7	2803.4	6.1253
225	0.08027	2605.6	2806.3	6.2639								
250	0.08700	2662.6	2880.1	6.4085	0.07058	2644.0	2855.8	6.2872	0.05872	2623.7	2829.2	6.1749
300	0.09890	2761.6	3008.8	6.6438	0.08114	2750.1	2993.5	6.5390	0.06842	2738.0	2977.5	6.4461
350	0.10976	2851.9	3126.3	6.8403	0.09053	2843.7	3115.3	6.7428	0.07678	2835.3	3104.0	6.6579
400	0.12010	2939.1	3239.3	7.0148	0.09936	2932.8	3230.9	6.9212	0.08453	2926.4	3222.3	6.8405
450	0.13014	3025.5	3350.8	7.1746	0.10787	3020.4	3344.0	7.0834	0.09196	3015.3	3337.2	7.0052
500	0.13998	3112.1	3462.1	7.3234	0.11619	3108.0	3456.5	7.2338	0.09918	3103.0	3450.9	7.1572
600	0.15930	3288.0	3686.3	7.5960	0.13243	3285.0	3682.3	7.5085	0.11324	3282.1	3678.4	7.4339
700	0.17832	3468.7	3914.5	7.8435	0.14838	3466.5	3911.7	7.7571	0.12699	3464.3	3908.8	7.6837
800	0.19716	3655.3	4148.2	8.0720	0.16414	3653.5	4145.9	7.9862	0.14056	3651.8	4143.7	7.9134
900	0.2159	3847.9	4387.6	8.2853	0.17980	3846.5	4385.9	8.1999	0.15402	3845.0	4384.1	8.1276
1000	0.2346	4046.7	4633.1	8.4861	0.19541	4045.4	4631.6	8.4009	0.16743	4044.1	4630.1	8.3288
1100	0.2532	4251.5	4884.6	8.6762	0.21098	4250.3	4883.3	8.5912	0.18080	4249.2	4881.9	8.5192
1200	0.2718	4462.1	5141.7	8.8569	0.22652	4460.9	5140.5	8.7720	0.19415	4459.8	5139.3	8.7000
1300	0.2905	4677.8	5404.0	9.0291	0.24206	4676.6	5402.8	8.9442	0.20749	4675.5	5401.7	8.8723

$T = $ °C; $v = $ m^3/kg; $u = $ kJ/kg; $h = $ kJ/kg; $s = $ kg/kg°K.

From G.J. Van Wylen, R. E. Sonntag, and C. Borganakke,C. 1994. *Fundamentals of Classical Thermodynamics* New York: John Wiley & Sons.

Table G.3 Compressed liquid

T		P = 5.00 MPa (Sat. 263.99°C)				P = 10.00 MPa (Sat. 311.06°C)				P = 15.00 MPa (Sat. 342.24°C)		
	v	u	h	s	v	u	h	s	v	u	h	s
Sat.	0.0012859	1147.8	1154.2	2.9202	0.0014524	1393.0	1407.6	3.3596	0.0016581	1585.6	1610.5	3.6848
0	0.0009977	0.04	5.04	0.0001	0.0009952	0.09	10.04	0.0002	0.0009928	0.15	15.05	0.0004
20	0.0009995	83.65	88.65	0.2956	0.0009972	83.36	93.33	0.2945	0.0009950	83.06	97.99	0.2934
40	0.0010056	166.95	171.97	0.5705	0.0010034	166.35	176.38	0.5686	0.0010013	165.76	180.78	0.5666
60	0.0010149	250.23	255.30	0.8285	0.0010127	249.36	259.49	0.8258	0.0010105	248.51	263.67	0.8232
80	0.0010268	333.72	338.85	1.0720	0.0010245	332.59	342.83	1.0688	0.0010222	331.48	346.81	1.0656
100	0.0010410	417.52	422.72	1.3030	0.0010385	416.12	426.50	1.2992	0.0010361	414.74	430.28	1.2955
120	0.0010576	501.80	507.09	1.5233	0.0010549	500.08	510.64	1.5189	0.0010522	498.40	514.19	1.5145
140	0.0010768	586.76	592.15	1.7343	0.0010737	584.68	595.42	1.7292	0.0010707	582.66	598.72	1.7242
160	0.0010988	672.62	678.12	1.9375	0.0010953	670.13	681.08	1.9317	0.0010918	667.71	684.09	1.9260
180	0.0011240	759.63	765.25	2.1341	0.0011199	756.65	767.84	2.1275	0.0011159	753.76	770.50	2.1210
200	0.0011530	848.1	853.9	2.3255	0.0011480	844.5	856.0	2.3178	0.0011433	841.0	858.2	2.3104
220	0.0011866	938.4	944.4	2.5128	0.0011805	934.1	945.9	2.5039	0.0011748	929.9	947.5	2.4953
240	0.0012264	1031.4	1037.5	2.6979	0.0012187	1026.0	1038.1	2.6872	0.0012114	1020.8	1039.0	2.6771
260	0.0012749	1127.9	1134.3	2.8830	0.0012645	1121.1	1133.7	2.8699	0.0012550	1114.6	1133.4	2.8576
280					0.0013216	1220.9	1234.1	3.0548	0.0013084	1212.5	1232.1	3.0393
300					0.0013972	1328.4	1342.3	3.2469	0.0013770	1316.6	1337.3	3.2260
320									0.0014724	1431.1	1453.2	3.4247
340									0.0016311	1567.5	1591.9	3.6546

$T = °C$; $v = m^3/kg$; $u = kJ/kg$; $h = kJ/kg$; $s = kJ/kg°K$.
From G.J. Van Wylen, R. E. Sonntag, and C. Borganakke,C. 1994. *Fundamentals of Classical Thermodynamics* New York: John Wiley & Sons.

Index